Selected Titles in This Series

269 **Mikhail Lyubich, John W. Milnor, and Yair N. Minsky, Editors,** Laminations and foliations in dynamics, geometry and topology, 2001

268 **Robert Gulliver, Walter Littman, and Roberto Triggiani, Editors,** Differential geometric methods in the control of partial differential equations, 2000

267 **Nicolás Andruskiewitsch, Walter Ricardo Ferrer Santos, and Hans-Jürgen Schneider, Editors,** New trends in Hopf algebra theory, 2000

266 **Caroline Grant Melles and Ruth I. Michler, Editors,** Singularities in algebraic and analytic geometry, 2000

265 **Dominique Arlettaz and Kathryn Hess, Editors,** Une dégustation topologique: Homotopy theory in the Swiss Alps, 2000

264 **Kai Yuen Chan, Alexander A. Mikhalev, Man-Keung Siu, Jie-Tai Yu, and Efim I. Zelmanov, Editors,** Combinatorial and computational algebra, 2000

263 **Yan Guo, Editor,** Nonlinear wave equations, 2000

262 **Paul Igodt, Herbert Abels, Yves Félix, and Fritz Grunewald, Editors,** Crystallographic groups and their generalizations, 2000

261 **Gregory Budzban, Philip Feinsilver, and Arun Mukherjea, Editors,** Probability on algebraic structures, 2000

260 **Salvador Pérez-Esteva and Carlos Villegas-Blas, Editors,** First summer school in analysis and mathematical physics: Quantization, the Segal-Bargmann transform and semiclassical analysis, 2000

259 **D. V. Huynh, S. K. Jain, and S. R. López-Permouth, Editors,** Algebra and its applications, 2000

258 **Karsten Grove, Ib Henning Madsen, and Erik Kjær Pedersen, Editors,** Geometry and topology: Aarhus, 2000

257 **Peter A. Cholak, Steffen Lempp, Manuel Lerman, and Richard A. Shore, Editors,** Computability theory and its applications: Current trends and open problems, 2000

256 **Irwin Kra and Bernard Maskit, Editors,** In the tradition of Ahlfors and Bers: Proceedings of the first Ahlfors-Bers colloquium, 2000

255 **Jerry Bona, Katarzyna Saxton, and Ralph Saxton, Editors,** Nonlinear PDE's, dynamics and continuum physics, 2000

254 **Mourad E. H. Ismail and Dennis W. Stanton, Editors,** q-series from a contemporary perspective, 2000

253 **Charles N. Delzell and James J. Madden, Editors,** Real algebraic geometry and ordered structures, 2000

252 **Nathaniel Dean, Cassandra M. McZeal, and Pamela J. Williams, Editors,** African Americans in Mathematics II, 1999

251 **Eric L. Grinberg, Shiferaw Berhanu, Marvin I. Knopp, Gerardo A. Mendoza, and Eric Todd Quinto, Editors,** Analysis, geometry, number theory: The Mathematics of Leon Ehrenpreis, 2000

250 **Robert H. Gilman, Editor,** Groups, languages and geometry, 1999

249 **Myung-Hwan Kim, John S. Hsia, Yoshiyuki Kitaoka, and Rainer Schulze-Pillot, Editors,** Integral quadratic forms and lattices, 1999

248 **Naihuan Jing and Kailash C. Misra, Editors,** Recent developments in quantum affine algebras and related topics, 1999

247 **Lawrence Wasson Baggett and David Royal Larson, Editors,** The functional and harmonic analysis of wavelets and frames, 1999

246 **Marcy Barge and Krystyna Kuperberg, Editors,** Geometry and topology in dynamics, 1999

245 **Michael D. Fried, Editor,** Applications of curves over finite fields, 1999

(*Continued in the back of this publication*)

Laminations and Foliations in Dynamics, Geometry and Topology

CONTEMPORARY
MATHEMATICS

269

Laminations and Foliations in Dynamics, Geometry and Topology

Proceedings of the Conference on
Laminations and Foliations in
Dynamics, Geometry and Topology
May 18–24, 1998
SUNY at Stony Brook

Mikhail Lyubich
John W. Milnor
Yair N. Minsky
Editors

American Mathematical Society
Providence, Rhode Island

This volume contains the proceedings of the Conference on Laminations and Foliations in Dynamics, Geometry and Topology held at SUNY at Stony Brook, May 18–24, 1998, with support from the National Science Foundation, grant DMS-9805524, and the Rosenbaum Foundation.

2000 *Mathematics Subject Classification.* Primary 53–06, 57–06, 37–06, 32–06, 53C12, 57R30, 57Mxx, 57M25, 37Fxx, 37F10, 37F75.

Library of Congress Cataloging-in-Publication Data

Laminations and foliations in dynamics, geometry and topology : laminations and foliations, May 18–24, 1998, Stony Brook / Mikhail Lyubich, John W. Milnor, Yair N. Minsky, editors.

p. cm. — (Contemporary mathematics, ISSN 0271-4132 ; 269)

Includes bibliographical references.

ISBN 0-8218-1985-2 (alk. paper)

1. Foliations (Mathematics)—Congresses. 2. Geometry, Differential—Congresses. I. Lyubich, Mikhail, 1959– II. Milnor, John Willard, 1931– III. Minsky, Yair Nathan. IV. Contemporary mathematics (American Mathematical Society) ; v. 269.

~~QA613.62 .L36 2000~~

514′.72—dc21 00-052559

∞ The paper used in this book is acid-free and falls within the guidelines
established to ensure permanence and durability.
Visit the AMS home page at URL: http://www.ams.org/

10 9 8 7 6 5 4 3 2 1 06 05 04 03 02 01

Contents

Introduction

The concepts of lamination and foliation have recently gained fresh strength. They have enabled advances in long-standing problems in hyperbolic geometry, led to new concepts in topology, witnessed solution of classical problems on analytic differential equations, contributed to the renormalization theory of one-dimensional dynamical systems, and played a crucial role in the foundations of higher dimensional holomorphic dynamics.

In spite of their deep relations, these different topics have developed as distinct research fields, with little interaction between their practitioners. In an effort to overcome this separation and bring these diverse points of view closer together, we held a conference at Stony Brook in May of 1998. The conference brought together experts in the different aspects of lamination and foliation theory, as well as many young researchers and graduate students.

The Present Volume

All but the last of the papers in this volume are based on lectures given at the Stony Brook conference. The first two are based on minicourses from the conference:

- **Geodesic laminations,** by Francis Bonahon.
  One-dimensional geodesic laminations on hyperbolic surfaces arose in Thurston's work on Teichmüller spaces and hyperbolic 3-manifolds, where they play a central role. For example, in Teichmüller theory they allow one to describe asymptotic properties of degenerating hyperbolic metrics on the surface. In hyperbolic 3-manifolds they describe the bending of convex hulls and the asymptotic geometry of ends. This presentation sets out the basic theory of geodesic laminations, discussing both its topological aspects and the more analytical ones arising from consideration of transverse measures and distributions. Some applications are given to the geometry of convex-hull boundaries in hyperbolic 3-manifolds.
- **Laminations, Foliations and the Topology of 3-Manifolds,** by David Gabai.
  The notion of (two-dimensional) taut foliation in a 3-manifold, and the related notion of essential lamination, generalize in some ways the properties of incompressible surfaces and have been much developed and applied by Gabai, Hatcher, Oertel, Thurston and others. A recurring theme in this minicourse is the notion of "minimal position" of curves or surfaces with respect to a taut foliation or essential lamination, and the applications of these ideas to answering deep questions about knot theory, covering spaces and algebraic topology of 3-manifolds.

- **Dicritical singularities of holomorphic vector fields,** by Cesar Camacho.
  This paper discusses the local structure of singularities of holomorphic vector fields in $\mathbb{C}^2$ and of the associated foliations. It describes the resolution procedure and the monodromy group, and then gives criteria for existence of meromorphic and Liouvillian integrals, and introduces two new invariants of the singularity.
- **Dynamics of $\mathbb{P}^2$: examples,** by John Eric Fornaess and Nessim Sibony.

  The study of holomorphic dynamics in two or more complex dimensions is a relatively new but rapidly developing field. This paper explores the dynamics of holomorphic maps from the complex projective plane to itself, providing a number of surprising and instructive examples.

- **Rational laminations of complex polynomials,** by Jan Kiwi.
  If f is a monic polynomial map with connected Julia set, then the associated *rational lamination* is an equivalence relation on $\mathbb{Q}/\mathbb{Z}$, where two rational angles are equivalent if and only if the associated external rays land at a common point of the Julia set. This concept was perhaps first introduced by McMullen, although it is based on earlier work by Thurston, Douady and Hubbard. Two polynomials with the same rational lamination are sometimes said to be *combinatorially equivalent.* The present paper provides a big step towards understanding polynomials of higher degree by giving a complete characterization of those equivalence relations which can arise as rational laminations. The results are easy to state, but quite difficult to prove.

- **Actions of discrete groups on complex projective spaces,** by José Seade and Alberto Verjovsky.
  By definition, a higher dimensional "*Complex Kleinian Group*" is a discrete subgroup of $\mathrm{PSL}(n+1, \mathbb{C})$ which act on the projective space $P^n_{\mathbb{C}}$, $n > 1$ in such a way that the domain of discontinuity is non-empty. This paper provides a survey of the field. In particular, it discusses higher dimensional analogues of Fuchsian groups. Using twistor theory, it shows that every Kleinian group of conformal automorphisms of S^4 gives rise to a complex Kleinian group of automorphisms of $P^3_{\mathbb{C}}$.

- **Dynamics of singular holomorphic foliations on the complex projective plane,** by Saeed Zakeri.
  Any holomorphic foliation of the complex projective plane by curves, with only mild singularities, is induced by a polynomial vector field on the plane. This is an outline of the theory, including discussion of concepts of 'degree', of monodromy, density of leaves, and ergodicity, as well as the possible existence of foliations with a non-trivial minimal set.

The Stony Brook Conference

A number of the lectures at the conference are not represented in this volume. First there were two further minicourses:

- **Dynamics and currents in $\mathbb{C}^2$,** by E. Bedford and J. Smillie.
Classically, laminations arise in the theory of hyperbolic dynamical systems as families of stable and unstable manifolds. There are natural "Ruelle-Sullivan currents" associated with these laminations. Remarkably, in two-dimensional holomorphic dynamics (iteration theory of polynomial diffeomorphisms of $\mathbb{C}^2$), currents come first as distributional $\partial\bar{\partial}$-derivatives of certain pluripotential Green's functions. These can then be interpreted as geometric currents supported on (1-complex-dimensional) laminations, which provide deep insight into the topological dynamics of the map.
[For much of this material, see for example the series of papers "*Polynomial diffeomorphisms of* $\mathbb{C}^2$" by Bedford and Smillie, in Invent. Math. **103** (1991) and **112** (1993) (joint with Lyubich), J. Amer. Math. Soc. **4** (1991), Math. Ann. **294** (1992), J. Geom. Anal. **8** (1998), Ann. Math. **148** (1998), and Ann. Sci. Éc. Norm. Sup. **32** (1999).]

- **Riemann surface laminations: uniformization and meromorphic functions,** by É. Ghys.
This minicourse focused on two foundational problems in the theory of Riemann surface laminations: the problem of embedding a Riemann surface lamination into a projective space, and the uniformization problem for Riemann surface laminations. These are motivated by the classical theory of Riemann surfaces, the Poincaré-Bendixon theory, and a conjecture of Camacho. The lectures developed a conceptual background for the discussion (foliated cycles, harmonic currents, divisors...) and presented many fascinating examples (e.g., an example of a non-flat lamination with all leaves parabolic) and several nice results (e.g., some necessary conditions for the embedding into a projective space) in this direction.
[An exposition of these ideas can be found in "*Laminations par surfaces de Riemann*", Panoramas et Synthèses **8** (1999), 49–95.]

In addition to the minicourses, there were a number of one-hour talks on current research topics:

- C. Camacho. *Complex foliations near a singularity.*
- A. Connes. *The Riemann flow and the Zeros of Zeta.*
- Ya. Eliashberg. *Foliations and contact structures on 3-manifolds.*
- S. Fenley. *Foliations with good geometry.*
- J.-E. Fornaess. *Remarks on dynamics on* $\mathbb{P}^2$.
- X. Gomez-Mont. *The attractor of a Riccati equation and the foliated geodesic flow.*
- A. Hatcher. *Kontsevich's conjecture on diffeomorphism groups of 3-manifolds.*
- J. H. Hubbard. *Foliations of domains in* $\mathbb{C}^2$ *with pluri-harmonic functions.*
- S. Hurder. *The Global Geometry of Riemannian Foliations.*
- Yu. Ilyashenko. *Covering manifolds for analytic families of leaves of foliations by analytic curves.*
- V. Kaimanovich. *Brownian motion on foliations: entropy, invariant measures, mixing.*
- S. Kerkhoff. *Seifert fibered spaces in the orbifold theorem.*
- J. Kiwi. *Rational laminations of complex polynomials.*

– L. Mosher. *Laminations and solvable groups.*
– D. Sullivan. *Some remarks on quantum topology.*
– A. Verjovsky. *Complex Kleinian groups in higher dimensions.*

Finally, two parallel evening sessions gave many active researchers an opportunity to present their recent results in the field.

Acknowledgments. The conference was funded by the Rosenbaum Foundation, and NSF grant DMS-9805524. The other members of the organizing committee – Étienne Ghys, Yakov Eliashberg, Tony Phillips, Dennis Sullivan, and Alberto Verjovsky– were instrumental in setting the framework of the conference, selecting speakers and obtaining funds. Nothing of value could have happened at the conference were it not for the tireless effort of the staff at the mathematics department. Gerri Sciulli and Lucille Meci kept everything together starting many months before the meeting itself. Amy Dellorusso, Grace Hunt, and Barbara Wichard helped with many details and long hours.

Mikhail Lyubich, John Milnor, Yair Minsky
Stony Brook, July 2000

Contemporary Mathematics
Volume **269**, 2001

Geodesic laminations on surfaces

Francis Bonahon

Contents

1991 *Mathematics Subject Classification.* 53C25, 30F40, 57N05.
Key words and phrases. geodesic laminations, transverse structures,.
This research was partially supported by N.S.F. grants DMS-9504282 and DMS-9803445.

Geodesic laminations on surfaces were introduced by W.P. Thurston a little over 20 years ago, and have since been a very powerful tool in hyperbolic geometry, low-dimensional topology and dynamical systems. In particular, there are several different contexts where geodesic laminations now routinely occur. Geodesic laminations can be considered as:

- topological objects, occurring as generalizations of simple closed curves on surfaces;
- geometric objects, such as bending laminations of hyperbolic convex cores, shearing loci of earthquakes on hyperbolic surfaces, stable laminations of pseudo-Anosov diffeomorphisms, maximal stretch laminations between hyperbolic surfaces;
- interesting dynamical objects, in particular because of their connection with interval exchange maps.

This versatility can be somewhat confusing for the non-expert. To the expert, it provides interesting challenges when these very different points of view need to interact with each other. The minicourse given at the Workshop was devoted to illustrations of the above three aspects of geodesic laminations and of their interactions. Each of the three lectures focused on one of these viewpoints.

This article is similarly divided into three parts.

The first part is devoted to generalities on geodesic laminations, gives some examples, and discusses dynamically interesting transverse structures for geodesic laminations. This includes the analytic notion of a transverse Hölder distribution, and the more combinatorial notion of a transverse cocycle; these two transverse structures are later shown to be equivalent.

The second part discusses topological applications of geodesic laminations. In particular, we consider the space of measured geodesic laminations, as a completion of the space of simple closed curves on the surface. We mention the piecewise linear structure $\mathcal{ML}(S)$, and indicate how the combinatorial tangent vectors of this piecewise linear manifold have a geometric interpretation as geodesic laminations with transverse Hölder distributions.

Finally, the third part is devoted to some geometric applications of geodesic laminations. We chose to focus on geodesic laminations as bending loci of boundaries of convex cores of hyperbolic 3–dimensional manifolds, and as pleating loci of pleated surfaces. In particular, we show how the bending of a pleated surface along its pleating locus can be measured by a transverse cocycle. We also connect the bending of pleated surfaces to the rotation angle of closed geodesics in the hyperbolic 3–manifold.

We made the deliberate effort of closely following the style and structure of the minicourse. Anybody who has attended a talk where the lecturer directly reads from a preprint knows that the translation between mathematical writing and mathematical lecturing often requires a serious reorganization. However, we decided to try the experiment, hoping that this informal and pictorial[1] style will be more effective at communicating the main ideas, while the reader can always turn to the published references in the bibliography for a more rigorous treatment of these topics.

We are grateful to the referee for a careful reading of the manuscript, and for the suggestion of many improvements.

[1]Many pictures are taken from the monograph [**Bo6**] in preparation.

PART I. THE DYNAMICAL VIEWPOINT

Definitions and first properties

One of the usual frameworks for geodesic laminations is that of a surface S with no boundary, endowed with a complete metric whose curvature is constant equal to -1 (namely a *hyperbolic metric*) and whose area is finite. There are possible variations of this setting, where we can allow the curvature of the metric to vary between two negative constants, and/or where the surface S can be required to be compact with convex boundary. However, these variations only lead to minor technical differences, and we will restrict ourselves to this first setting.

An important consequence of the existence of such a hyperbolic metric is that S is a surface of finite topological type, namely is homeomorphic to the complement of finitely many points in a compact surface. In addition, the Gauss-Bonnet formula implies that the Euler characteristic $\chi(S)$ is strictly negative. Conversely, it is well known that a surface S of finite topological type and negative Euler characteristic always admits a hyperbolic metric of this type.

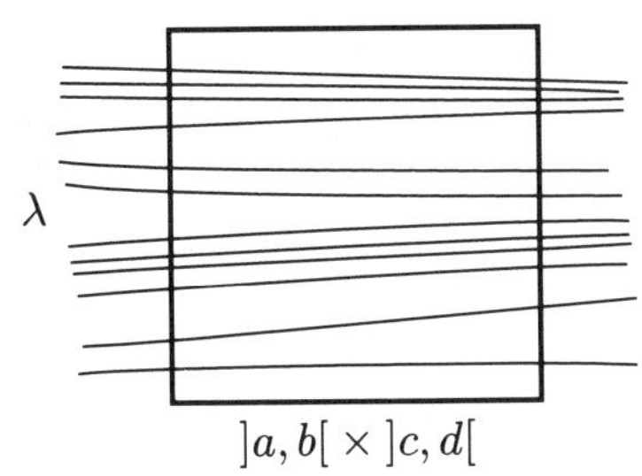

FIG. 1. The local type of a geodesic lamination

A *geodesic lamination* on the surface S is a lamination λ of S whose leaves are geodesic. Recall that the fact that λ is a lamination means that λ is a closed subset of S and is decomposed into subsets called its *leaves*, so that locally the situation is as follows: Every point of λ has a neighborhood U homeomorphic to the product $]a,b[\times]c,d[$ of two open intervals in such a way that $U\cap\lambda$ corresponds to $K\times]c,d[$ for some compact subset K of $]a,b[$ and, for every leaf g of λ, $U\cap g$ corresponds to $A_g\times]c,d[$ for some subset A_g of $]a,b[$. Note that this local description implies that, for a geodesic lamination, the leaves are *complete* geodesics in the sense that each leaf is either closed or has infinite length in both of its ends.

In practice, checking whether a subset of S is a geodesic lamination is much easier than one might think from this elaborate description of its local type. Indeed, the following property holds.

PROPOSITION 1. *If λ is a closed subset of S which is a disjoint union of simple geodesics, then λ is a geodesic lamination.*

Here, a geodesic is *simple* if it does not cross itself. It may be closed or infinite.

The key point in the proof of Proposition 1 is the following estimate in hyperbolic geometry, whose proof can be found in [**CEG, §5.2.6**].

LEMMA 2. *In the hyperbolic plane $\mathbb{H}^2$, there is a constant C with the following property: If the geodesic g has unit tangent vector v at x, if the geodesic h has unit*

FIG. 2. A geodesic lamination

tangent vector w at y, and if the geodesics g and h are disjoint, then,

$$d(v, \pm w) \leqslant C d(x, y).$$

In particular, if λ is a closed subset of S which is a disjoint union of simple geodesics, the tangent directions of these geodesics form a Lipschitz direction field on λ. Extending it to a Lipschitz direction field over a neighborhood of λ and using the standard existence results for solutions to ordinary differential equations, one obtains local charts which show that λ is a lamination, in the sense indicated above.

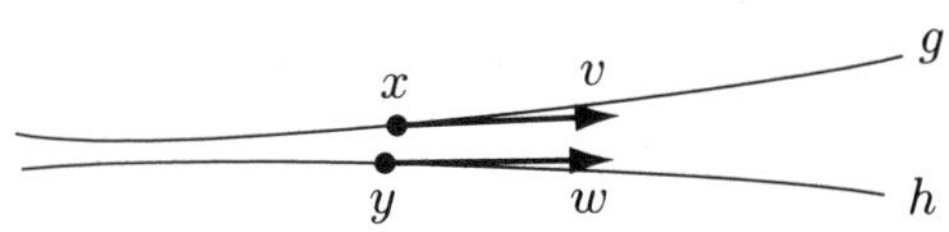

FIG. 3. Disjoint geodesics have nearby directions at nearby points

Examples of geodesic laminations

Geodesic laminations with finitely many leaves. The first example is provided by a family of disjoint simple closed geodesics $\gamma_1, \gamma_2, \dots, \gamma_n$.

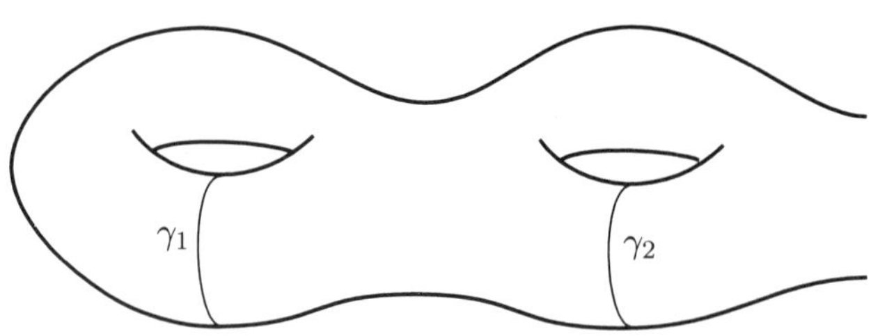

FIG. 4. A geodesic lamination with all leaves closed.

This example can be enlarged by adding a finite family of infinite geodesics which spiral along these closed geodesics, as in Figure 5.

Interval exchange maps. More complex examples are associated to interval exchange maps, which we now define. Consider a compact interval I in $\mathbb{R}$, which we decompose as a finite union $I = I_1 \cup I_2 \cup \dots \cup I_n$ of intervals I_i with disjoint

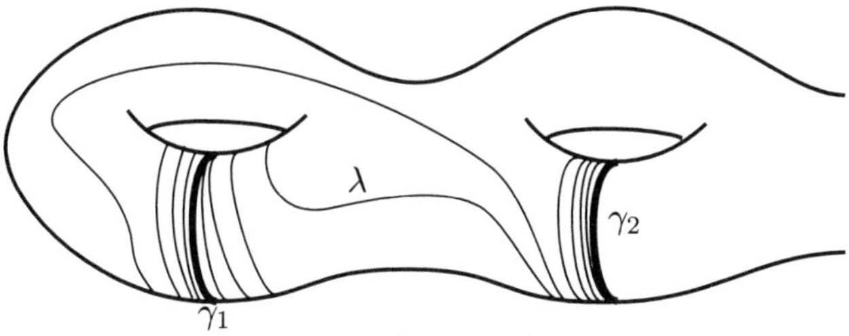

FIG. 5. A geodesic lamination with finitely many leaves.

interiors. Choose another decomposition of $I = J_1 \cup J_2 \cup \ldots \cup J_n$ into intervals J_j with disjoint interiors such that, for every i, there is an isometry $\varphi_i : I_i \to J_i$. The collection of the φ_i defines a 'map' $\varphi : I \to I$. This map is in general 1-to-2 at the end points of the I_i, but is well-defined everywhere else. Such a φ is an *interval exchange map*.

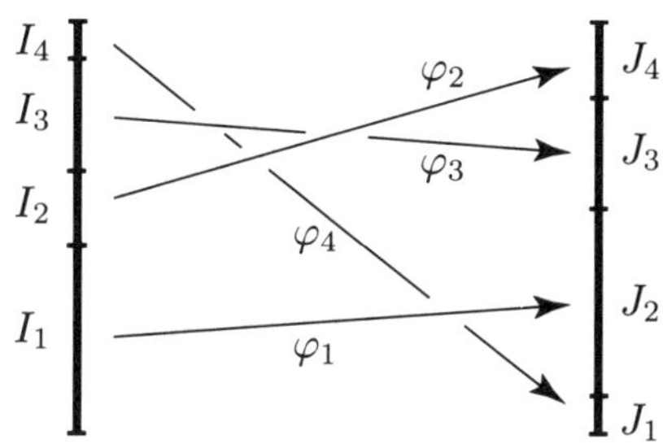

FIG. 6. An interval exchange map.

Let us restrict attention to the case where $\varphi(\partial I) \cap \partial I \neq \emptyset$. Otherwise, φ restricts to an exchange of fewer intervals.

To the interval exchange map $\varphi : I \to I$ is associated its *suspension* Σ_φ, obtained by gluing n rectangles $I_i \times [0,1]$ to the interval I as follows: $I_i \times \{0\}$ is identified to the interval $I_i \subset I$ by the obvious map, and $I_i \times \{1\}$ to the interval $J_i \subset I$ by the (isometric) restriction $\varphi_i : I_i \to J_i$ of φ. In particular, Σ_φ looks like a freeway interchange without entry or exit ramps.

Foliate each rectangle $I_i \times [0,1]$ by the arcs $\{*\} \times [0,1]$. This defines a foliation of the suspension Σ_φ by lines and simple closed curves. The leaves of this foliation are in one-to-one correspondence with the 2–sided orbits of φ, namely with the bi-infinite sequences $\{\ldots, x_{-n}, \ldots, x_{-1}, x_0, x_1, \ldots, x_n, \ldots\}$ where $x_{n+1} = \varphi(x_n)$ for every $n \in \mathbb{Z}$ and where the indexing is only defined modulo translation in $\mathbb{Z}$. Let a leaf be *regular* if it does not pass through any of the end points of the I_i.

Note that the boundary of the suspension Σ_φ consists of a certain number of cycles. Gluing a semi-open annulus along each of these cycles, we obtain a surface S of finite type, whose Euler characteristic $\chi(S) = 1 - n$ is negative. Endow S with a finite area hyperbolic metric.

Pulling tight each leaf l of Σ_φ defines a geodesic g_l of S. The key technical property here is that each leaf of Σ_φ is quasi-geodesic in Σ_φ and (by a small argument using the fact that $\varphi(\partial I) \cap \partial I \neq \emptyset$) in S, so that there is a unique geodesic g_l

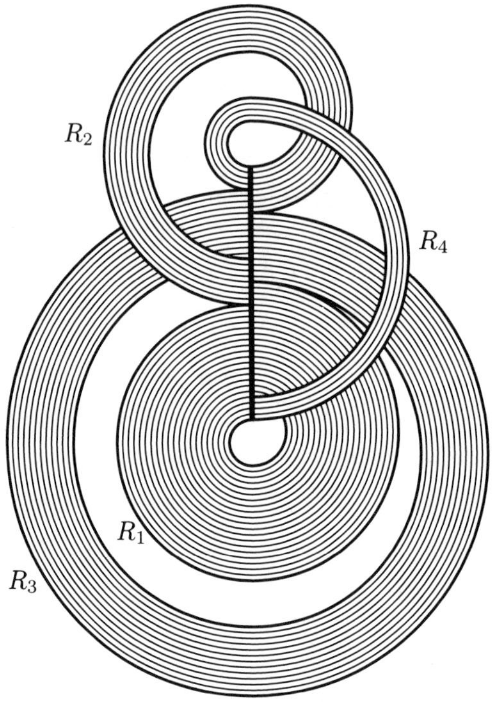

FIG. 7. The suspension of an interval exchange map.

of S for which there is a homotopy from l to g_l which moves points by a uniformly bounded distance.

Consider the family of those geodesics g_l which are associated in this way to the regular leaves of Σ_φ. Since distinct regular leaves l and l' are disjoint, the boundedness of the homotopies from l to g_l and from l' to $g_{l'}$ shows that the associated geodesics cannot cross each other, and therefore are either disjoint or equal. Similarly, since a regular leaf does not cross itself, the geodesic g_l must be simple (closed or infinite). Let λ' be the union of the geodesics g_l associated to the regular leaves l of Σ_φ, and let λ' be the closure of λ'. An application of Lemma 2 and Proposition 1 shows that λ is a geodesic lamination.

If the lengths of all the intervals I_i are rational, then every regular leaf of Σ_φ is closed. Consequently, λ consists of finitely many simple closed geodesics, and we are back to our first example.

The situation is much more interesting in the other extreme case where the lengths of the intervals I_i are linearly independent over $\mathbb{Q}$. Then, assuming in addition that φ does not restrict to an exchange of fewer intervals, all the regular leaves of Σ_φ are infinite. In addition, two distinct regular leaves l, l' must diverge at some point and follow different routes on the 'freeway' Σ_φ; it easily follows that the associated geodesics g_l and $g_{l'}$ must be distinct. As a consequence, the geodesic lamination λ so constructed now has uncountably many leaves, all infinite.

The topology of geodesic laminations

Consider the complement $S - \lambda$ of a geodesic lamination λ in the hyperbolic surface S. On the open subset $S - \lambda$, we can consider the path metric induced by the metric of S, for which the distance between x and $y \in S - \lambda$ is the infimum of the lengths of all paths going from x to y in $S - \lambda$ (and is infinite if x and y are in different components of $S - \lambda$).

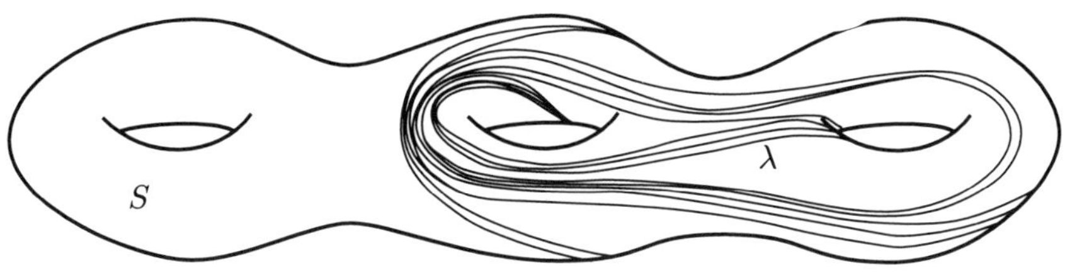

FIG. 8. A geodesic lamination λ.

Let $\widehat{S - \lambda}$ be the completion of $S - \lambda$ for this metric. From the local picture of λ, we see that $\widehat{S - \lambda}$ locally is a hyperbolic surface with totally geodesic boundary. More precisely, $\widehat{S - \lambda}$ is obtained by abstractly adding to $S - \lambda$ a boundary made up of leaves of λ.

FIG. 9. The completion $\widehat{S - \lambda}$ of the complement $S - \lambda$.

Since S has finite area, it follows from the above description that $\widehat{S - \lambda}$ also has finite area. From this, we conclude that $\widehat{S - \lambda}$ is the union of a compact part, of a finite number of cusps corresponding to cusps of S, and of finitely many spikes, each delimited by two (possibly equal) boundary components of $\widehat{S - \lambda}$. In particular, the topological type of $\widehat{S - \lambda}$ is finite and its boundary consists of finitely many geodesics.

We can be a little more precise. Using Lemma 2, we can construct a direction field which is non-singular and is transverse to the leaves on λ, and has finitely many isolated singularities on $S - \lambda$. Using the Poincaré-Hopf formula to express the Euler characteristic $\chi(\ )$ in terms of the indices of these singularities, we conclude that

$$\chi\left(\widehat{S - \lambda}\right) = \chi(S) + \tfrac{1}{2}s$$

where s is the number of spikes of $\widehat{S - \lambda}$.

A corollary of this formula is that the number of components of λ is uniformly bounded. In particular, λ can contain at most finitely many *sub-laminations*, namely closed subsets which are the union of leaves of λ. As a consequence, λ

contains finitely many sub-laminations which are minimal for inclusion. Analyzing the leaves of λ which are not contained in minimal sub-laminations and using the above Euler characteristic formula to guarantee finiteness, one can prove (see [**Th1, §8**], [**CEG, §4.2.8**] or [**CaB, §4**]):

PROPOSITION 3. *A geodesic lamination is the union of finitely many minimal sub-laminations and of finitely many infinite isolated leaves, whose ends spiral along the minimal sub-laminations or converge to a cusp.*

Note that a sub-lamination λ' is minimal if and only if every half-leaf of λ' is dense in λ'. A simple example consists of a closed leaf of λ. However, interval exchange maps provide many examples where λ is minimal and has uncountably many leaves.

A leaf g of λ is *isolated* if every point of g has a neighborhood whose intersection with λ is a single arc (contained in g).

As a corollary of the Proposition 3, we have:

PROPOSITION 4. *Each cusp of S has a neighborhood whose intersection with λ consists of finitely many isolated half-leaves converging towards the cusp.*

The higher dimensional case

Before going any further, we might wonder if we would not get more insight by considering geodesic laminations in all dimensions, namely laminations whose leaves are totally geodesic. Compact totally geodesic submanifolds provide examples of such laminations, although somewhat trivial. However, Abdelghani Zeghib essentially showed that this is the only example:

THEOREM 5. (Zeghib [**Ze**]) *Let M be a compact Riemannian manifold of dimension $n \geqslant 3$ and of negative curvature. If λ is a codimension 1 lamination of M whose leaves are totally geodesic, then all the leaves of λ are compact.*

The idea of the proof is based on a volume argument: If a boundary component F of $\widehat{S-\lambda}$ has infinite $(n-1)$–volume, then the other boundary components cannot be packed close enough to F to enclose a finite n–dimensional volume; in particular, $\widehat{S-\lambda}$ will then have infinite volume, contradicting the compactness of M. Namely, the existence of finite area spikes in $\widehat{S-\lambda}$ is a phenomenon which is typical of the dimension 2.

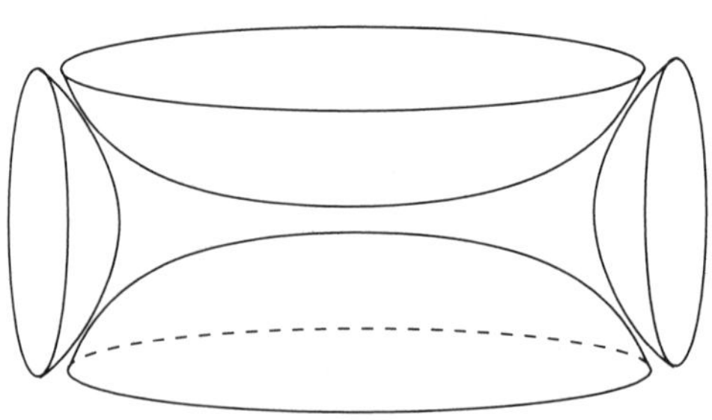

FIG. 10. In dimension $\geqslant 3$, totally geodesic surfaces cannot be packed close together.

A more explicit example

Traditionally, geodesic laminations are often represented by rough pictures analogous to the ones which have appeared above, such as Figures 2 or 8. These pictures are convenient and are topologically correct, but they often are metrically incorrect. This fact had been overlooked for a long time by the author until, recently, he had the curiosity of drawing computer pictures of explicit geodesic laminations. Such an example appears on Figure 11. Here, the underlying surface S is a once punctured torus, obtained by identifying opposite sides in a quadrilateron in (the Poincaré model for) the hyperbolic plane $\mathbb{H}^2$ with its vertices on the circle at infinity. The identifications between sides respect the intersection of the geodesic lamination λ with these sides.

In this example, λ is the geodesic lamination associated to the suspension of an exchange of two intervals, whose length are rationally independent. In particular, λ has uncountably many leaves.

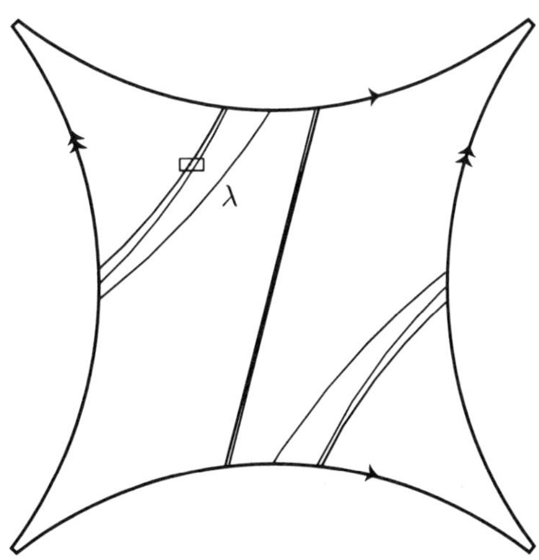

FIG.11. A geodesic lamination on the punctured torus.

Two striking features of this picture are that λ occupies remarkably little space on the surface S, and that there is not much of the structure of λ which is really visible. Actually, drawing this picture required a careful choice of λ to make it more interesting: Most geodesic laminations on S appear only as the union of three fat lines on the quadrilateron.

However, if we zoom on the little rectangle shown on Figure 11, the enlarged rectangle is the one represented on Figure 12(a). Similarly, Figure 12(b) is en enlargement of the small rectangle indicated on Figure 12(a), and Figure 12(c) represents the small rectangle of Figure 12(b). These pictures suggest that the leaves of λ are grouped in patterns reminiscent of Cantor sets.

(a)

(b)

(c)

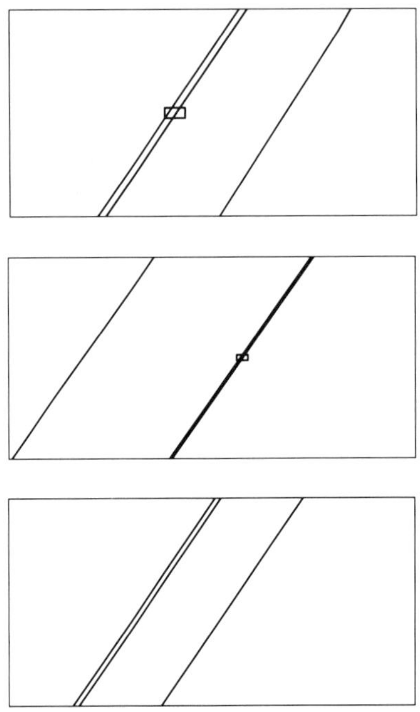

FIG.12. Zooming in on Figure 11.

Local properties of geodesic laminations

These heuristic observations can be justified as follows.

First, a surprising result of Joan Birman and Caroline Series [**BiS**] (see also [**Th3, §10**] for a proof which is technically equivalent, but perhaps more conceptual) shows that very few geodesics of S are simple.

THEOREM 6. (J. Birman, C. Series) *The union of all simple geodesics of S has Hausdorff dimension* 1.

Note that the union of all simple geodesics is also the union of all geodesic laminations.

An immediate corollary of Theorem 6 is that an individual geodesic lamination λ has Hausdorff dimension 1. In particular, if k is a differentiable arc which is transverse to the leaves of λ, then $k\cap\lambda$ has Hausdorff dimension 0, and in particular is totally disconnected. This proves:

PROPOSITION 7. *Suppose that the geodesic lamination λ has no isolated leaves. Then, for every arc k transverse to the leaves of λ, the intersection $k\cap\lambda$ is a Cantor set.*

However, metrically, $k \cap \lambda$ does not look like the typical middle third Cantor set at all.

PROPOSITION 8. *If the arc k is transverse to the geodesic lamination λ and if we index the components $d_1, d_2, \ldots, d_n, \ldots$ of the complement $k - \lambda$ by order of non-increasing size, then*

$$|\log \text{length}\,(d_n)| \asymp n$$

Here the symbol $a_n \asymp b_n$ means that the ratio a_n/b_n is bounded between two positive constants for n large enough. It should be noted that

$$|\log \text{length}\,(d_n)| \asymp \log n$$

for the middle third Cantor set.

To explain why Proposition 8 holds, we should observe that all but finitely many d_n are in spikes of $\widehat{S - \lambda}$. For a given spike, list the d_n contained in this spike as $d'_1, d'_2, \ldots, d'_p, \ldots$ in the order of their progression towards the end of the spike.

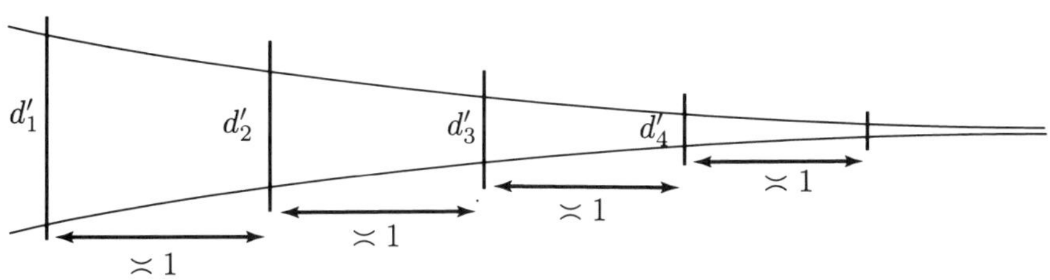

FIG.13. Components of $k - \lambda$ in a spike of $\widehat{S - \lambda}$.

For any two consecutive components d'_p, d'_{p+1}, an estimate using hyperbolic geometry gives that the ratio

$$\frac{\text{length}\,(d'_{p+1})}{\text{length}\,(d'_p)} \asymp e^{-\left(\text{distance from } d'_p \text{ to } d'_{p+1}\right)}.$$

By compactness of $k \cap \lambda$, the distance between d'_p and d'_{p+1} is bounded between two constants. It follows that

$$\left|\log \text{length}\,(d'_p)\right| \asymp p.$$

Since $\widehat{S - \lambda}$ has only finitely many spikes, the estimate of Proposition 8 follows.

Transverse structures

Transverse measures. The first type of transverse structure which we can consider for a geodesic lamination is provided by the classical notion of transverse measure.

A transverse measure for a lamination is locally a measure on the space of leaves of the lamination. For a geodesic lamination λ, the leaves of λ are locally determined by their intersection points with a differentiable arc transverse to λ. We can therefore define a *transverse measure* for the geodesic lamination λ as the assignment of a measure on each arc k transverse to λ, subject to the following two conditions: If the arc k' is contained in the transverse arc k, then the measure assigned to k' is the restriction of the measure assigned to k; if the two arcs k and k' are homotopic through a family of arcs transverse to λ, the homotopy sends the measure assigned to k to the measure assigned to k'.

We are here using the traditional geometric analysis convention that a *measure* actually is a *Radon measure*. Namely, it assigns a finite non-negative mass to each relatively compact Borelian subset, in a countably additive manner.

An immediate consequence of the condition of invariance under homotopy of arcs transverse to λ is that the support of the measure assigned to the transverse arc k must be contained in $k \cap \lambda$.

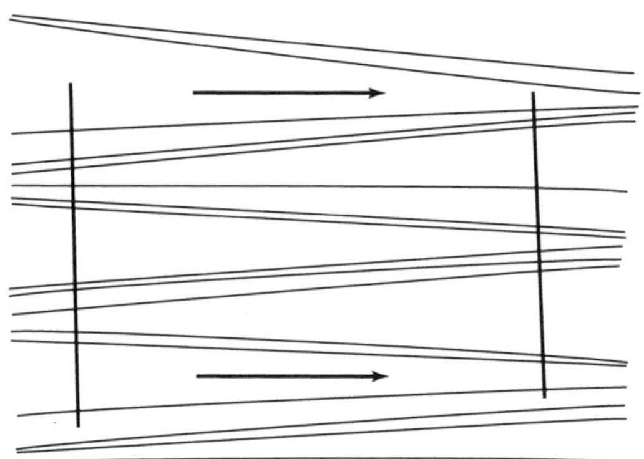

FIG.14. A homotopy of arcs transverse to λ.

The simplest example of such a transverse measure occurs for a geodesic lamination λ which admits a certain number of closed leaves $l_1, \dots, l_n$. Pick positive numbers $a_1, \dots, a_n$. For each arc k transverse to λ, consider the measure α defined by the property that

$$\alpha(A) = \sum_{i=1}^{n} a_i \# A \cap l_i$$

for every subset A of k, where $\#X$ denotes the cardinal of X. This α is clearly invariant under homotopy of arcs transverse to λ, and consequently defines a transverse measure for λ.

A more elaborate construction uses the fact that the leaves of λ are 1–dimensional, and therefore have polynomial growth. Then, if we apply a construction similar to the above ones to arbitrary large pieces of leaves of λ, an averaging process provides a non-trivial transverse measure for λ; compare [**Pl**].

PROPOSITION 9. *Every geodesic lamination λ admits a transverse measure whose support consists of all the minimal sub-laminations of λ.*

It is easy to check that infinite isolated leaves of λ cannot be in the support of any transverse measure α for λ, as they would otherwise provide arcs transverse to λ and with infinite α–mass. Therefore, Proposition 9 is optimal.

Interval exchange maps also provide interesting examples of geodesic laminations with transverse measures. Indeed, the regular 2-sided orbits of an interval exchange map $\varphi : I \to I$ defines leaves of the associated geodesic lamination λ.

Since φ preserves the Lebesgue measure of I, the space of 2–sided orbits of φ inherits a natural measure from the Lebesgue measure of I. This measure pushes forward to locally define a measure on the space of leaves of λ, and therefore defines a transverse measure for λ.

Transverse distributions(?). Transverse measures are classical objects in the theory of foliations and laminations. A natural generalization, with a wider analytical scope, is to consider transverse distributions.

At this point, it is probably worth recalling that, by the Riesz Representation Theorem, a Radon measure on a locally compact space X is the same thing as a positive linear functional on the space of all continuous functions $\varphi : X \to \mathbb{R}$ with compact support. A *distribution* on a manifold X is a continuous linear form on the space of infinitely differentiable functions $\varphi : X \to \mathbb{R}$ with compact support. See for instance [**Sc**].

In the case of a geodesic lamination λ, one can try to define a transverse distribution as a distribution defined on each arc transverse to λ, such that, when the arcs k and k' are homotopic through a family of arcs transverse to λ, the homotopy sends the distribution defined on k to the distribution defined on k'. However, there is a serious technical problem here. To be able to compare in full generality distributions on k to distributions on k', the homotopy from k to k' must be differentiable. This almost never happens.

Indeed, if k is an arc transverse to λ, for all but finitely many components of $k - \lambda$, the leaves of λ passing through the end points of this component of $k - \lambda$ are asymptotic and bound a spike of the completed complement $\widehat{S - \lambda}$. If we homotop k to k' through a family of arcs transverse to λ, this homotopy establishes a one-to-one correspondence between components of $k - \lambda$ and components of $k' - \lambda$. However, if the homotopy moves each point of k by a distance of approximately $D > 0$, we notice two possible behaviors for a component d of $k - \lambda$: if the two leaves passing through the end points of d are asymptotic in the direction of k', then the corresponding component d' of $k' - \lambda$ has length $l\,(d')$ which is approximately $\mathrm{e}^{-D} l\,(d)$, by an easy hyperbolic geometry estimate; on the other hand, if these two leaves are asymptotic in the direction opposite to k', then $l\,(d')$ is approximately $\mathrm{e}^{+D} l\,(d)$. For a typical geodesic lamination, every neighborhood of a point of $k \cap \lambda$ contains components of $k - \lambda$ which are of each of these two types. This clearly prevents the homotopy from k to k' from being differentiable. See Figure 15.

In particular, a typical geodesic lamination admits no transverse differentiable structure. As a consequence, it does not in general make sense to talk of transverse distributions for a geodesic lamination λ, which explains the question mark in the heading of this subsection.

However the above argument, which shows that a homotopy of arcs transverse to λ cannot be differentiable, also shows that the homotopy can be chosen to be Lipschitz, and in particular can be chosen to be Hölder continuous.

Recall that a function φ between metric spaces is *Hölder continuous* with *Hölder exponent* $\nu \leqslant 1$ if

$$d\,(\varphi\,(x)\,, \varphi\,(y)) \leqslant \text{constant}\;\; d\,(x, y)^{\nu}$$

for all x, y. It is *Lipschitz continuous* if it is Hölder continuous with Hölder exponent $\nu = 1$.

Having thus showed that geodesic laminations admit transverse Hölder structures, this leads us to consider:

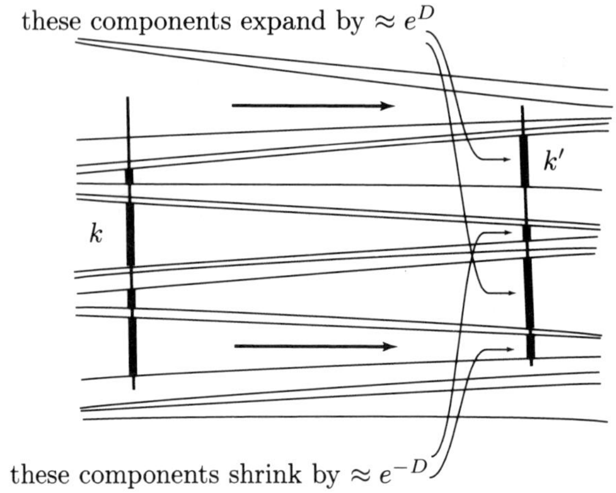

FIG.15. A homotopy of arcs transverse to λ cannot be differentiable.

Transverse Hölder distributions. Let us define a *Hölder distribution* on a compact metric space X as a continuous linear functional on the space of all Hölder continuous functions $\varphi : X \to \mathbb{R}$ with compact support.

A *transverse Hölder distribution* α for a geodesic lamination λ is a Hölder distribution defined on each arc k transverse to λ, which is invariant under every Hölder continuous homotopy of arcs transverse to λ. A consequence of the invariance property is that, for a transverse arc k, the Hölder distribution induced on k by α has support contained in $k \cap \lambda$ (namely $\alpha(\varphi) = 0$ for every Hölder continuous function whose support is contained in $k - \lambda$).

Here is an example of such a transverse Hölder distribution. Consider the geodesic lamination λ of Figure 16, consisting of one closed geodesic γ and of one infinite leaf whose ends spiral along γ.

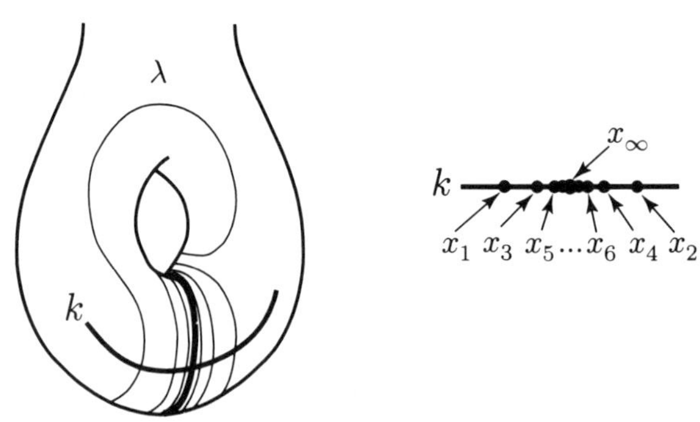

FIG. 16. A geodesic lamination with a transverse Hölder distribution

For the arc k shown on Figure 16,

$$k \cap \lambda = \{x_\infty, x_1, x_2, \ldots, x_n, \ldots\}$$

where x_∞ is the point $k \cap \gamma$. Note that the other points $x_1, x_2, \ldots, x_n$ accumulate on x_∞. For $a, b \in \mathbb{R}$, define a Hölder distribution α on k by

$$\alpha(\varphi) = a\varphi(x_\infty) + b\sum_{1}^{\infty} (\varphi(x_n) - \varphi(x_\infty)).$$

for every Hölder continuous function $\varphi : k \to \mathbb{R}$. To check that the infinite sum does converge note that, if we index the x_n so that the distance $d(x_n, x_\infty)$ monotonically converges to 0, then $d(x_n, x_\infty) \asymp (e^{-L})^n$ where L is the length of γ; indeed, this easily follows from an estimate in hyperbolic geometry. This implies that $|\varphi(x_n) - \varphi(x_\infty)| \prec (e^{-\nu L})^n$, where ν is the Hölder exponent of φ, so that the infinite sum is absolutely bounded by a converging geometric series.

Note that this formula is easily transported by a homotopy of arcs transverse to λ. In general, every transverse arc k' can be decomposed into smaller arcs, each homotopic to a subarc of k through a family of transverse arcs. The formula for k then enables us to define a Hölder distribution on this transverse arc k'.

In this way, we define a transverse Hölder distribution for the geodesic lamination of Figure 16. This construction depends on the choice of the two reals a, $b \in \mathbb{R}$, so that we actually have a 2–parameter family of such transverse Hölder distributions.

Transverse cocycles. Transverse measures and transverse Hölder distributions are analytic objects. We now consider simpler transverse structures, of a more combinatorial nature.

A *transverse cocycle* α for the geodesic lamination λ associates to each arc k transverse to λ a number $\alpha(k)$ in a way which is:

- invariant under homotopy of arcs transverse to λ;
- finitely additive, in the sense that, if we split k as the union of two subarcs k_1 and k_2 with disjoint interiors, then $\alpha(k) = \alpha(k_1) + \alpha(k_2)$.

We can give a more analytic flavor to this definition by observing that a transverse cocycle is equivalent to the data, for each arc k transverse to λ, of a linear function on the space of all locally constant functions $\varphi : K \cap \lambda \to \mathbb{R}$, in a way which is invariant under homotopy of arcs transverse to λ.

The combinatorial nature of transverse cocycles makes it very easy to analyze them. In particular, using train tracks, one easily proves [**Bo2**]:

PROPOSITION 10. *For a geodesic lamination λ, the space $\mathcal{H}(\lambda; \mathbb{R})$ of transverse cocycles for λ is a vector space of dimension*

$$n_O(\lambda) + \tfrac{1}{2}\#\text{spikes of } \widehat{S - \lambda},$$

where $n_O(\lambda)$ is the number of orientable components of λ.

These transverse cocycles look very different from transverse Hölder distributions. However, it turns out that the two notions are actually equivalent. Indeed, a transverse Hölder distribution defines a transverse cocycle α by the property that

$$\alpha(k) = \alpha(\text{constant function } 1 \text{ on } k).$$

THEOREM 11. [**Bo2**] *This establishes as one-to-one correspondence between transverse Hölder distributions and transverse cocycles for the geodesic lamination* λ.

The key part of the proof is to show that two Hölder distributions which induce the same transverse cocycle are actually equal. This is provided by the following lemma, which, for a transverse Hölder distribution α and a Hölder continuous function $\varphi : k \to \mathbb{R}$ defined on an arc transverse to λ, expresses $\alpha(\varphi)$ in terms of the transverse cocycle associated to α and of the values of the function φ on $k \cap \lambda$.

PROPOSITION 12. (Gap Formula) *Let α be a transverse Hölder distribution for the geodesic lamination λ. For every Hölder continuous function $\varphi : k \to \mathbb{R}$ defined on an oriented arc k transverse to λ,*

$$\alpha(\varphi) = \alpha(k)\,\varphi\left(x_k^+\right) + \sum_d \alpha(k_d)\left(\varphi\left(x_d^-\right) - \varphi\left(x_d^+\right)\right)$$

where: x_k^+ is the positive end point of the oriented arc k; d ranges over all components of $k - \lambda$ (= gaps); x_d^+ and x_d^- are the positive and negative end points of d; k_d is the sub-arc of k joining the negative end point x_k^- of k to an arbitrary point in d.

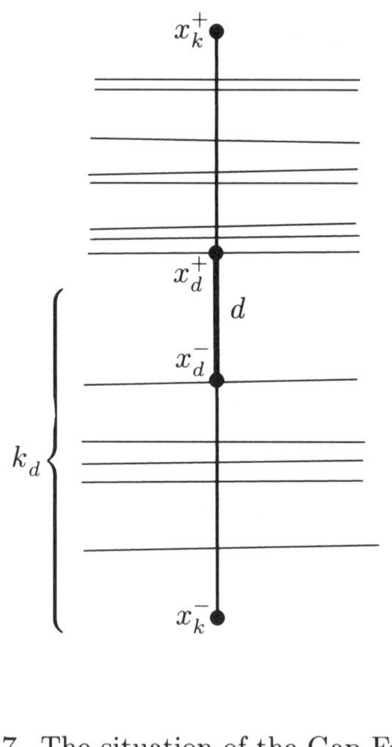

FIG. 17. The situation of the Gap Formula

Note that this formula holds in particular if α is a transverse measure, but that it is necessary for φ to be Hölder continuous even in this case (when $k \cap \lambda$ is a Cantor set, think of the Lebesgue function $\varphi : k \to \mathbb{R}$ which is continuous, non-constant, but constant on each component of $k - \lambda$).

Although the result is apparently local, the proof of Proposition 12 uses the global structure of λ in a crucial way. We can illustrate this by showing why the infinite sum in the Gap Formula converges for a Hölder continuous function

$\varphi : k \to \mathbb{R}$. Recall from Proposition 8 that we can index the components d_1, d_2, $\dots$, d_n, $\dots$ of $k - \lambda$ so that $|\log \text{length}\,(d_n)| \asymp n$. In particular,

$$\text{length}\,(d_n) \prec e^{-An}$$

for some constant $A > 0$, and

$$\left|\varphi\left(x_{d_n}^-\right) - \varphi\left(x_{d_n}^+\right)\right| \prec e^{-A\nu n}$$

if φ has Hölder exponent ν. In addition, the index n measures how deep d_n converges towards the spikes of $\widehat{S - \lambda}$. By a careful analysis of the combinatorics of transverse cocycles, one can show that

$$\left|\alpha\left(k_{d_n}\right)\right| \prec n.$$

It follows that the infinite sum in the Gap Formula is bounded by a geometric series, and therefore converges.

What the Gap Formula shows is that the map which associates a transverse cocycle to each transverse Hölder distribution is injective. Conversely, given a transverse cocycle for λ, we can use the Gap Formula to *define* a Hölder distribution transverse to λ. A little argument shows the invariance under homotopy of arcs transverse to λ, and concludes the proof of Theorem 11.

As an aside, we should mention another important ingredient of the proof of Proposition 12, namely the following general (and elementary) lemma. Recall that the *support* of a distribution α on the manifold X is the set of those points $x \in X$ which satisfy the following property: For every neighborhood U of x, there is a differentiable function $\varphi : X \to \mathbb{R}$ whose support is contained in U and such that $\alpha\,(\varphi) \neq 0$; see for instance [**Sc**]. Replacing "differentiable" by "Hölder continuous", this definition immediately extends to the case of a Hölder distribution α on a metric space X.

PROPOSITION 13. (Support Lemma) *If α is a Hölder distribution in a metric space X and if $\varphi : X \to \mathbb{R}$ is Hölder continuous, then $\alpha\,(\varphi)$ depends only on the restriction of φ to the support of α.*

This property is of course satisfied by measures. However, it is false for arbitrary distributions. For instance, the distribution on $\mathbb{R}$ which to a function $\varphi : \mathbb{R} \to \mathbb{R}$ associates its derivative at 0 does not satisfy this property. In some sense the Support Lemma shows that, although more general, Hölder distributions still have much in common with measures.

We can conclude this first part with an open question. We observed that a geodesic lamination λ admits, not only a Hölder transverse structure, but also a Lipschitz transverse structure. We can therefore generalize transverse Hölder distributions to transverse Lipschitz distributions for λ. A natural question is to ask whether this is a generalization at all, namely if a geodesic lamination can admit a transverse Lipschitz distribution which is not a transverse Hölder distribution. (Conjecture: No). The main obstacle to answering this question is that we do not know if the Support Lemma of Proposition 13 also applies to Lipschitz distributions. Of course, this leads to another question, which is to ask why anybody should care about this first question. Indeed, we will see in the second part that geodesic laminations can be defined as purely topological objects, and that their transverse Hölder structures are independent of the choice of a negatively curved metric on the surface S. However, the transverse Lipschitz structure of a geodesic lamination strongly depends on the metric; for instance, for the geodesic lamination of Figure 5, the lengths of the closed leaves of the geodesic lamination are completely determined by the transverse Lipschitz structure (compare Figure 15). Therefore, for someone who was hardwired as a topologist, such as the author of these notes, only geodesic laminations with transverse Hölder distributions really make sense, and transverse Lipschitz distributions are just a curiosity. On the other hand, for a geometer, if there is indeed a difference between Hölder and Lipschitz distributions, this could be a way to introduce very subtle invariants of a negatively curved metric on the surface S.

Part II. The topological viewpoint

From the point of view of the topologist, one of the motivations for the study of geodesic laminations on a surface S (with negative Euler characteristic $\chi(S)$) is to analyze the set of simple closed curves on S, considered up to deformation. By a deformation, we mean here a homotopy or an isotopy, since these two notions are equivalent for simple closed curves on surfaces [**Ba**][**Ep**][**Bu, Appendix**].

The idea is to endow S with an arbitrary metric of constant negative curvature -1 and of finite area, which always exists since $\chi(S)$ is negative.

Then, every simple closed curve can be deformed to a unique simple closed geodesic, unless it can be shrunk to a point or pushed to one of the cusps of the surface. So, essentially, this establishes a one-to-one correspondence between the set of deformation classes of simple closed curves and the set of closed geodesics, since we only loose finitely many uninteresting isotopy classes of simple closed curves. Let a simple closed curve in S be *essential* if it cannot be shrunk to a point or pushed to a cusp.

As observed in Part I, a simple closed geodesic can be interpreted as a geodesic lamination. This embeds the set

$$\mathcal{S}(S) = \{\text{essential simple closed curves in } S\}/\text{deformation}$$

in the set

$$\mathcal{L}(S) = \{\text{geodesic laminations in } S\}.$$

The set $\mathcal{L}(S)$ has a natural topology, coming from the Hausdorff metric on the space of closed non-empty subsets of S. Recall that the space of closed non-empty subsets of a metric space is compact in the topology induced by the Hausdorff metric, and it is relatively easy to check that $\mathcal{L}(S)$ is closed in this space. As a consequence, $\mathcal{L}(S)$ is compact. In particular, every sequence of simple closed curves has a subsequence which converges in $\mathcal{L}(S)$.

However, the following result, originally announced by Thurston in [**Th3, §10**] and proved by Xiadong Zhu and the author in [**Zh**][**ZhB**], shows that the space $\mathcal{L}(S)$ with this topology is totally disconnected, and in particular is not very convenient.

THEOREM 14. (Thurston, Zhu-Bonahon) *The space $\mathcal{L}(S)$ has Hausdorff dimension 0 with respect to the Hausdorff metric. In particular, the space $\mathcal{L}(S)$ is totally disconnected.*

A much better idea is to use geodesic laminations with transverse measures.

As a geodesic lamination, a simple closed geodesic γ admits a natural transverse measure α defined as follows: If A is a subset of an arc k transverse to γ, $\alpha(A)$ is equal to the number of points of $A \cap \gamma$. This embeds the set

$$\mathcal{S}(S) = \{\text{essential simple closed curves in } S\}/\text{deformation}$$

in

$$\mathcal{ML}(S) = \{\text{measured geodesic laminations in } S\}$$

where a *measured geodesic lamination* α consists of a compact geodesic lamination λ_α endowed with a transverse measure α whose support is equal to all of λ_α. It easily follows from the existence of α that λ_α has no infinite isolated leaf, namely that every connected component of λ_α is minimal.

The topology and piecewise linear structure of $\mathcal{ML}(S)$

An arc k is *generic (with respect to simple geodesics)* if it is transverse to every simple geodesic of S. We saw in Theorem 6 that the union of all simple geodesics has Hausdorff dimension 1. It follows that almost every geodesic arc is generic, and that every arc can be approximated by a generic arc.

If the arc k is generic, then every measured geodesic lamination $\alpha \in \mathcal{ML}(S)$ defines a measure on k, which we will still denote by α. In particular, for each continuous function $\varphi : k \to \mathbb{R}$, we can consider its integral $\alpha(\varphi) = \displaystyle\int \varphi \, d\alpha$ with respect to α.

Endow $\mathcal{ML}(S)$ with the topology defined by the family of semi-norms $\alpha \mapsto |\alpha(\varphi)|$, where $\varphi : k \to \mathbb{R}$ ranges over all continuous functions defined on transverse arcs. In particular, a sequence $\alpha_n \in \mathcal{ML}(S)$ converges to α if and only if $\alpha_n(\varphi)$ converges to $\alpha(\varphi)$ for every continuous function φ defined on a generic arc k. (By the standard approximation techniques, we can restrict attention to countably many functions φ to define the topology of $\mathcal{ML}(S)$, so that the topology is metrizable.)

PROPOSITION 15. (Thurston) *The positive real multiples of simple closed geodesics are dense in* $\mathcal{ML}(S)$.

In some sense, we can think of the passage from the set $\mathcal{S}(S)$ of simple closed curves to the topological space $\mathcal{ML}(S)$ as analogous to the passage from the lattice $\mathbb{Z}^n$ to the euclidean space $\mathbb{R}^n$. In particular, one of the motivations to introduce the space $\mathcal{ML}(S)$ is to analyze asymptotic directions when going to infinity in $\mathcal{S}(S)$.

If k is a generic arc and if $\alpha \in \mathcal{ML}(S)$, let $\alpha(k) \in \mathbb{R}$ be the α–integral of the constant function 1 on k.

THEOREM 16. (Thurston) *There exists a finite family* $k_1, k_2, \dots, k_n$ *such that the map* $\mathcal{ML}(S) \to \mathbb{R}^n$ *defined by* $\alpha \mapsto (\alpha(k_i))_{i=1,\dots,n}$ *induces a homeomorphism between* $\mathcal{ML}(S)$ *and a piecewise linear submanifold of* $\mathbb{R}^n$. *In addition, for every generic arc* k, *there exists a piecewise linear function which expresses* $\alpha(k)$ *in terms of the* $\alpha(k_i)$ *for every measured lamination* $\alpha \in \mathcal{ML}(S)$.

Variations of this result can be found in [**FLP**] or [**PeH**]. Recall that a map $\varphi : \mathbb{R}^p \to \mathbb{R}^q$ is *piecewise linear* if one can decompose $\mathbb{R}^p$ as the union of finitely many polyhedra so that the restriction of φ to each polyhedron is affine. A *piecewise linear submanifold* of $\mathbb{R}^n$ is a subset M such that, for every point $x \in M$, there is a piecewise linear homeomorphism from a neighborhood U of M to an open subset V of $\mathbb{R}^n$ which sends $U \cap M$ to $V \cap \mathbb{R}^p$. A *piecewise linear manifold* is a space M which is endowed with a family of charts which locally model M over open subsets of $\mathbb{R}^n$, such that the changes of charts are restrictions of piecewise linear maps $\mathbb{R}^n \to \mathbb{R}^n$.

An immediate consequence of Theorem 16 is that $\mathcal{ML}(S)$ admits the structure of a piecewise linear manifold.

THEOREM 17. (Thurston) *As a piecewise linear manifold, $\mathcal{ML}(S)$ is isomorphic to $\mathbb{R}^{3|\chi(S)|-b}$, where b is the number of ends of S and where $\chi(S)$ is its Euler characteristic.*

Again, references include [**FLP**] or [**PeH**].

Change of metric

To define measured geodesic laminations, we used a finite area hyperbolic metric m on the surface S. A natural question is to wonder what will happen if we use a different finite area hyperbolic metric m'.

LEMMA 18. *For every m–geodesic g, there is a unique m'–geodesic g' such that g can be deformed to g' by a homotopy which moves points by a uniformly bounded distance. In addition, if the two m–geodesics g, h are simple and disjoint, then the corresponding m'–geodesics g', h' are also simple and disjoint.*

See [**CEG, §4.1.4**]. The key property here is that an m–geodesic is quasi-geodesic for the metric m' which, because the curvature of m' is negative, implies that it stays within bounded distance of an m–geodesic.

Lemma 18 establishes a one-to-one correspondence between m–geodesic laminations and m'–geodesic laminations. In addition, if we interpret a transverse measure for a geodesic lamination λ as a locally defined measure on the space of leaves of λ, this correspondence also induces a one-to-one correspondence between measured m–geodesic laminations and measured m'–geodesic laminations.

With somewhat more work, it can also be shown that the piecewise linear structure of $\mathcal{ML}(S)$ is independent of the metric m.

As an addendum which will be important later on, we should mention that a consequence of Lemma 18 and of an estimate of hyperbolic geometry is that the correspondence between m–geodesics and m'–geodesics is Hölder continuous. In particular, each transverse Hölder distribution for the m–geodesic lamination λ gets transported to a transverse Hölder distribution for the corresponding m'–geodesic lamination λ'. In other words, geodesic laminations with transverse Hölder distributions are metric independent objects.

The length function

Many notions defined for deformation classes of simple closed curves extend to measured geodesic laminations. A typical example is the length function l_m associated to a hyperbolic metric m.

Fix a hyperbolic metric m on S. (This hypothesis can be relaxed to allow m to be a complete metric of negative curvature). For a simple closed curve γ, define

$$l_m(\gamma) = \inf_{\gamma' \sim \gamma} m - \text{length}(\gamma'),$$

where the infimum is taken over all simple closed curves γ' homotopic to γ.

PROPOSITION 19. (Thurston) *The length function uniquely extends to a continuous function*

$$l_m : \mathcal{ML}(S) \to \mathbb{R}^+$$

that is positively homogeneous, namely such that $l_m(a\alpha) = a l_m(\alpha)$ *for every* $a \in \mathbb{R}^+$ *and* $\alpha \in \mathcal{ML}(S)$, *where* $a\alpha$ *is obtained by multiplying the transverse measure of* α *by* a.

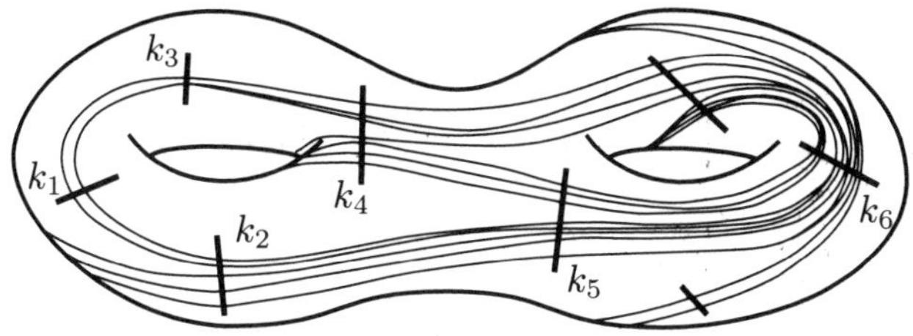

FIG. 18. Measuring the length $l_m(\alpha)$ of a measured geodesic lamination α

The length $l_m(\alpha)$ of a measured geodesic lamination $\alpha \in \mathcal{ML}(S)$ is defined as follows. Pick a finite family of arcs $k_1, \ldots, k_n$ transverse to the geodesic lamination λ_α underlying α, so that $\lambda_\alpha - \bigcup_{i=1}^n k_i$ consists of arcs of finite length. Each component of $\lambda_\alpha - \bigcup_{i=1}^n k_i$ is characterized by (any one of) its end points in $\bigcup_{i=1}^n k_i$, and the measure defined by α on the k_i consequently induces a measure on the space of these finite length components of $\lambda_\alpha - \bigcup_{i=1}^n k_i$. Then $l_m(\alpha)$ is defined as the integral of the lengths of these arcs with respect to the measure defined by α. The invariance of α under homotopy of arcs transverse to λ_α easily implies that $l_m(\alpha)$ is independent on the choice of the k_i. If we choose the k_i generic with respect to simple geodesics and so that every simple geodesic meets at least one of them, the continuity of $l_m : \mathcal{ML}(S) \to \mathbb{R}^+$ immediately follows from the definition of the topology of $\mathcal{ML}(S)$.

This length function l_m plays an important rôle in many branches of low-dimensional topology and geometry.

In this context, a natural question is to ask whether the function $l_m : \mathcal{ML}(S) \to \mathbb{R}^+$ might be differentiable. If l_m is differentiable at $\alpha \in \mathcal{ML}(S)$, it should have a differential $T_\alpha\alpha : T_\alpha\mathcal{ML}(S) \to \mathbb{R}$ defined on the tangent space $T_\alpha\mathcal{ML}(S)$ of $\mathcal{ML}(S)$ at α.

However, we now encounter a conceptual problem in defining this tangent space $T_\alpha\mathcal{ML}(S)$. Indeed, we only saw that $\mathcal{ML}(S)$ is a piecewise linear manifold, not a differentiable manifold. In fact, it can be shown that, except when S is the three times punctured sphere, there is no homeomorphism between $\mathcal{ML}(S)$ and a differentiable manifold for which the natural action of the diffeomorphism group of S corresponds to a differentiable action.

Tangent vectors to $\mathcal{ML}(S)$

Fortunately, a piecewise linear manifold classically admits a well-defined tangent space at each point, although its properties are weaker than that of the tangent space of a differentiable manifold. The definition mimics the differentiable context. A *tangent vector* of $\mathcal{ML}(S)$ at α is an equivalence class of curves

$\gamma : [0, \varepsilon[\to \mathcal{ML}(S)$ with $\gamma(0) = \alpha$ and with the following property: If $\varphi : U \to V$ is a chart of the piecewise linear structure of $\mathcal{ML}(S)$, identifying a neighborhood U of α in $\mathcal{ML}(S)$ to an open subset V of $\mathbb{R}^n$, then the curve $\varphi \circ \gamma$ admits a tangent vector at 0; two such curves γ and γ' are equivalent if and only if $\varphi \circ \gamma$ and $\varphi \circ \gamma'$ have the same tangent vector at 0 in $\mathbb{R}^n$. Since the change of charts are piecewise linear, these properties are clearly independent of the choice of the chart φ.

The space $T_\alpha \mathcal{ML}(S)$ of tangent vectors of $\mathcal{ML}(S)$ at α is homeomorphic to $\mathbb{R}^n$, and admits a natural multiplicative action of the group $\mathbb{R}^+$ of positive integers. However, in general, it admits no natural structure of vector space, or even of multiplication by -1.

This combinatorial definition of tangent vectors is easy to introduce but, for applications, we would like to have a more geometric interpretation of them. Namely, given a curve $t \mapsto \alpha_t \in \mathcal{ML}(S)$, we want to analyze the tangent vector $\dot{\alpha}_0 = d\alpha_t/dt^+{}_{|t=0}$ in a very geometric way.

For this, we can heuristically study a very specific example. Let α_t be the geodesic lamination which is associated to the suspension of the exchange of two intervals of length 1 and t, namely the interval exchange $[0, 1+t] \to [0, 1+t]$ which sends $[0, 1]$ to $[t, 1+t]$ and $[1, 1+t]$ to $[0, t]$. The suspension of this interval exchange is a punctured torus.

The structure of α_t and of its underlying geodesic lamination λ_t is in general relatively complex. However, it is easy to describe for $t = 1/n$ with n a positive integer. Indeed, all orbits of the interval exchange map are then closed, and $\alpha_{1/n}$ consists of a single closed geodesic $\lambda_{1/n}$ going n times around one direction of the punctured torus and once around the other one, and endowed with the transverse measure which is $1/n$ times the Dirac transverse measure. For $t = 0$, α_0 consists of the closed geodesic λ_0 going once around the first direction of the punctured torus. However, as n tends to ∞ (so that $\alpha_{1/n}$ tends to α_0), the geodesic lamination $\lambda_{1/n}$ converges for the Hausdorff topology to the geodesic lamination λ_0^+ consisting of the closed geodesic λ_0 and of an additional infinite geodesic whose two ends spiral around λ_0. See Figure 20. Note that we already encountered the geodesic lamination λ_0^+ in Figure 16.

Consider the arc k represented on Figure 19. Its intersection with λ_0 consists of a single point x_∞^∞. Its intersection with λ_0^+ consists of x_∞^∞ and of infinitely many points $x_1^\infty, x_2^\infty, \ldots, x_i^\infty, \ldots$ which accumulate on x_∞^∞. For every n, $k \cap \lambda_{1/n}$ consists of n points $x_1^n, x_2^n, \ldots, x_n^n$. In addition, it is possible to choose the indexings so that, as n tends to ∞, the point x_i^n converges to x_i^∞.

Let us perform a formal computation. For a continuous function $\varphi : k \to \mathbb{R}$, its integrals with respect to the measures defined by α_0 and $\alpha_{1/n}$ on k are

$$\alpha_0(\varphi) = \varphi(x_\infty^\infty)$$

and

$$\alpha_{1/n}(\varphi) = \frac{1}{n} \sum_{i=1}^{n} \varphi(x_i^n).$$

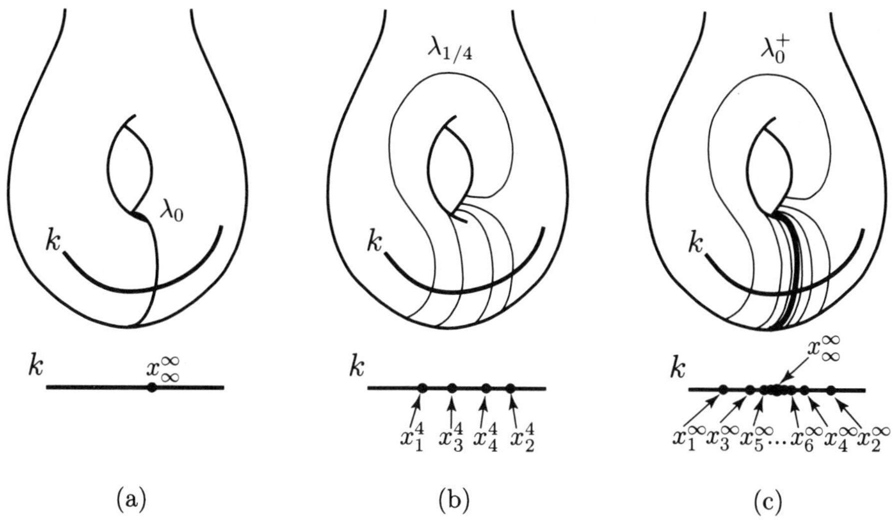

FIG. 19. A formal computation

If the derivative $\dot{\alpha}_0 = d\alpha_t/dt^+{}_{|t=0}$ exists, it should associate to φ the quantity

$$\begin{aligned}
\dot{\alpha}_0(\varphi) &= \lim_{t\to 0^+} \frac{\alpha_t(\varphi) - \alpha_0(\varphi)}{t} \\
&= \lim_{n\to\infty} n\left(\alpha_{1/n}(\varphi) - \alpha_0(\varphi)\right) \\
&= \lim_{n\to\infty} \sum_{i=1}^{n} \varphi(x_i^n) - n\varphi(x_\infty^\infty) \\
&= \lim_{n\to\infty} \sum_{i=1}^{n} \left(\varphi(x_i^n) - \varphi(x_\infty^\infty)\right) \\
&= \sum_{i=1}^{\infty} \left(\varphi(x_i^\infty) - \varphi(x_\infty^\infty)\right)
\end{aligned}$$

since x_i^n tends to x_i^∞ as n tends to ∞. Of course, this is only a formal computation. In particular, for a general continuous φ, there is no reason why the infinite sum of the last expression would even converge. However, a geometric estimate shows that the distances $d(x_i^\infty, x_\infty^\infty)$ are asymptotic to e^{-Li}, where $L > 0$ is the length of the closed geodesic λ_0. Compare Proposition 8. If we impose in addition that $\varphi : k \to \mathbb{R}$ is Hölder continuous, then the above infinite sum is consequently bounded by a converging geometric series. A little more work enables one to rigorously justify the above computation, and to prove that

$$\dot{\alpha}_0(\varphi) = \lim_{t\to 0^+} \frac{\alpha_t(\varphi) - \alpha_0(\varphi)}{t} = \sum_{i=1}^{\infty} \left(\varphi(x_i^\infty) - \varphi(x_\infty^\infty)\right)$$

when $\varphi : k \to \mathbb{R}$ is Hölder continuous. Note that we already encountered this transverse Hölder distribution for the geodesic lamination of Figure 16.

A similar argument can be made on every arc transverse to λ_0^+, and provides a transverse Hölder distribution for λ_0^+.

This heuristic argument in one specific example can actually be extended to the full setting of tangent vectors to $\mathcal{ML}(S)$.

THEOREM 20. [**Bo1**] *Let the curve $t \mapsto \alpha_t \in \mathcal{ML}(S)$ have a tangent vector $\dot\alpha_0 \in T_{\alpha_0}\mathcal{ML}(S)$ at α_0. Then, for every Hölder continuous function $\varphi : k \to \mathbb{R}$ defined on a generic arc k, the right derivative $d\alpha_t(\varphi)/dt^+_{|t=0}$ exists and defines a Hölder distribution on k.*

In addition, as k ranges over all generic arcs, the support of the corresponding Hölder distributions form a geodesic lamination λ_0^+, and $\dot\alpha_0$ defines a transverse Hölder distribution for λ_0^+.

In this way, we have associated to each tangent vector $\dot\alpha_0 \in T_{\alpha_0}\mathcal{ML}(S)$ a geodesic lamination λ_0^+ endowed with a transverse Hölder distribution. The following proposition provides a converse, by characterizing which geodesic laminations with transverse distributions correspond to tangent vectors to $\mathcal{ML}(S)$.

PROPOSITION 21. [**Bo1**] *Let α be a transverse Hölder distribution for the geodesic lamination λ, and assume that the support of α is all of λ. This data corresponds to a tangent vector $\dot\alpha_0 \in T_{\alpha_0}\mathcal{ML}(S)$ at $\alpha_0 \in \mathcal{ML}(S)$ if and only if:*

(i) *no leaf of λ crosses the geodesic lamination λ_0 underlying α_0, namely their union is a geodesic lamination;*

(ii) *every infinite isolated half-leaf g of λ is asymptotic to a component of λ_0, and $\alpha(k) \geqslant 0$ for an (arbitrary) transverse arc k with $k \cap \lambda = k \cap g = 1$ point;*

(iii) *for every minimal sublamination λ_1 of λ that is not in the geodesic lamination λ_0 underlying α_0, the restriction of α to λ_1 is a transverse measure.*

The conditions of Proposition 21 just reflect the fact that, if $t \mapsto \alpha_t$ is tangent to α at α_0, then $\alpha_t(k) \geqslant 0$ for every transverse arc k and every t.

For almost every $\alpha_0 \in \mathcal{ML}(S)$, the geodesic lamination λ_0 underlying α_0 is maximal for inclusion among all geodesic laminations. In this case, the conditions of Proposition 21 are automatically satisfied, and it follows that $T_{\alpha_0}\mathcal{ML}(S)$ is equal to the vector space $\mathcal{H}(\lambda_0)$ of all transverse Hölder distributions or, equivalently, of all transverse cocycles for λ_0.

On the other hand, when the geodesic lamination λ_0 underlying α_0 is not maximal, every chain recurrent geodesic lamination λ containing λ_0 determines a linear face of $T_{\alpha_0}\mathcal{ML}(S)$, for the piecewise linear structure of $\mathcal{ML}(S)$. This face is naturally isomorphic to a convex cone in $\mathcal{H}(\lambda)$, bounded by finitely many hyperplanes.

The derivative of the length function

We are now ready to compute the tangent map

$$T_{\alpha_0} l_m : T_{\alpha_0}\mathcal{ML}(S) \to \mathbb{R}$$

of the length function

$$l_m : \mathcal{ML}(S) \to \mathbb{R}^+$$

at $\alpha_0 \in \mathcal{ML}(S)$. By definition, if this tangent map exists, $T_{\alpha_0} l_m(\dot{\alpha}_0)$ should be equal to $dl_m(\alpha_t)/dt^+{}_{|t=0}$ for every curve $t \mapsto \alpha_t$ tangent to the vector $\dot{\alpha}_0 \in T_{\alpha_0}\mathcal{ML}(S)$ at α_0.

When we defined the length of a measured geodesic lamination α, we cut the underlying geodesic lamination λ_α along a family of transverse arcs $k_1, \ldots, k_n$ so that $\lambda_\alpha - \bigcup_i k_i$ consisted of arcs of finite length, noted that each component of $\lambda_\alpha - \bigcup_i k_i$ is characterized by (any one of) its end points in $\bigcup_i k_i$, and defined the length $\lambda_m(\alpha)$ as the integral of the lengths of the components with respect to the measure defined by α on the k_i. However, not only is the length of a component of $\lambda_\alpha - \bigcup_i k_i$ a continuous function of each of its end points, Lemma 2 actually shows that this length is a Lipschitz continuous function of this end points. In particular, the same process can be used to define the length $l_m(\alpha)$ when α is only a Hölder distribution for the geodesic lamination λ_α.

Because the construction is so natural, the following immediately follows from Theorem 20.

THEOREM 22. [**Bo1**] *Let the curve $t \mapsto \alpha_t \in \mathcal{ML}(S)$ be tangent to the vector $\dot{\alpha}_0 \in T_{\alpha_0}\mathcal{ML}(S)$ at α_0. If we interpret the tangent vector $\dot{\alpha}_0$ as a geodesic lamination with a transverse Hölder distribution, then*

$$dl_m(\alpha_t)/dt^+{}_{|t=0} = l_m(\dot{\alpha}_0).$$

In other words, the tangent map $T_{\alpha_0} l_m$ of the length function

$$l_m : \mathcal{ML}(S) \to \mathbb{R}^+$$

at α_0 coincides with the length function

$$l_m : T_{\alpha_0}\mathcal{ML}(S) \to \mathbb{R}$$

defined by interpreting the vectors of $T_{\alpha_0}\mathcal{ML}(S)$ as geodesic laminations with transverse Hölder distributions.

PART III. THE GEOMETRIC VIEWPOINT

As indicated in the introduction, geodesic laminations occur in many geometric contexts. We will focus on only one example: pleated surfaces and convex cores of hyperbolic 3–manifolds.

The convex core of a hyperbolic 3–manifold

Let M be a hyperbolic 3–manifold, namely a 3–dimensional manifold with a complete Riemannian metric of constant curvature -1. In addition, we assume that M is non-compact.

The *convex core* $C(M)$ of M is the smallest closed convex subset of M. Here a subset A of M is *convex* if every geodesic arc of M whose end points are contained in A is completely contained in A; in particular, a single point is in general not convex. The convex core $C(M)$ exists and is unique, except in a few degenerate cases which we will exclude; see for instance [**Th1**][**EpM**].

Here is another description of the convex core $C(M)$. Endowed with the lift of the metric of M, the universal covering $\widetilde{M}$ of M must be isometric to the hyperbolic

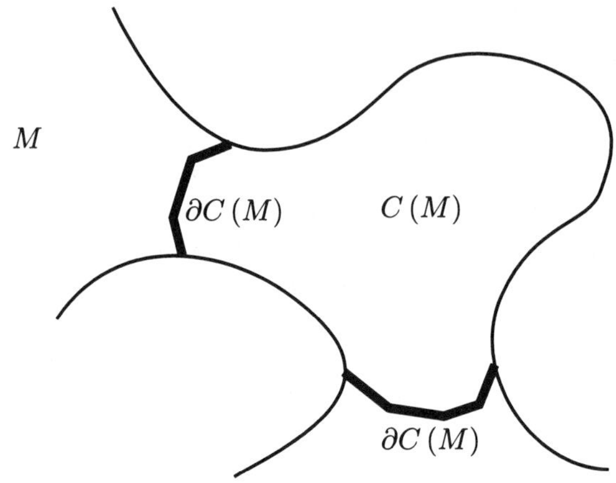

FIG. 20. The convex core $C(M)$ of a hyperbolic 3–manifold M

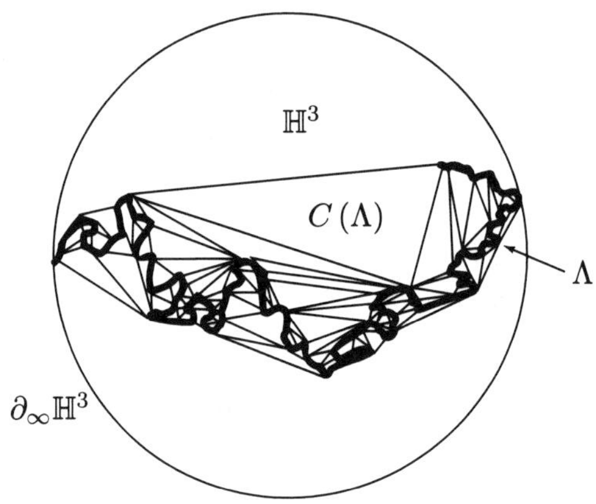

FIG. 21. The convex hull $C(\Lambda)$ of the limit set Λ

space $\mathbb{H}^3$, as this is the only simply connected hyperbolic 3–manifold. It follows that M is isometric to the quotient of $\mathbb{H}^3$ under an isometric action of the fundamental group $\pi_1(M)$.

In the union of $\mathbb{H}^3$ and of its sphere at infinity $\partial_\infty \mathbb{H}^3$, the *limit set* $\Lambda \subset \partial_\infty \mathbb{H}^3$ of M is defined as the set of cluster points of any orbit of the action of $\pi_1(M)$ on $\mathbb{H}^3$. We can then consider the *convex hull* $C(\Lambda)$ of Λ, namely the smallest convex subset of $\mathbb{H}^3 \cup \partial_\infty \mathbb{H}^3$ containing Λ. If we use the projective model for $\mathbb{H}^3$, in which geodesics correspond to euclidean line segments of $\mathbb{R}^3$, then $C(\Lambda)$ just corresponds to the euclidean convex hull of Λ. It is now easy to see that

$$C(M) = \left(C(\Lambda) - \partial_\infty \mathbb{H}^3\right) / \pi_1(M)$$

in $M = \mathbb{H}^3/\pi_1(M)$, except in the very degenerate cases where Λ consists of 0, 1 or 2 points, which we henceforth exclude.

One of the advantages of this construction is that it provides some insight on the geometry of the boundary $\partial C(M)$, which is locally the same as the boundary of $C(\Lambda)$.

First of all $\partial C(M)$ topologically is a surface.

Its local geometry is reminiscent of that of the boundary of a convex polyhedron, in the sense that it is a union of disjoint embedded geodesics of M (the 'edges') and of embedded totally geodesic pieces bounded by geodesics (the 'faces'). However, there are no points corresponding to the vertices of a polyhedron, as these points are now 'at infinity', in Λ.

In addition, exactly as on the boundary of a hyperbolic polyhedron away from the vertices, the path metric induced by the metric of M on $\partial C(M)$ is a (2–dimensional) hyperbolic metric. If, in addition, the fundamental group $\pi_1(M)$ is finitely generated, a theorem of Ahlfors implies that $\partial C(M)$ has finite area.

In sharp contrast with the case of a finite polyhedron, the family of 'edges' along which $\partial C(M)$ is not totally geodesic is not necessarily locally finite. However, what we know about the structure of geodesic laminations provides us with some information. Indeed, the union of these edges forms a geodesic lamination on the hyperbolic surface $\partial C(M)$, called the *bending locus* of $\partial C(M)$. From Theorem 6, we conclude that the bending locus has Hausdorff dimension 1, in spite of the fact that it may consist of uncountably many geodesics. We should emphasize that this local property of $\partial C(M)$ depends on global information, and that it is false if we do not require the fundamental group $\pi_1(M)$ to be finitely generated.

The amount by which $\partial C(M)$ actually bends along the bending locus is measured by a transverse measure for the bending locus, as explained below. The bending locus together with this transverse measure is the *bending measured lamination* of $\partial C(M)$.

Again, [**Th1,** §**8**] and [**EpM**] are good references for these facts.

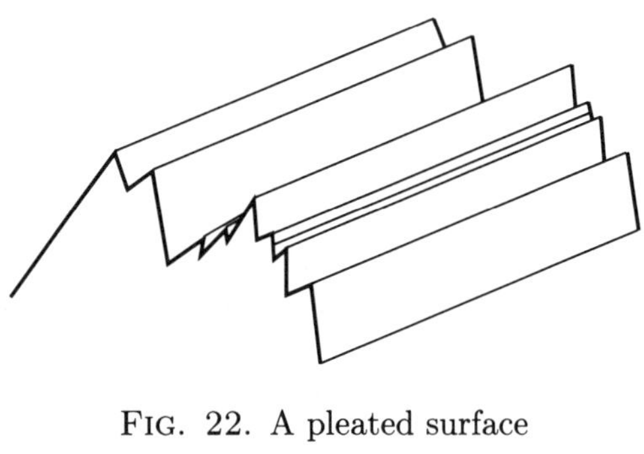

FIG. 22. A pleated surface

Pleated surfaces in hyperbolic 3–manifolds

The boundary of the convex core is a special example of a pleated surface. A *pleated surface* in a hyperbolic 3–manifold M consists of a hyperbolic surface S and of a map $f : S \to M$ such that

1. f sends each continuous arc in S to an arc of the same length in M;
2. there is a geodesic lamination $\lambda \subset S$ (called a *pleating locus* for f) such that f sends each leaf of λ to a geodesic of M and is totally geodesic on $S - \lambda$;

3. when S is non-compact, f is *cusp-preserving* in the sense that it sends each small neighborhood of a cusp of S to a small neighborhood of a cusp of M.

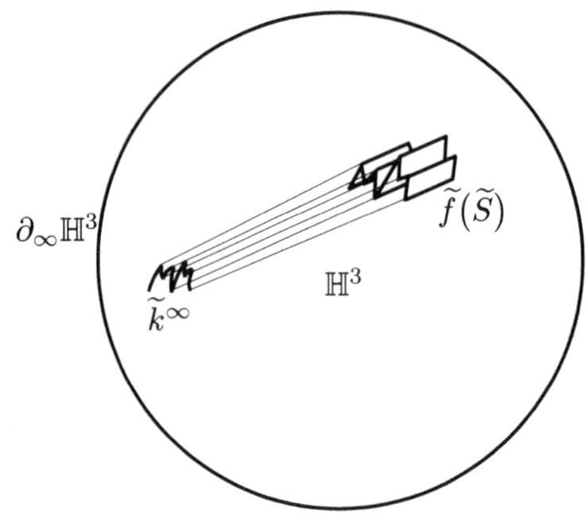

FIG. 23. Measuring the bending of a pleated surface

We now explain how to measure the bending of the pleated surface $f : S \to M$ along λ.

Consider an arc k transverse to λ in S. We want to measure the amount of bending which occurs along k. It is convenient to lift the situation to a map $\widetilde{f} : \widetilde{S} \to \widetilde{M}$ between the universal coverings of S and M, and to isometrically identify $\widetilde{M}$ to the hyperbolic space $\mathbb{H}^3$. Lift k to an arc $\widetilde{k}$ in $\widetilde{S}$, and let $\widetilde{\lambda} \subset \widetilde{S}$ be the pre-image of λ in $\widetilde{S}$.

We can "project $\widetilde{k}$ to $\partial_\infty \mathbb{H}^3$ parallel to $\widetilde{f}(\widetilde{S})$ " as follows. Pick an orientation for $\widetilde{k}$, and orient the leaves of $\widetilde{\lambda}$ meeting $\widetilde{k}$ from left to right (using the orientations of k and S). To each $x \in \widetilde{k} \cap \widetilde{\lambda}$ we can then associate the positive end point $x^\infty \in \partial_\infty \mathbb{H}^3$ of the oriented geodesic $\widetilde{f}(g)$ of $\mathbb{H}^3$, where g is the leaf of $\widetilde{\lambda}$ passing through x. To extend this set of points to a curve in $\partial_\infty \mathbb{H}^3$ we need, for every component d of $\widetilde{k} - \widetilde{\lambda}$, to connect by an arc the two points x_d^∞, $y_d^\infty \in \partial_\infty \mathbb{H}^3$ associated to the two points x_d, $y_d \in \widetilde{k} \cap \widetilde{\lambda}$ of ∂d. Such an arc is provided by the totally geodesic plane of $\mathbb{H}^3$ which coincides with $\widetilde{f}(\widetilde{S})$ on a neighborhood of each point of $\widetilde{f}(d)$. This plane intersects the sphere at infinity along a round circle, and consequently specifies a circle arc d^∞ joining x_d^∞ to y_d^∞. Note that d^∞ may consist of a single point, when $x_d^\infty = y_d^\infty$; this will happen very often, for instance roughly 50% of the time when λ is maximal for inclusion, namely when its complement $S - \lambda$ consists only of infinite triangles. So far, we have implicitly assumed that d does not contain one of the end points of $\widetilde{k}$, so that the end points x_d^∞, $y_d^\infty \in \widetilde{k} \cap \widetilde{\lambda}$ are well-defined; for the two components containing one of the end points of $\widetilde{d}$, the same construction only associates a germ d^∞ of circle arc (of arbitrary length) associated to each such component d.

Let $\widetilde{k}^\infty$ be the curve in $\partial_\infty \mathbb{H}^3$ which is made up of the points x^∞ associated to all $x \in \widetilde{k} \cap \widetilde{\lambda}$ and of the circle arcs d^∞ associated to all components d of $\widetilde{k} - \widetilde{\lambda}$. Intuitively, the total amount by which $f(S)$ bends at the points of $k \cap \lambda$ is equal to the amount by which $\widetilde{k}^\infty$ turns at its 'corners', corresponding to points of $\widetilde{k} \cap \widetilde{\lambda}$. Since there may

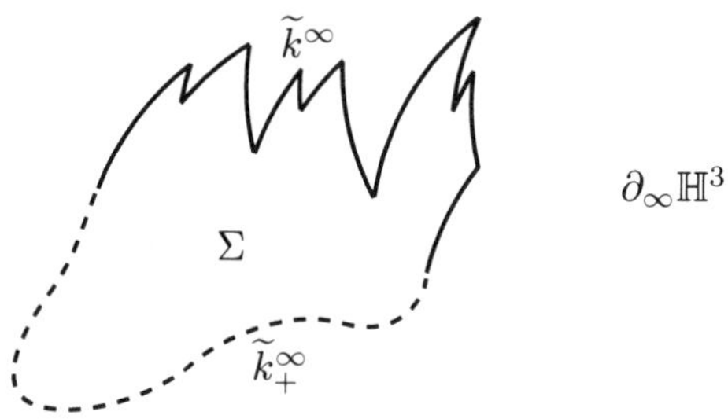

FIG. 24. Measuring the turning of the arc $\widetilde{k}^\infty$ at its corners

be uncountably many such corners, we will use an indirect arguments to measure this amount of turning.

Extend the arc $\widetilde{k}^\infty$ to a closed curve $\widetilde{k}_+^\infty$ in $\partial_\infty \mathbb{H}^3$, in such a way that $\widetilde{k}_+^\infty$ is smooth near the two end points of $\widetilde{k}^\infty$. The closed curve $\widetilde{k}_+^\infty$ bounds a cycle Σ in the sphere $\partial_\infty \mathbb{H}^3$, and we can consider the oriented area of this cycle for the standard metric of $\partial_\infty \mathbb{H}^3$ considered as the unit sphere in $\mathbb{R}^3$. Then, define

$$\beta(k) = \text{area}(\Sigma) - \int_{\widetilde{k}_+^\infty} \text{geodesic curvature of the smooth pieces} \\ \in \mathbb{R}/2\pi\mathbb{Z}.$$

The finiteness of the integral is easily proved. Since the area of the sphere is 4π, this $\beta(k) \in \mathbb{R}/2\pi\mathbb{Z}$ is clearly independent of the choice of the cycle Σ. Also, it easily follows from the Gauss-Bonnet formula that $\beta(k)$ is independent of the closed curve extension $\widetilde{k}_+^\infty$ of the arc $\widetilde{k}^\infty$.

When $\widetilde{k}^\infty$ has only finitely many corners, the Gauss-Bonnet formula shows that $\beta(k)$ is the sum of the turning angles at these corners. In the general case, we *define* the sum of the turning angles at the corners of $\widetilde{k}^\infty$ to be equal to $\beta(k)$.

One can check that $\beta(k)$ is independent of the orientation of k and of the choice of the lift $\widetilde{k}$ of k. Also, $k \mapsto \beta(k)$ is clearly finitely additive, and invariant under homotopy of arcs transverse to the pleating locus λ. As a consequence, this defines an $\mathbb{R}/2\pi\mathbb{Z}$–valued transverse cocycle $\beta \in \mathcal{H}(\lambda; \mathbb{R}/2\pi\mathbb{Z})$, called the *bending transverse cocycle* of the pleated surface $f : S \to M$. By definition, an $\mathbb{R}/2\pi\mathbb{Z}$–valued transverse cocycle associates an element of $\mathbb{R}/2\pi\mathbb{Z}$ to each arc transverse to λ, in a way which is finitely additive and invariant under homotopy of arcs transverse to λ. When the surface is non-compact, the bending transverse cocycle β satisfies an additional condition for every cusp c of S: Recall that c admits a neighborhood which meets λ only along finitely many isolated half-leaves $l_1, \dots, l_n$ converging towards c; then $\sum_{i=1}^n \beta(l_i) = 0 \in \mathbb{R}/2\pi\mathbb{Z}$, where $\beta(l_i)$ is the element of $\mathbb{R}/2\pi\mathbb{Z}$ associated by β to an arbitrary transverse arc k_i with $k_i \cap \lambda = k_i \cap l_i = 1$ point. Let $\mathcal{H}_0(\lambda; \mathbb{R}/2\pi\mathbb{Z})$ denote the space of those $\mathbb{R}/2\pi\mathbb{Z}$–valued transverse cocycles which satisfy this cusp relation for every cusp c of S.

Conversely, if we are given the hyperbolic metric m of S, the bending locus λ and the bending transverse cocycle in $\mathcal{H}_0(\lambda; \mathbb{R}/2\pi\mathbb{Z})$, we can completely reconstruct the local geometry of the pleated surface $f : S \to M$, the lift $\widetilde{f} : \widetilde{S} \to \widetilde{M} = \mathbb{H}^3$, and in particular the monodromy representation

$$\rho : \pi_1(S) \to \mathrm{Isom}\left(\mathbb{H}^3\right)$$

with respect to which $\widetilde{f}$ is equivariant. The proof is comparatively easy when β is a transverse measure (see [**EpM,** §**3**]), but technically harder in the general case [**Bo3**]. It uses an analog of the Gap Formula in the (non-commutative) group $\mathrm{Isom}\left(\mathbb{H}^3\right)$.

We can use this to locally parametrize near ρ the space of representations

$$\rho : \pi_1(S) \to \mathrm{Isom}\left(\mathbb{H}^3\right)$$

which are *cusp-preserving*, in the sense that ρ' sends the elements of $\pi_1(S)$ corresponding to cusps of S to parabolic elements of $\mathrm{Isom}\left(\mathbb{H}^3\right)$. For this, assume in addition (and without loss of generality) that λ is *maximal*, namely that it is maximal for inclusion among all geodesic laminations. This is equivalent to the property that its complement consists only of infinite triangles.

Deform the representation ρ to a nearby cusp-preserving representation

$$\rho' : \pi_1(S) \to \mathrm{Isom}\left(\mathbb{H}^3\right).$$

If ρ' is sufficiently close to ρ, there is a pleated surface $\widetilde{f}' : \widetilde{S} \to \mathbb{H}^3$ which still has pleating locus λ but is now equivariant with respect to ρ', and this pleated surface is unique modulo pre-composition by the lift of an isotopy of S. This pleated surface has hyperbolic metric

$$m' \in \mathcal{T}(S) = \{\text{hyperbolic metrics on } S\} / \text{isotopy}$$

and bending cocycle

$$\beta' \in \mathcal{H}_0(l; \mathbb{R}/2\pi\mathbb{Z}).$$

PROPOSITION 23. *The map $\rho' \mapsto (m', \beta')$ defines a diffeomorphism between a neighborhood of ρ in*

$$\mathcal{R} = \left\{\text{cusp} - \text{preserving representations } \pi_1(S) \to \mathrm{Isom}\left(\mathbb{H}^3\right)\right\} / \mathrm{Isom}\left(\mathbb{H}^3\right)$$

and a neighborhood of (m, β) in

$$\mathcal{T}(S) \times \mathcal{H}_0(\lambda; \mathbb{R}/2\pi\mathbb{Z}).$$

When λ is maximal, it can be shown that $\mathcal{H}_0(\lambda; \mathbb{R}/2\pi\mathbb{Z})$ consist of two copies of $(\mathbb{R}/2\pi\mathbb{Z})^{3|\chi(S)|-c}$, where c is the number of cusps of S.

The main point of Proposition 23 is that it provides a local parametrization of $\mathcal{R}$ by $\mathcal{T}(S) \times \mathcal{H}_0(\lambda; \mathbb{R}/2\pi\mathbb{Z})$ which is very well adapted to the geodesic lamination λ. We now discuss an application of this machinery to convex cores of hyperbolic 3–manifolds.

Variations of the geometry of convex cores

Let M_0 be a hyperbolic 3–manifold, and deform the metric of M_0 to M.

Under relatively mild hypotheses (for instance if M_0 is convex co-compact, or if M_0 is geometrically finite and we restrict to cusp-preserving deformations, or if we restrict to holomorphic deformations. etc ...), the convex core boundary $\partial C(M)$ has a natural topological identification with $\partial C(M_0)$, which is well-defined up to isotopy. If S is a component of $\partial C(M_0)$, the geometry of the corresponding component of $\partial C(M)$ is then described by its hyperbolic metric $m \in \mathcal{T}(S)$ and its bending measured lamination $\beta \in \mathcal{ML}(S)$.

The set

$$\mathcal{D} = \{\text{deformations of the hyperbolic metric of } M\}$$

suitably restricted as above so that the topology of $\partial C(M)$ does not change, is a differentiable manifold. Recall that the Teichmüller space $\mathcal{T}(S)$ is a differentiable manifold, and that the space $\mathcal{ML}(S)$ of measured geodesic lamination is a piecewise linear manifold. This leads to the following natural questions:

QUESTION 1. Is the map $\mathcal{D} \to \mathcal{T}(S)$ defined by $M \mapsto m$ differentiable

QUESTION 2. Is the map $\mathcal{D} \to \mathcal{ML}(S)$ defined by $M \mapsto \beta$ differentiable, in the appropriate sense?

To answer these questions, we take the problem backwards, mixing the differentiable and piecewise linear context.

If U is open in $\mathbb{R}^m$, the *tangent map* of $\varphi : U \to \mathbb{R}^n$ is, if it exists, the map $T_x\varphi : \mathbb{R}^m \to \mathbb{R}^n$ such that

$$T_x\varphi(v) = (\varphi \circ \alpha)'(0)$$

for every curve $\alpha : [0, \varepsilon[\to U$ with $\alpha(0) = x$ and $\alpha'(0) = v$. The map φ is *tangentiable* if it admits a tangent map everywhere. Typical examples of tangent maps include differentiable maps and piecewise linear maps, and these are exactly the examples which are relevant here.

This immediately leads to a notion of *tangentiable manifolds*, defined by atlases of charts where the changes of charts are all tangentiable. Typical examples are differentiable manifolds, piecewise linear manifolds or, as we will encounter later on, products of differentiable and piecewise linear manifolds. By an immediate translation of the corresponding definitions in the differentiable case, one can introduce notions of tangent vectors to a tangentiable manifold X at the point x, of tangent spaces T_xX, of tangentiable maps $f : X \to Y$ between tangentiable manifolds, and of tangent maps $T_xf : T_xX \to T_{f(x)}Y$ of tangentiable maps. We already encountered these notions when discussing the piecewise linear structure of $\mathcal{ML}(S)$. In particular, the tangent space T_xX is homeomorphic to $\mathbb{R}^n$, where n is the dimension of X, but does not necessarily admit a natural vector space structure.

The key to computing tangent maps is the following very elementary lemma.

LEMMA 24. *If a local homeomorphism φ is tangentiable at x and if the tangent map $T_x\varphi$ is injective, then φ^{-1} is tangentiable at $\varphi(x)$ and $T_{\varphi(x)}\varphi^{-1} = (T_x\varphi)^{-1}$. (The tangent map of a local homeomorphism is always surjective, by a degree argument.)*

We want to show that the map $\mathcal{D} \to \mathcal{T}(S) \times \mathcal{ML}(S)$ defined by $M \mapsto (m, \beta)$ is tangentiable at m_0.

For this, it is convenient to consider the space

$$\mathcal{R} = \left\{\text{cusp} - \text{preserving representations } \pi_1(S) \to \text{Isom}\left(\mathbb{H}^3\right)\right\} / \text{Isom}\left(\mathbb{H}^3\right)$$

and the map $\mathcal{D} \to \mathcal{R}$ which to a hyperbolic structure M associates $\rho \circ i$, where $\rho : \pi_1(S) \to \text{Isom}\left(\mathbb{H}^3\right)$ is the monodromy of M and where $i : \pi_1(S) \to \pi_1(M)$ is the homomorphism induced by the inclusion map. When discussing Proposition 23, we also explained how, given a hyperbolic metric $m \in \mathcal{T}(S)$ and a measured geodesic lamination $\beta \in \mathcal{ML}(S)$, we can reconstruct a pleated surface $\widetilde{f} : \widetilde{S} \to \mathbb{H}^3$ which is equivariant with respect to a representation $\rho : \pi_1(S) \to \text{Isom}\left(\mathbb{H}^3\right)$, and which induces on S the hyperbolic metric m and the bending measured lamination β. This defines a map

$$\varphi : \mathcal{T}(S) \times \mathcal{ML}(S) \to \mathcal{R}.$$

In his study of complex projective structures on surfaces (see [**KaT**]), Thurston showed that φ is a local homeomorphism. It is then immediate that the map $\mathcal{D} \to \mathcal{T}(S) \times \mathcal{ML}(S)$ we are interested in coincides near M_0 with the composition of the map $\mathcal{D} \to \mathcal{R}$ and of a local inverse φ^{-1} of φ.

By Proposition 23, for instance, the space $\mathcal{R}$ is a differentiable manifold near ρ_0. Also, the map $\mathcal{D} \to \mathcal{R}$ is a differentiable map between differentiable manifolds, by construction. Therefore, to prove that the map $\mathcal{D} \to \mathcal{T}(S) \times \mathcal{ML}(S)$ is tangentiable at M_0, it suffices to show that the local inverse φ^{-1} is tangentiable at ρ_0. By Lemma 24, we only need to show that φ is tangentiable at the point $(m_0, \beta_0) \in \mathcal{ML}(S)$ corresponding to M_0, and that the tangent map $T_{(m_0,\beta_0)}\varphi$ is injective.

These properties are proved in [**Bo4**], by carefully analyzing the proof of Proposition 23. It follows that the map $\mathcal{D} \to \mathcal{T}(S) \times \mathcal{ML}(S)$ is tangentiable. Composing with the appropriate projections, we conclude that the maps $\mathcal{D} \to \mathcal{T}(S)$ and $\mathcal{D} \to \mathcal{ML}(S)$ are tangentiable at M_0. A close examination of the proof actually shows that the tangent map $T_{M_0}\mathcal{D} \to T_{m_0}\mathcal{T}(S)$ is linear (which makes sense since $\mathcal{D}$ and $\mathcal{T}(S)$ are differentiable manifolds), and varies continuously with M_0; in other words, the map is continuously differentiable in the usual sense.

Letting S range over all components of $\partial C(M_0)$, this proves:

THEOREM 25. [**Bo4**] *The map $\mathcal{D} \to \mathcal{T}(\partial C(M_0))$, which associates to a deformation $M \in \mathcal{D}$ of the hyperbolic metric M_0 the hyperbolic metric $m \in \mathcal{T}(\partial C(M_0))$ of $\partial C(M) \cong \partial C(M_0)$, is continuously differentiable.*

THEOREM 26. [**Bo4**] *The map $\mathcal{D} \to \mathcal{ML}(\partial C(M_0))$, which associates to a deformation $M \in \mathcal{D}$ of the hyperbolic metric M_0 the bending measured lamination $\beta \in \mathcal{ML}(\partial C(M_0))$ of $\partial C(M) \cong \partial C(M_0)$, is tangentiable.*

Explicit examples show that the map $\mathcal{D} \to \mathcal{T}(S)$ is in general not C^2. This low differentiability can actually be attributed to the 'corners' of the piecewise linear structure of $\mathcal{ML}(S)$.

This machinery has a nice application to the function vol : $\mathcal{D} \to \mathbb{R}$ which associates to M the volume of its convex core $C(M)$.

Consider the bending map $b : \mathcal{D} \to \mathcal{ML}(\partial C(M_0))$. We saw that b is tangentiable. If $v \in T_{M_0}\mathcal{D}$ is a tangent vector of $\mathcal{D}$, we can therefore consider its image $T_{M_0}b(v) \in T_{b(M_0)}\mathcal{ML}(S)$ under the tangent map of b. By Theorem 20, this tangent vector $T_{M_0}b(v)$ can be interpreted as a geodesic lamination with a transverse

Hölder distribution, and consequently has a well-defined length $l_{m_0}(T_{M_0}b(v))$ for the hyperbolic metric m_0 of the boundary $\partial C(M_0)$.

THEOREM 27. **[Bo5]** *The convex core volume function* $\mathrm{vol} : \mathcal{D} \to \mathbb{R}$ *is tangentiable at* M_0 *and, for every tangent vector* $v \in T_{M_0}\mathcal{D}$,

$$T_{M_0}\mathrm{vol}(v) = \frac{1}{2} l_{m_0}(T_{M_0}b(v)).$$

This is the natural generalization of the well-known Schläfli formula for the variation of volumes of polyhedra in hyperbolic spaces.

Rotation angles, bending cocycles, and Thurston's intersection form

We conclude with a last application of the bending transverse cocycle of a pleated surface.

Let M be an oriented hyperbolic 3–manifold, and let γ be a closed geodesic of M. If we start with a vector v_x orthogonal to γ at x, we can extend the vector v_x to a parallel vector field v (orthogonal to γ) along γ. However, as we go around γ, there is no reason why v should return to the value v_x at x. Let $\mathrm{rot}_M(\gamma) \in \mathbb{R}/2\pi\mathbb{Z}$ be the angle between the new value v'_x of v at x and the original vector v_x, measured in the plane orthogonal to γ at x. It is immediate that $\mathrm{rot}_M(\gamma)$ is independent of the choices of x, v_x and of an orientation for γ.

By definition, $\mathrm{rot}_M(\gamma) \in \mathbb{R}/2\pi\mathbb{Z}$ is the *rotation angle* of the closed geodesic γ.

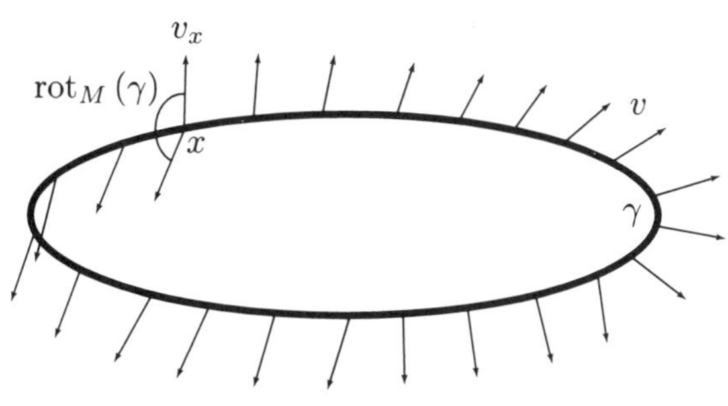

FIG. 25. Measuring the rotation angle of a closed geodesic

This definition can be extended to the wider context of measured geodesic laminations. Suppose that we are given a finite type surface S and a cusp-preserving map $f : S \to M$. A geodesic lamination λ in S is *realizable* if there exists a pleated surface $f' : S \to M$, homotopic to f through a family of cusp-preserving maps, which admits a pleating locus containing λ. In particular, f' sends each leaf of λ to a complete geodesic of M.

Let $\mathcal{O}_f \subset \mathcal{ML}(S)$ denote the set of those measured geodesic laminations whose underlying geodesic lamination is realizable. Under relatively mild hypotheses on f or M, the subset $\mathcal{O}_f$ is open in $\mathcal{ML}(S)$. For instance, $\mathcal{O}_f$ is equal to all of $\mathcal{ML}(S)$ if the convex core $C(M)$ is compact.

By re-interpreting the rotation angles of closed geodesics in the appropriate framework, one can show:

PROPOSITION 28. **[Bo3]** *There is a continuous function:*

$$\widetilde{\mathrm{rot}}_M : \mathcal{O}_f \to \mathbb{R}$$

such that

$$\widetilde{\mathrm{rot}}_M(\gamma) = \mathrm{rot}_M(\gamma^*) \mod 2\pi$$

for every simple closed curve $\gamma \in \mathcal{O}_f \subset \mathcal{ML}(S)$, *where* γ^* *denotes the closed geodesic of* M *that is homotopic to* $f(\gamma)$, *and such that* $\widetilde{\mathrm{rot}}_M(a\alpha) = a\widetilde{\mathrm{rot}}_M(\alpha)$ *for every* $\alpha \in \mathcal{O}_f$, $a \in \mathbb{R}^+$.

Note that the last property obliges us to replace $\mathbb{R}/2\pi\mathbb{Z}$ by $\mathbb{R}$. The function $\widetilde{\mathrm{rot}}_M : \mathcal{O}_f \to \mathbb{R}$ is not unique and depends on certain choices, as there are several ways to lift rotation angles from $\mathbb{R}/2\pi\mathbb{Z}$ to $\mathbb{R}$ in a continuous way. However, compare Theorem 30 below and the remark thereafter.

The same construction, which enabled us to define a rotation angles for transverse measures for a realizable geodesic lamination λ, automatically extends to transverse Hölder distributions for λ, and defines a linear map

$$\widetilde{\mathrm{rot}}_M : \mathcal{H}(\lambda) \to \mathbb{R}$$

when the geodesic lamination λ stays away from the cusps of S. Note that this is always the case when λ is the underlying geodesic lamination of a measured geodesic lamination, or when λ is the underlying geodesic lamination of a tangent vector to $\mathcal{ML}(S)$ interpreted as a geodesic lamination with a transverse Hölder distribution.

As in the case of Theorem 22, the naturality of the construction gives in a quasi-automatic way the following differentiability result.

THEOREM 29. *The function* $\widetilde{\mathrm{rot}}_M : \mathcal{O}_f \to \mathbb{R}$ *of Proposition 28 is tangentiable at each* $\alpha \in \mathcal{O}_f$, *and the tangent map*

$$T_\alpha \widetilde{\mathrm{rot}}_M : T_\alpha \mathcal{ML}(S) \to \mathbb{R}$$

just associates to $v \in T_\alpha \mathcal{ML}(S)$ *the rotation angle* $\widetilde{\mathrm{rot}}_M(v)$, *if we interpret the tangent vector* v *as a geodesic lamination with a transverse Hölder distribution.*

When the geodesic lamination λ has leaves converging towards a cusp of S, one can still define a rotation angle $\widetilde{\mathrm{rot}}_M(\alpha)$ for a transverse cocycle $\alpha \in \mathcal{H}(\lambda)$ provided that α satisfies an additional condition: for every cusp c of S, the weights assigned by α to the finitely many half-leaves converging to c must add up to 0. The same construction as above, subject to the same choices, then defines a linear function

$$\widetilde{\mathrm{rot}}_M : \mathcal{H}_0(\lambda) \to \mathbb{R}$$

where $\mathcal{H}_0(\lambda)$ denotes the vector space of all transverse cocycles for λ which satisfy the above cusp condition.

It turns out that these rotation angles are related to the bending cocycle of a pleated surface through the Thurston intersection form, which has an algebro-topological flavor.

Intuitively, if α and β are two transverse cocycles for the same geodesic lamination λ, their Thurston intersection number $\tau(\alpha, \beta) \in \mathbb{R}$ is the algebraic intersection number of λ with multiplicity α and of λ with multiplicity β.

In general, to compute an algebraic intersection number between two 1–dimensional objects a and b on the surface S, one needs an orientation for a and b; after perturbing b to some b' in transverse position with respect to a, the

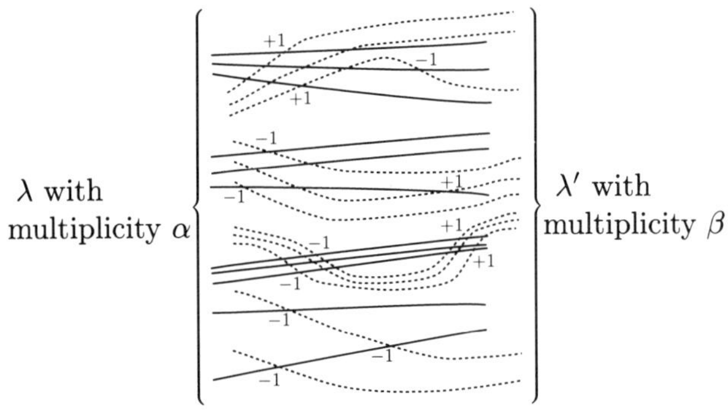

FIG. 26. Measuring the Thurston intersection number of α, $\beta \in \mathcal{H}_0(\lambda)$

algebraic intersection of a and b is then defined by adding up the signs of the intersection points of a and b', taking into account whatever multiplicities may have been assigned to a and b.

In our case, we do not have any orientation assigned to the geodesic lamination λ. However, we can proceed as follows. Choose a small C^1 perturbation from λ to a lamination λ', so that the leaves of λ' are transverse to the leaves of λ. The key property is that at each $x \in \lambda \cap \lambda'$, the leaf l' of λ' passing through x makes a very small angle with the leaf l of λ passing through x. If we choose an orientation for the leaf l at x, this uniquely determines an orientation for l' at x, namely the one for which their oriented tangent vectors make a small angle; this assigns a sign to the intersection of l and l' at x. If we reverse the orientation of l, the orientation of l' also gets reversed and the sign of the intersection at x is unchanged. This assign a well-defined sign ± 1 to each intersection point $x \in \lambda \cap \lambda'$. Then, if α and β are two transverse cocycles for λ, the Thurston intersection number $\tau(\alpha, \beta) \in \mathbb{R}$ is defined by taking the 'sum' of the signs of the intersection points of λ with multiplicity α and λ' with multiplicity β; since $\lambda \cap \lambda'$ may be uncountable, one needs to be a little careful when computing this sum, by grouping these intersection points in finitely many boxes in S. The correct framework to formalize this, and to make sure that the output does not depend on the choice of the perturbation of λ to λ', is to lift the situation to the orientation covering of the geodesic lamination λ; see [**PeH, §3.2**] or [**Bo3, §3**], for instance.

Actually, when λ has leaves converging towards a cusp of S, one needs an additional condition for α and β for $\tau(\alpha, \beta)$ to be well-defined: They need to satisfy the *cusp condition* that, for every cusp c of S, the weights assigned by α (resp. β) to the finitely many half-leaves converging to c add up to 0.

If $\mathcal{H}_0(\lambda)$ denotes the vector space of all transverse cocycles for λ which satisfy the above cusp condition, this defines the *Thurston intersection form*

$$\tau : \mathcal{H}_0(\lambda) \times \mathcal{H}_0(\lambda) \to \mathbb{R}.$$

This is an antisymmetric bilinear form. An easy exercise shows that τ is non-degenerate if and only if every end of $S - \lambda$ that is not a whole cusp of S has an odd number of spikes. In particular, τ is non-degenerate when λ is maximal.

THEOREM 30. [**Bo3**] *Let $f : S \to M$ be a pleated surface with bending locus λ and with bending transverse cocycle $\beta \in \mathcal{H}_0(\lambda; \mathbb{R}/2\pi\mathbb{Z})$. Then, for every choice of a rotation angle function*

$$\widetilde{\mathrm{rot}}_M : \mathcal{H}_0(\lambda) \to \mathbb{R}$$

there is a lift of β to an $\mathbb{R}$-valued transverse cocycle $\widetilde{\beta} \in \mathcal{H}_0(\lambda)$ such that

$$\widetilde{\mathrm{rot}}_M(\alpha) = \tau(\alpha, \beta)$$

for every $\alpha \in \mathcal{H}_0(\lambda)$.

Conversely, choosing a lift $\widetilde{\beta} \in \mathcal{H}_0(\lambda)$ for the bending transverse cocycle $\beta \in \mathcal{H}_0(\lambda; \mathbb{R}/2\pi\mathbb{Z})$ defines a rotation angle function $\widetilde{\mathrm{rot}}_M : \mathcal{H}_0(\lambda) \to \mathbb{R}$ by the property that $\widetilde{\mathrm{rot}}_M(\alpha) = \tau(\alpha, \beta)$. Making such a choice at the pleated surface realizing a measured geodesic lamination $\alpha \in \mathcal{O}_f$ in general determines in a unique way a function $\widetilde{\mathrm{rot}}_M : \mathcal{O}_f \to \mathbb{R}$ as in Proposition 28.

References

[Ba] R. Baer, *Isotopie von Kurven auf orientierbaren, geschlossenen Flächen und ihr Zusammenhang mit der topologischen Deformation der Flächen*, J. Reine Angew. Math. **159** (1928), 101–111.

[BiS] J.S. Birman, C. Series, *Geodesics with bounded intersection numbers on surfaces are sparsely distributed*, Topology **24** (1985), 217–225.

[Bo1] F. Bonahon, *Geodesic laminations with transverse Hölder distributions*, Ann. Sci. École Norm. Sup. **30** (1997), 205–240.

[Bo2] F. Bonahon, *Transverse Hölder distributions for geodesic laminations*, Topology **36** (1997), 103–122.

[Bo3] F. Bonahon, *Shearing hyperbolic surfaces, bending pleated surfaces, and Thurston's symplectic form*, Ann. Fac. Sci. Toulouse **5** (1996), 233-297.

[Bo4] F. Bonahon, *Variations of the boundary geometry of 3-dimensional hyperbolic convex cores*, J. Differential Geom. **50** (1998), 1–24.

[Bo5] F. Bonahon, *A Schläfli-type formula for convex cores of hyperbolic 3-manifolds*, J. Differential Geom. **50** (1998), 25–58.

[Bo6] F. Bonahon, *Closed curves on surfaces*, manuscript in progress.

[Bu] P. Buser, *Geometry and spectra of compact Riemann surfaces*, Progr. Math. vol. 106, Birkhäuser, Boston, 1992.

[CEG] R.D. Canary, D.B.A. Epstein, P. Green, *Notes on notes of Thurston*, Analytical and Geometrical aspects of Hyperbolic space (D.B.A. Epstein, ed.), L.M.S. Lecture Notes Series vol. 111, Cambridge Univ. Press, 1987, pp. 3–92.

[CaB] A. Casson, S.A. Bleiler, *Automorphisms of surfaces after Thurston and Nielsen*, Cambridge University Press, 1988.

[Eps] D.B.A. Epstein, *Curves on 2-manifolds and isotopies*, Acta Math. **115** (1966), 83–107.

[EpM] D.B.A. Epstein, A. Marden, *Convex hulls in hyperbolic spaces, a theorem of Sullivan, and measured pleated surfaces*, Analytical and geometric aspects of hyperbolic space (D.B.A. Epstein, ed.), L.M.S. Lecture Note Series vol. 111, Cambridge University Press, 1986, pp. 113–253.

[FLP] A. Fathi, F. Laudenbach, V. Poenaru, *Travaux de Thurston sur les surfaces*, Astérisque 66–67, Société Mathématique de France, 1979.

[KaT] Y. Kamishima, S. Tan, *Deformations spaces of geometric structures*, Aspects of low-dimensional manifolds (Y. Matsumo, S. Morita, eds.), Advanced Studies in Pure Math. vol. 20, Kinokuniya Company Ltd., Tokyo, Japan, 1992, pp. 263–299.

[Ke] M. Keane, *Interval exchange transformations*, Math. Zeit. **141** (1975), 25–31.

[PeH] R.C. Penner, J.L. Harer, *Combinatorics of train tracks*, Ann. Math. Studies vol. 125, Princeton Univ. Press, 1992.

[Pl] J. Plante, *Foliations with measure preserving holonomy*, Ann. Math. **127** (1975), 327–362.

[Sc] L. Schwartz, *Théorie des distributions*, Nouvelle édition, Hermann, Paris, 1966.

[Th1] W.P. Thurston, *The topology and geometry of 3–manifolds*, Lecture notes, Princeton University, 1976–79.

[Th2] W.P. Thurston, *On the geometry and dynamics of diffeomorphisms of surfaces*, Bull. Amer. Math. Soc. **19** (1988), 417–431.

[Th3] W.P. Thurston, *Minimal stretch maps between hyperbolic surfaces*, unpublished manuscript.

[Ze] A. Zeghib, *Laminations et hypersurfaces géodésiques des variétés hyperboliques*, Ann. Sci. Ec. Norm. Sup. **24** (1991), 171–188.

[Zh] X. Zhu, *Fractal dimensions of spaces of geodesic laminations*, PhD Thesis, Univ. of Southern California, 2000.

[ZhB] X. Zhu, F. Bonahon, in preparation.

Index

DEPARTMENT OF MATHEMATICS, UNIVERSITY OF SOUTHERN CALIFORNIA, LOS ANGELES CA 90089-1113, U.S.A.

E-mail address: fbonahon@math.usc.edu

Contemporary Mathematics
Volume **269**, 2001

Dicritical Singularities of Holomorphic Vector Fields

César Camacho

A differential equation of the form

$$w = a(x,y)dy - b(x,y)dx = 0$$

where $a(x,y)$ and $b(x,y)$ are holomorphic functions defined in an open subset $\mathcal{U} \subset \mathbb{C}^2$, induces a foliation $\mathcal{F}$ on $\mathcal{U}$ whose leaves are Riemann surfaces tangent at each point of $\mathcal{U}$ to the kernel of w. A singular point of $\mathcal{F}$ is a point $(x_0, y_0) \in \mathcal{U}$ where $a(x_0,y_0) = b(x_0,y_0) = 0$. Here we will be concerned with local properties of $\mathcal{F}$ near an isolated singular point that we assume to be $0 \in \mathbb{C}^2$.

The main tool in this study is the resolution theorem of Seidenberg [Sei] that establishes the existence of a finite sequence of maps (blow-up's)

$$\left(M^{(n)}, D^{(n)}\right) \xrightarrow{\pi^{(n)}} \dots \longrightarrow \left(M^{(1)}, D^{(1)}\right) \xrightarrow{\pi^{(1)}} (\mathbb{C}^2, 0)$$

with the following properties.

$M^{(1)}$ is a complex 2-manifold defined by two charts: $(x,t) \in \mathbb{C}^2$ and $(u,y) \in \mathbb{C}^2$ glued according to the rule $y = tx$, $t \neq 0$ and $x = uy$, $u \neq 0$. The map $\pi^{(1)}\colon M^{(1)} \to \mathbb{C}^2$ is defined by $\pi^{(1)}(x,t) = (x,y)$, $y = tx$, in the chart $(x,t) \in \mathbb{C}^2$ and by $\pi^{(1)}(u,y) = (uy,y)$ in the chart $(u,y) \in \mathbb{C}^2$.
The *divisor* $D^{(1)}$ is the subset of $M^{(1)}$ defined as $D^{(1)} := {\pi^{(1)}}^{-1}(0)$. Since $ut = 1$, $D^{(1)}$ is in fact an embedded Riemann sphere with self intersection number -1. Moreover $M^{(1)} \to D^{(1)}$ is a fiber space with projection $(x,t) \mapsto (0,t)$ in one chart and $(u,y) \mapsto (u,0)$ in the other. Clearly the restriction of $\pi^{(1)}$ to $M^{(1)} \backslash D^{(1)}$ is a complex diffeomorphism onto $\mathbb{C}^2 \backslash \{0\}$. The foliation $\mathcal{F}^{(1)}$ is then defined as the pull back by $\pi^{(1)}$ of $\mathcal{F}$; it is well defined in ${\pi^{(1)}}^{-1}(\mathcal{U})$. The foliation $\mathcal{F}^{(1)}$ is given by the pull back of w, ${\pi^{(1)}}^{*} w$, as follows. Near $0 \in \mathbb{C}^2$ w can be written as $w = w_\nu + w_{\nu+1} + \dots$, where w_j are 1-forms whose coefficients are homogeneous polynomials of degree j. In the chart (x,t), ${\pi^{(1)}}^{*} w$ is written as ${\pi^{(1)}}^{*} w = x^d w^{(1)}$ and in the chart (u,y), $\mathcal{F}^{(1)}$ is given by ${\pi^{(1)}}^{*} w = y^d w^{(2)}$ where d is such that w_1 and w_2 are holomorphic 1-forms with isolated singularities. Then the foliation $\mathcal{F}^{(1)}$ is given by $w^{(1)} = 0$ in chart (x,t) and $w^{(2)} = 0$ in chart (u,y) with the identification $w_1 = t^d w_2$, $t \neq 0$, in the intersection of these charts. It is easy to see that there are two possibilities: 1) The *dicritical* case, when $d = \nu + 1$. It is characterized by the fact that, with the exception of $\nu - 1$ points in $D^{(1)}$, all leaves of $\mathcal{F}^{(1)}$ are transverse to $D^{(1)}$. These exceptional points can be either tangency

---

2000 *Mathematics Subject Classification.* Primary 32S65, 37F75.

points of $\mathcal{F}^{(1)}$ with $D^{(1)}$ or singularities of $\mathcal{F}^{(1)}$ both counted with multiplicity. 2) The *nondicritical* case, when $d = \nu$. In this case there are $\nu + 1$ singularities of $\mathcal{F}^{(1)}$ in $D^{(1)}$, counted with multiplicities, and $D^{(1)}$ is tangent to $\mathcal{F}^{(1)}$.

The map $\pi^{(2)}: M^{(2)} \to M^{(1)}$ is the blow-up at a point $p \in D^{(1)}$. This point $p \in D^{(1)}$ is either a singular point of $\mathcal{F}^{(1)}$ or a point of tangency of $\mathcal{F}^{(1)}$ with $D^{(1)}$. Again $D^{(2)} = {\pi^{(2)}}^{-1}(p)$ is a projective line with selfintersection number -1 in $M^{(2)}$, crossing normally the embedded projective line $D^{(1)} = \overline{{\pi^{(2)}}^{-1}(D^{(1)} \setminus \{p\})}$ that we denote again by $D^{(1)}$. The effect of ${\pi^{(2)}}^{-1}$ on $D^{(1)}$ is to decrease by one its selfintersection number in $M^{(2)}$.

The foliation $\mathcal{F}^{(2)}$ is defined as the pull back of $\mathcal{F}^{(1)}$ by $\pi^{(2)}$ and, as before, $D^{(2)}$ can be either a dicritical component or a nondicritical component.

The resolution theorem states that after a finite number of these blow-up's, say n, at each singular point $q \in M^{(n)}$ of $\mathcal{F}^{(n)}$ there are coordinate charts (z, w) such that $z(q) = w(q) = 0$, $\mathcal{F}^{(n)}$ is given locally by the expression $B(z,w)dz - A(z,w)dw = 0$ and the Jacobian $\dfrac{\partial(A,B)}{\partial(z,w)}(0,0)$ has at least one nonzero eigenvalue. Moreover if $\lambda_1 \neq 0 \neq \lambda_2$ are eigenvalues of this matrix, then $\lambda_1/\lambda_2 \in \mathbb{Q}_+$. Singularities with exactly one nonzero eigenvalue of its Jacobian are called *saddle-nodes*. The divisor of the resolution is $D := \bigcup_{s=1}^{n} D^{(s)}$, a union of projective lines with normal crossing at points called *corners*. By [La] Theorem 4.4 the intersection matrix $\left(D^{(i)} \cdot D^{(j)}\right)$ is negative definite. The divisor D can be written also as $D = \bigcup_{i=1}^{k} \mathcal{D}_i \bigcup_{j=1}^{\ell} \mathcal{N}_j$, $k+\ell = n$, where the $\mathcal{D}_i$ are dicritical components and the $\mathcal{N}_j$ are nondicritical components. We can assume, by means of further blow-up's if necessary, that $\mathcal{F}^{(n)}$ has no singularities at the dicritical components and that any two dicritical components have empty intersection. Any connected component of $\bigcup_{j=1}^{\ell} \mathcal{N}_j$ is called a *nondicritical branch* of the resolution of $\mathcal{F}$. It follows from [La] see also [Mo] that

Lemma. The intersection matrix $(\mathcal{N}_i \cdot \mathcal{N}_j)$ of each nondicritical branch is negative definite.

The main invariant associated to each component $\mathcal{N}_j$ of a nondicritical branch is its *holonomy*. It is a representation $\varphi: \pi_1(\mathcal{N}_j \setminus \operatorname{sing} \mathcal{F}^{(n)}) \to \operatorname{Diff}(\mathbb{C}, 0)$ of the fundamental group of $\mathcal{N}_j \setminus \operatorname{sing} \mathcal{F}^{(n)}$ into the group of germs of diffeomorphisms of $\mathbb{C}$ fixing $0 \in \mathbb{C}$. Here $(\mathbb{C}, 0)$ can be realized as a local cross section to $\mathcal{N}_j$ at a regular point of $\mathcal{F}^{(n)}$, $q \in \mathcal{N}_j$. To each closed path $\alpha: [0,1] \to \mathcal{N}_j \setminus \operatorname{sing} \mathcal{F}^{(n)}$, $\alpha(0) = \alpha(1) = q$, a diffeomorphism $f_\alpha: (\mathbb{C}, 0) \hookleftarrow$ is associated as follows. Given $p \in (\mathbb{C}, 0)$ close enough to q, there is a unique path $\tilde{\alpha}$, $\tilde{\alpha}(0) = p$, with image contained in the leaf of $\mathcal{F}^{(n)}$ passing through p, and over α, i.e. $\tilde{\alpha}(t)$ and $\alpha(t)$ are in the same fiber over $\mathcal{N}_j$. Then $f_\alpha(p) = \tilde{\alpha}(1)$, and $\varphi([\alpha]) =$ germ f_α at $q \in \mathcal{N}_j$. The notion of holonomy of a nondicritical component can be extended to the holonomy of a component in a nondicritical branch. This notion is called *singular holonomy* was introduced in [Ce-Sc] and used in [Ca-Sc]. Essentially it is the holonomy over closed paths that are allowed to pass through corners of a branch. It is defined as follows, Suppose $B = \bigcup_{i=1}^{m} \mathcal{N}_i$ is a branch of $\mathcal{F}^{(n)}$ and assume that all the corners in the branch are singular points of the foliation $\mathcal{F}^{(n)}$, with a local *linear* expression. In other words, if $p = \mathcal{N}_i \cap \mathcal{N}_j$, then near p there are coordinates (x, y) such that $x(p) = y(p) = 0$, $\mathcal{N}_i = (x = 0)$, $\mathcal{N}_j = (y = 0)$ and $\mathcal{F}^{(n)}$: $\lambda y\, dx + x\, dy = 0$. We take points $q_i \in \mathcal{N}_i$, $q_j \in \mathcal{N}_j$ close to p and such that $q_i \neq p \neq q_j$. Let $(\Sigma_i, q_i), (\Sigma_j, q_j)$ be small cross sections to $\mathcal{F}^{(n)}$ such that $\Sigma_i \cap \mathcal{N}_i = q_i$ and $\Sigma_j \cap \mathcal{N}_j = q_j$. Suppose moreover that G_i (resp. G_j) is the holonomy group of $\mathcal{N}_i$ (resp. $\mathcal{N}_j$) at q_i (resp. q_j). The Dulac correspondence $\mathcal{D}$: $(\Sigma_i, q_i) \to (\Sigma_j, q_j)$ is obtained by following the local leaves of $\mathcal{F}^{(n)}$. In our situation it is given by $\mathcal{D}(x) = x^\lambda$. To an element $h \in G_i$ we associate an element $h^{\mathcal{D}} \in \mathrm{Diff}(\Sigma_j, q_j)$ verifying the relation $h^{\mathcal{D}} \circ \mathcal{D} = \mathcal{D} \circ h$, called the *adjunction equation.* This induces a subgroup $G_j * (\mathcal{D}_* G_i)$ generated by G_j and the elements $h^{\mathcal{D}}$, $h \in G_i$. Thus if h is, for instance, linear, i.e., $h(x) = \mu x$, then $h^{\mathcal{D}}(y) = \mu^\lambda \cdot y$ where μ^λ is any solution of $z^{1/\lambda} = \mu$. Other situations are treated in [Ca-Sc]. The singular holonomy group of an element in the branch B is the maximal subgroup of $\mathrm{Diff}(\Sigma_j, q_j)$ obtained by iteration of the adjunction process made at all the corners in the branch B, see [Ca-Sc] for a detailed description of this construction.

The following theorem is a restatement of the theorem of Mattei and Moussu [Ma-Mou]:

Theorem. *The singularity $0 \in \mathbb{C}^2$ of $\mathcal{F}$ admits a local holomorphic first integral if and only if all components of the resolution divisor are nondicritical and the singular holonomy of D is finite.*

It is well known [Ce-Ma] that it is not possible to give a topological characterization of foliations admitting a meromorphic first integral. A necessary condition of the existence of a meromorphic first integral is that the singular holonomy of the nondicritical components of the resolution is finite. Thus we have the following

Theorem A. *The foliation $\mathcal{F}$ admits a meromorphic first integral near $0 \in \mathbb{C}^2$ if and only if*

(i) *The singular holonomy of the nondicritical branches is finite,*

(ii) *For each dicritical component $\mathcal{D}_i$ of the resolution of $\mathcal{F}$, there is a meromorphic function $f_i: \mathcal{D}_i \to \overline{\mathbb{C}}$ such that it is invariant by the singular holonomy of the nondricritical branches intersecting $\mathcal{D}_i$.*

(iii) *Given any two dicritical components $\mathcal{D}_i$ and $\mathcal{D}_j$ and a branch B_{ij} between them, i.e. $q_i = B_{ij} \cap \mathcal{D}_i \neq \phi \neq B_{ij} \cap \mathcal{D}_j = q_j$. Then the germs of f_i at q_i and f_j at q_j are compatible, i.e. for any leaf L near B_{ij}, the value of f_i at $L \cap D_i$ is the same as the value of f_j at $L \cap D_j$.*

Proof. Suppose that $\mathcal{F}$ admits a meromorphic first integral f near $0 \in \mathbb{C}^2$. Let

$\pi = \pi^{(n)} \circ \pi^{(n-1)} \circ \cdots \circ \pi^{(1)}$ be the resolution map of $\mathcal{F}$ at $0 \in \mathbb{C}^2$. Then π maps $(M^{(n)}, D)$ to $(\mathbb{C}^2, 0)$. We fix a dicritical component $\mathcal{D}_i$ of this resolution. Then the map $\tilde{f} = \pi \circ f \colon M^{(n)} \backslash D \to \mathbb{C}^2 \backslash \{0\}$ is clearly meromorphic and is trivially extended to the points of D. We denote still with $\tilde{f}$ this extension. Let B be any nondicritical branch. By the Lemma and the main theorem of [Ca] there is a singular point $q \in B$, which is not a corner, such that through q passes a separatrix s transverse to B. The value of the function $\tilde{f}$ at B is the value of f at s. Clearly the singular holonomy of B is finite and $f_i = \tilde{f} \mid \mathcal{D}_i$ is a meromorphic function invariant by the singular holonomy of B near $B \cap \mathcal{D}_i$. Conversely, suppose that (i)-(iii) hold. Then we proceed to construct a meromorphic first integral $\tilde{f}$ for $\mathcal{F}^{(n)}$. Since the singular holonomy of each nondicritical branch is finite, Lemma 2 of [Ma-Mou] allows to define $\tilde{f}$ in a neighborhood of the nondicritical branches, as extension of the f_i's, invariantly along the leaves of $\mathcal{F}^{(n)}$. For each dicritical component $\mathcal{D}_i$ and any leaf L_p of $\mathcal{F}^{(n)}$ passing through a point p near $\mathcal{D}_i$, we have $L_p \cap \mathcal{D}_i = p_i$. Then $\tilde{f}(p) = f_i(p_i)$ defines $\tilde{f}$ in a neighborhood of $\mathcal{D}_i$. The blow down of $\tilde{f}$ yields a meromorphic first integral for $\mathcal{F}$, near $0 \in \mathbb{C}^2$.

The theorem above puts in evidence the role played by the dicritical components, as vertices in the resolution tree, where the global holonomy of the nondicritical branches is concentrated. Thus the classification of the invariants of the dicritical components appears as a fundamental problem for the classification of more general singularities. In what follows we will introduce two invariants. The first invariant is an index theorem associated to dicritical singularities, which complements the index theorem of [Ca-S] for the nondicritical situation. This goes as follows.

Suppose that $0 \in \mathbb{C}^2$ is a singular point of the differential equation

$$\mathcal{F} : a(x,y)dy - b(x,y)dx = 0$$

such that it is 1-dicritical in the sense that the foliation $\mathcal{F}^{(1)}$ obtained by blowing up $0 \in \mathbb{C}^2$ is transverse to the divisor $D^{(1)}$, except at a finite number points, which are either points of tangency of $\mathcal{F}^{(1)}$ with $D^{(1)}$ or singular points of $\mathcal{F}^{(1)}$. Let $w = w_\nu + w_{\nu+1} + \dots$ be the Taylor development at $0 \in \mathbb{C}^2$, where $w_\nu \neq 0$, and $w_j = a_j(x,y)dy - b_j(x,y)dx$ is, for each $j \geq \nu$ a homogeneous 1-form of degree j. The foliation $\mathcal{F}^{(1)}$ is given by the 1-forms

$$\begin{aligned} w^{(1)} &= P(x,t)dx + Q(x,t)dt \\ &= \frac{1}{x^{\nu+1}} \left[(t\,a(x,tx) - b(x,tx))dx + x\,a(x,tx)dt \right] \\ w^{(2)} &= \frac{1}{y^{\nu+1}} \left[(a(uy,y) - ub(uy,y))dy - yb(uy,y)du \right]. \end{aligned}$$

Clearly, in the intersection of the charts (x,t) and (u,y) we have $w^{(1)} = t^{\nu+1}\, w^{(2)}$, and to be 1-dicritical means that $t\,a_\nu(1,t) - b_\nu(1,t) \equiv 0$.

The fibers of the fibration $M^{(1)} \to D^{(1)}$ are, in the (x,t)-chart, the sets $t =$ constant, and in the (u,y)-chart the sets $u =$ constant. We assume that the set (curve) of tangency points of $\mathcal{F}^{(1)}$ with these fibers cuts $D^{(1)}$ in a finite number of points.

This means that $t\,a_{\nu+1}(1,t) - b_{\nu+1}(1,t) \not\equiv 0$. Suppose that $p_0 = (0,t_0)$ is one of these points. We define the index:

$$\tau_{p_0}(\mathcal{F}^{(1)}, D^{(1)}) := \operatorname*{Res}_{t=t_0} \frac{\partial}{\partial x}\Big(-\frac{Q}{P}\Big)(0,t)dt.$$

This index is invariant by local automorphisms of the line bundle, i.e., maps of the form $\varphi(x,t) = (\xi(t)\cdot x, \psi(t))$, where $\xi(0) \neq 0$ and $\psi'(0) \neq 0$.

Theorem. $\sum_{p\in D^{(1)}} \tau_p(\mathcal{F}^{(1)}, D^{(1)}) = -1.$

Proof. Suppose for the sake of simplicity that $\mathcal{F}^{(1)}$ is transverse to the fiber of $M^{(1)} \to D^{(1)}$ passing through the point $(u,y) = (0,0)$. This means that $a_{\nu+1}(0,1) \neq 0$.
Define $P_j(x,y) := y\,a_j(x,y) - x\,b_j(x,y)$ for $j \geq \nu$. Then $w^{(1)} = \big(P_{\nu+1}(1.t) + xP_{\nu+2}(1,t) + \dots\big)dx + (a_\nu(1,t) + x\,a_{\nu+1}(1,t) + \dots)dt$. At the generic point $p_{t_0} = (0,t_0) \in D^{(1)}$, the index of $\mathcal{F}^{(1)}$, relative to $D^{(1)}$, is:

$$\begin{aligned}\tau_{p_0}(\mathcal{F}^{(1)}, D^{(1)}) &= \operatorname*{Res}_{t=t_0} -\frac{\partial}{\partial x}\left(\frac{a_\nu(1,t) + xa_{\nu+1}(1,t)+\dots}{P_{\nu+1}(1,t) + xP_{\nu+2}(1,t)+\dots}\right)(0,t)dt\\ &= -\operatorname*{Res}_{t=t_0}\left(\frac{a_{\nu+1}(1,t)}{P_{\nu+1}(1,t)} - \frac{a_\nu(1,t)\cdot P_{\nu+2}(1,t)}{P_{\nu+1}(1,t)^2}\right)dt.\end{aligned}$$

By the Residue Theorem, we have

$$\begin{aligned}&\sum_{p\in D^{(1)}} \tau_p(\mathcal{F}^{(1)}, D^{(1)}) = \sum_{p\in\mathbb{C}} \tau_p(\mathcal{F}^{(1)}, D^{(1)})\\ &= +\operatorname*{Res}_{t=\infty}\left(\frac{a_{\nu+1}(1,t)}{P_{\nu+1}(1,t)} - \frac{a_\nu(1,t)P_{\nu+2}(1,t)}{P_{\nu+1}(1,t)^2}\right)dt\\ &= -\operatorname*{Res}_{u=0}\left(\frac{a_{\nu+1}(u,1)}{a_{\nu+1}(u,1) - ub_{\nu+1}(u,1)} + a_\nu(u,1)\,\frac{a_{\nu+2}(u,1) - ub_{\nu+2}(u,1)}{(a_{\nu+1}(u,1) - ub_{\nu+1}(u,1))^2}\right)\frac{du}{u}\\ &= -1.\end{aligned}$$

The above theorem is, in fact, more general and can be applied to a foliation $\widetilde{\mathcal{F}}$ on a line bundle over a compact Riemann surface S. The foliation is assumed to be transverse to the zero section (S) of the line bundle except at a finite number of points and tangent to the fibers only at a finite number of points in S. Then

Theorem B.

$$\sum_{p\in S} \tau_p(\widetilde{\mathcal{F}}, S) = c(S)$$

where $c(S)$ is the Chern class of the line bundle.

Another invariant associated to the dicritical situation can be derived by the slope "functions" which in the (x,t)-chart is defined as $\dfrac{dx}{dt}$ and in the (u,y)-chart as $\dfrac{dy}{du}\cdot$ The Blow-up at $0 \in \mathbb{C}^2$ is given by the change of coordinates $ut = 1$ and $y = tx$. Since

$$\frac{dy}{du} = t \cdot \frac{dx}{dt} \cdot \frac{dt}{dy} + x\,\frac{dt}{du}$$

we obtain that

$$\left.\frac{dy}{du}\right|_{y=0} = t \cdot \frac{dt}{du} \cdot \left.\frac{dx}{dt}\right|_{x=0}$$

in the intersection of the two charts. This means that the slope functions glue as a meromorphic section of the bundle $M^{(1)} \otimes \kappa$, where κ is the canonical bundle associated to $D^{(1)}$. Since the algebraic sum of zeros and poles of a meromorphic section is the Chern class of the line bundle, we obtain that the algebraic sum of zeros and poles of the "slope section" is equal to -3.

More generally, suppose that $\widetilde{\mathcal{F}}$ is a foliation on a line bundle $E \to S$ over a compact Riemann surface S. The foliation is assumed to be in general position with the zero section (S) of the line bundle. Moreover, it is supposed to be tangent to the fibers only at a finite number of points in S. Suppose that the local coordinates of E are $(x_\alpha, y_\alpha) \in U_\alpha \times \mathbb{C}$, where (U_α) is an covering of S by open subsets U_α and $y_\alpha = \xi_{\alpha\beta}(x_\alpha) \cdot y_\beta$ in $(U_\alpha \cap U_\beta) \times \mathbb{C}$ whenever $U_\alpha \cap U_\beta \neq \phi$. The *slope section* is defined as $\left(\left.\dfrac{dy_\alpha}{dx_\alpha}\right|_{y_\alpha=0}\right)_\alpha$. Then the zeros of this section are the points of tangency of $\widetilde{\mathcal{F}}$ with S. The sum of the orders of all these zeros is denoted $Z(\widetilde{\mathcal{F}}, S)$. The poles of this section are the points of tangency (in S) of $\widetilde{\mathcal{F}}$ with the fibers. The sum of the orders of these tangencies is denoted $P(\widetilde{\mathcal{F}}, S)$. Then

Theorem D. *$Z(\widetilde{\mathcal{F}}, S) - P(\widetilde{\mathcal{F}}, S) = c(S) + 2g - 2$ where $c(S)$ is the Chern class of the line bundle E and g is the genus of the Riemann surface S.*

From Theorem B we obtain also that

$$\sum_{p \in S} \tau_p(\widetilde{\mathcal{F}}, S) = Z(\widetilde{\mathcal{F}}, S) - P(\widetilde{\mathcal{F}}, S) + 2 - 2g.$$

We show now that the invariants for $\widetilde{\mathcal{F}}$ introduced above can be realized in a "linear" foliation $\mathcal{L}$ defined on E and associated to $\widetilde{\mathcal{F}}$ as follows. Suppose that $\widetilde{\mathcal{F}}$ is defined by a collection $(\Omega_\alpha)_\alpha$ of holomorphic 1-forms $\Omega_\alpha = A_\alpha(x_\alpha, y_\alpha)dy_\alpha - B_\alpha(x_\alpha, y_\alpha)dx_\alpha$. Each Ω_α is defined on $U_\alpha \times \mathbb{C}$, such that whenever $U_\alpha \cap U_\beta \neq \phi$ we have $\Omega_\alpha = g_{\alpha\beta}\Omega_\beta$, $g_{\alpha\beta} \in \mathcal{O}^*(U_\alpha \cap U_\beta)$, a nonzero holomorphic function on $U_\alpha \cap U_\beta$. The 1-forms $\zeta_\alpha := \dfrac{B_\alpha(x_\alpha, 0)}{A_\alpha(x_\alpha, 0)}\, dx_\alpha$ are called the *local slope differentials* and the 1-forms $\eta_\alpha := \dfrac{\partial}{\partial y_\alpha}\left(\dfrac{B_\alpha(x_\alpha, y_\alpha)}{A_\alpha(x_\alpha, y_\alpha)}\right)(x_\alpha, 0)\, dx_\alpha$ are called the *first variation* of the local slope differentials. The invariants introduced above are the orders of the zeros and poles of ζ_α and the residues of the η_α. Clearly, we have

$$A_\alpha(x_\alpha,0)\xi_{\alpha\beta}(x_\alpha) = g_{\alpha\beta}(x)B_\beta(x_\beta,0) \quad \text{and}$$
$$B_\alpha(x_\alpha,0)dx_\alpha = g_{\alpha\beta}(x)B_\beta(x_\beta,0)dx_\beta\,.$$

Thus $\zeta_\alpha = \xi_{\alpha\beta}(x)\cdot\zeta_\beta$ and $\eta_\alpha - \eta_\beta = d\log\xi_{\alpha\beta}(x)$ for $x \in U_\alpha \cap U_\beta$. The foliation $\mathcal{L}$ is defined by the 1-forms: $w_\alpha = dy_\alpha - \zeta_\alpha - y_\alpha\eta_\alpha$. It is immediate to verify that if $U_\alpha \cap U_\beta \neq \phi$, $w_\beta = \xi_{\alpha\beta}w_\beta$; i.e. the w_α's glue as to define a foliaton on E. Indeed the $(w_\alpha)_\alpha$ define a section of the line bundle E with values meromorphic $(1,0)$-forms. The relation between $\widetilde{\mathcal{F}}$ and $\mathcal{L}$ will be studied in a forthcoming paper.

References

[Ca] C. Camacho, *Quadratic forms and holomorphic foliations on singular surfaces.* Mathematische Annalen **282**, (1988), 177-184.

[Ca-S] C. Camacho, P. Sad, *Invariant varieties through singularities of holomorphic vector fields.* Ann. of Math. **115** (1982), 579–595.

[Ca-Sc] C. Camacho, B. Scárdua, *Holomorphic foliations with Liouvillian first integrals.* Preprint IMPA, (1997).

[Ce-Sc] D. Cerveau, B. Scárdua, *On the integration of polynomial vector fields in dimension two.* Preprint IRMAR, (1996).

[La] H. Laufer, *Normal two dimensional singularities.* Ann. Math. Stud. nº **71** (1971).

[Ma-Mou] J.F. Mattei, R. Moussu, *Holonomie et intégrales premières.* Ann. Sc. Éc. Norm. Sup. t **13**, (1980), 469–523.

[Mo] R. Mol, *Foliations with meromorphic first integrals.* These IMPA, (1998).

[Sc] B. Scárdua, *Transversely affine and transversely projective holomorphic foliations.* Ann. Sc. Norm. Sup. **13** (1980), 469–523.

[Sci] A. Seidenberg, *Reduction of the singularities of the differential equation $Ady = Bdx$.* American Journal of Mathematics (1968), 248–269.

[Si] M.F. Singer, *Liouvillian first integrals of differential equations.* Trans. Amer. Math. Soc. **333** (1992).

Instituto de Matemática Pura e Aplicada - IMPA
Estrada Dona Castorina 110, Jardim Botânico
CEP 22460-320 Rio de Janeiro RJ, Brasil
e-mail: camacho@ impa.br

Contemporary Mathematics
Volume **269**, 2001

Dynamics of $\mathbb{P}^2$ (Examples)

by

John Erik Fornæss* and Nessim Sibony

Contents

* The first author was supported by an NSF grant.

2000 *Mathematics Subject Classification.* Primary 32F50, 37Fxx; Secondary 32H50.

1 Introduction

The Fatou-Julia theory of the dynamics of rational maps in $\mathbf{P}^1$ has received a lot of attention in the past 20 years. The notes by J. Milnor ([Mi]) or the monograph by Carleson-Gamelin ([CG]) give an introduction to the classical theory. It is quite natural to extend the theory to holomorphic endomorphisms of $\mathbf{P}^k$, $k \geq 2$. Some progress has been made in this direction and we refer to the survey articles ([F]) and ([S1]) for a description of the main results obtained in this direction.

The purpose of the present article is to study some families of endomorphisms of $\mathbf{P}^2$ that exhibit interesting dynamical properties. Before discussing more precisely the content, we recall some basic properties, see ([F]), ([S1]) for the proofs.

Let f be a holomorphic endomorphism of $\mathbf{P}^k$. In homogeneous coordinates, $f = [f_0 : \cdots : f_k]$ where all the f_j are homogeneous polynomials of degree d. We will say that d is the algebraic degree of f. The topological degree is d^k. We will assume that $d \geq 2$.

The Fatou set for f is the maximal open set where the family of iterates (f^n) is locally equicontinuous. The complement of the Fatou set is the Julia set $\mathcal{J} = \mathcal{J}_f$. One has the following characterization for the Julia set (FS2]). Let ω denote the standard Kahler form in $\mathbf{P}^k$. The sequence of forms $\left(\frac{(f^n)^*\omega}{d^n}\right)$ converges in the sense of currents to a positive closed current T. The support of T is exactly the Julia set of f. The current T satisfies the functional equation $f^*T = dT$.

The saddle periodic points with their stable manifolds are contained in the Julia set. The periodic points which are repelling are also contained in the Julia set. So in contrast with the one dimensional theory, the Julia set is not contained in the non wandering set of f. The Julia set can be decomposed using the currents of bidegree (ℓ, ℓ), $T^\ell := T \wedge \cdots \wedge T$, ℓ times. We will restrict our attention here to the measure $\mu := T \wedge \cdots \wedge T$, k times. The measure μ is an invariant measure of maximal entropy $\log d^k$. The map f is mixing for μ. Moreover as was recently proved by Briend-Duval ([BT]) the Lyapunov exponents for μ are strictly positive. Hence the measure μ is a limit of $\mu_n := \frac{1}{d^{nk}} \sum_j \epsilon_{a_j}$ where the sum is over the repelling periodic points of order n.

A point p is in the support of μ if and only if for every $r > 0$ the set $\mathbf{P}^k \setminus \cup_{n>0} f^n(B(p,r))$ is pluripolar, in some sense it means that the set is "small", for example it is of Hausdorff dimension at most $2(k-1)$ ([FS2]). Since the map f is mixing for μ, the support S_μ of μ is contained in the non wandering set of f. It is in some sense the analogue of the Julia in $\mathbf{P}^1$, for example periodic repelling points are dense in S_μ.

Besides the Fatou components and S_μ, there are other pieces of $\mathbf{P}^k$ where interesting dynamics occurs, for example attractors, first investigated in this setting in ([FW]) . Recall that a closed set $A \subset \mathbf{P}^k$ is an *attracting* set if there is an open set $U \supset A$ such that $f(U) \subset\subset U$ and $A = \overline{\cap f^n(U)}$. The open set U is called a trapping region. The set A is an *attractor*, in Ruelle's sense ([Ru]), if A is a minimal equivalence class for the relation $x \succ y$ defined by $x \succ x$ for all x and $x \succ y$ for $x \neq y$ if there is for every $\epsilon > 0$ an ϵ chain from x to y, i.e. there are $\{x = x_0, \ldots, x_n = y\}$ such that for all $1 \leq j \leq n$, dist $(f(x_{j-1}), x_j)) < \epsilon$. For example let $f_0 = [P(z,w) : Q(z,w)]$ be a rational map on $\mathbf{P}^1$ with Julia set equal to $\mathbf{P}^1$. Then the map

$$\overline{f}([z : w : t]) := [P(z,w) : Q(z,w) : t^d]$$

has the hyperplane $(t = 0)$ as an attractor. We study in the second section perturbations of the map $\overline{f}$, we show that for some special perturbations f_ϵ there is an attractor A_ϵ where periodic points are dense and where the corresponding unstable manifolds are dense. The attractor A_ϵ is not algebraic in general, a first example of that phenomenon was given by Jonsson ([J1]), see also ([JW]).

In paragraph 3 we give examples in $\mathbf{P}^2$ where the Julia set is equal to $\mathbf{P}^2$. We also generalize to polynomial maps of $\mathbb{C}^k$ some standard facts for polynomials in $\mathbb{C}$. More precisely let f be a polynomial map of $\mathbb{C}^k$, which extends holomorphically to $\mathbf{P}^k$. Let

$$K := \{z; (f^n(z)) \text{ is bounded}\}.$$

If K does not intersect the critical set, then f is strictly expanding on K and $K = S_\mu$, so repelling periodic points are dense in K.

In paragraph 4 we study perturbations of generalized Hénon maps in $\mathbb{C}^2$. Consider the automorphisms of $\mathbb{C}^2$ defined by

$$f_0(z,w) = (p(z) + aw, z)$$

where p is a polynomial of degree d. We consider the perturbation

$$f_\epsilon([z:w:t]) = [t^d p(z/t) + awt^{d-1} : zt^{d-1} + \epsilon w^d : t^d].$$

For $\epsilon > 0$, f_ϵ is an endomorphism of $\mathbf{P}^2$. We show that the nonwandering set Ω_ϵ of a composition of such maps might contain a countable discrete set of repelling periodic orbits. When $f_0 = (z^2 + aw, z)$ and $|a| >> 1$ we give a more complete description of the dynamics of

$$f_\epsilon([z:w:t]) = [z^2 + awt : zt + \epsilon w^2 : t^2]$$

in P^2.

In paragraph 5 we give examples where $\mathrm{Supp}(\mu)$ equals the Julia set.

2 Attractors

The general form of maps on $\mathbf{P}^2$ preserving complex lines through $[0:0:1]$ is

$$f = [P(z,w) : Q(z,w) : R(z,w,t)].$$

We will consider the special case

$$f = f_\epsilon = [P(z,w) : Q(z,w) : t^d + \epsilon Q_1(z,w)].$$

We will assume in what follows that the map

$$f_0 = [P(z,w) : Q(z,w)]$$

has Julia set equal $\mathbf{P}^1$. In the first three subsections we will also normalize so that $[1:0]$ is a fixed point.

LEMMA 2.1 *For all small enough complex valued ϵ, the only Fatou component is the superattractive basin of* $[0:0:1]$.

Proof: If p is not in the basin of zero, let D be a disk through p that projects (via lines through the origin) onto an open subset of the line P^1 at infinity. Since the map preserves lines through the origin and its restriction to the line at infinity has Julia set equal to P^1, we see that forward images of D will eventually project onto the entire line at infinity. Hence iterates of f cannot be normal on D, unless $f^n(D)$ tends uniformly to zero, which our assumptions disallow. We conclude that p is in the Julia set.

■

LEMMA 2.2 *Let U be any neighborhood of $(t = 0)$. Then f_ϵ has an attracting set A contained in U for any small enough ϵ. The attracting set intersects all lines through $[0 : 0 : 1]$.*

Proof: When $\epsilon = 0$ the map f fixes $(t = 0)$ and this set is attracting. Hence there exists a small neighborhood $V \subset U$ of $(t = 0)$ so that for all small enough ϵ, $f_\epsilon(V) \subset\subset V$. Set $A = \cap f_\epsilon^n(V)$, then since $f_\epsilon^n(t = 0)$ intersects all lines, the same must be true for A.

■

We note that A is connected since we may take V to be connected.
The following is obvious.

THEOREM 2.3 *The periodic points in A are the following: Consider any periodic orbit in the space of lines. Next there is a unique periodic orbit in those lines belonging to A and this is attracting inside those lines. In particular, A does not contain any repelling periodic orbit.*

Note that on the fixed line and all its preimages the lines are divided in two by the preimages of the quasicircle in the fixed line. On the fixed line the boundary of the basin of 0 and the basin of A agree.

Remark 2.4 *If an attracting set A contains a repelling periodic point p of period ℓ, then A contains a nonempty open set. Indeed if r is small enough, $B(p, r) \subset A$ and so*
$\lim_{n\to\infty} f^{n\ell}(B(p,r)) \subset A$, *but $f^{n\ell}(B(p,r))$ increases, (at least after a linear coordinate change,) to an open set $V \subset A$. If f^ℓ is linearizable in a*

neighborhood of p*, then* V *is a non degenerate image of* $\mathbb{C}^2$*, using a map* $\lim_{n\to\infty} f^{n\ell} \circ \left[(f^\ell)'(0)\right]^{-n}$.

The following theorem is proved in the same way as Proposition 6.1 of ([FS2]).

THEOREM 2.5 *Take any forward invariant closed positive* $(1,1)$ *current* σ*, i.e.* $f_*\sigma = d\sigma$*,* d *being the algebraic degree of* f*, on* A*, for example a Cesaro mean of the forward orbits of the current* $(t=0)$*. Then* $\nu := \sigma \wedge T$ *is an invariant measure.*

Let B_0 be the basin of attraction of 0 and B_A the basin of attraction of A, i.e. $B_A = \cup_n f^{-n}(U)$ where U is any small neighborhood of A. Both are totally invariant connected open sets.

THEOREM 2.6 *We have that Supp*$(\mu) \subset \partial B_0 \subset \partial B_A$*. Moreover Supp*$(\mu)$ *projects onto* $\mathbf{P}^1$.

Proof: The basin B_A contains the line at infinity $(t=0)$. So $\mathbf{P}^2 \setminus \overline{B}_A$ is a bounded invariant open set in $\mathbb{C}^2$. As a consequence $\mathbf{P}^2 \setminus \overline{B}_A$ is a union of Fatou components. Since there is only one Fatou component, namely B_0, we have $\mathbf{P}^2 = \overline{B}_A \cup B_0$. It follows that $\partial B_0 \subset \partial B_A$.

Let $p \in$ Supp μ, and W any neighborhood of p. Then $\cup f^n(W)$ ([FS1]) is dense in $\mathbf{P}^2$. Hence W must contain points in B_0 and points in B_A, so Supp $\mu \subset \partial B_0 \cap \partial B_A = \partial B_0$.

Let p be a repelling periodic point in Supp(μ). By the Briend-Duval Theorem ([BD]) such p exists. It is a repelling fixed point for some iterate $g_n := f^n_{|L}$ where L is a line through p. The preimages of p under inverse iterates of g_n are dense in the Julia set of g_n. The preimages of p under f are in Supp(μ). Hence Supp(μ) projects onto $\mathbf{P}^1$.

∎

We consider the maps f_ϵ on the space of lines, i.e. the map f_0 on $\mathbf{P}^1 = (t=0)$. Let $\mathcal{F}$ be the compact set of histories $(x_{-n})_{n\leq 0}, (f_0(x_{-n-1}) = x_{-n})$ in the product topology of $\mathbb{Z}^{\mathbf{P}^1}$. We define a map $\Phi : \mathcal{F} \to A$. But we first need a more precise trapping region.

2.1 Trapping region

Let $\eta > 0$. We define the neighborhood W_η about $(t = 0)$ as follows:

$$\begin{aligned} W_\eta^1 &= \{[z : w : t]; |z| \leq |w|,\ |t| < \eta|w|\} \\ W_\eta^2 &= \{[z : w : t]; |w| \leq |z|,\ |t| < \eta|z|\} \\ W_\eta &= W_\eta^1 \cup W_\eta^2. \end{aligned}$$

So W_μ is the exterior of the bidisc of radius $1/\eta$ in the (z, w) coordinates. Clearly, if we fix any small $\eta_0 > 0$ then for all small enough ϵ and all $\eta < \eta_0$ we have $f(W_\eta) \subset W_1$. Then, for example, if $[z : 1 : t] \in W_\eta^1$ and $f([z : w : t]) \in W_1^1$ we can write

$$f([z : 1 : t]) = [\frac{P(z,1)}{Q(z,1)} : 1 : \frac{t^d + \epsilon Q_1(z,1)}{Q(z,1)}].$$

Here $|z| \leq 1$ and $|P(z,1)| \leq |Q(z,1)|$ so necessarily $Q(z,1)$ is bounded uniformly away from zero. So we see that we get:

LEMMA 2.7 *There exists a constant $C > 0$ independent of ϵ so that: If we restrict to any line and any $0 < \delta < \epsilon < \eta$, the image of the disc $\Delta(0, \delta)$ around $t = 0$ is contained in $\Delta(0, C(\delta^d + \epsilon))$ in the image-line and the derivative in the t direction is bounded by $C\delta^{d-1}$.*

Hence we get a trapping region of the form $W_{2C\epsilon}$ and on these the derivative in the t direction is bounded by $2^{d-1}C^d\epsilon^{d-1}$. It follows that this trapping region contains a nontrivial attracting set, intersecting all lines. In fact we see that the attracting set is an attractor by using the fact that in the space of lines there are arbitrarily long pseudo-orbits connecting two points.

2.2 The map Φ.

Let (x_{-n}) be any element of $\mathcal{F}$. For any n consider the image $f^n(\Delta(0, 2C\epsilon))$ where we take the disc in the line x_{-n}. It follows that the intersection of these images contains exactly one point $p \in A$. We set $\Phi((x_{-n})) = p$. So this map sends $\mathcal{F}$ to A. On the other hand, if $p \in A$, then there is a sequence (p_{-n}) of preimages in A (perhaps several), $p_0 = p$. Each of these p_{-n} are in some line x_{-n} and necessarily $(x_{-n}) \in \mathcal{F}$. Then $\Phi((x_{-n})) = p$. Hence Φ is onto.

The following Lemma is then obvious.

LEMMA 2.8 *$\Phi : \mathcal{F} \to A$ is continuous and onto. Also Φ is a semiconjugacy. Φ maps periodic points to periodic points and hence periodic points are dense in the attractor.*

Since $[0:0:1] \notin A$, we can define a projection $\pi : A \to \mathbf{P}^1, \pi([z:w:t]) = [z:w]$. Then π is also a semiconjugacy, $\pi \circ f = f_0 \circ \pi$.

Notice that this construction works even if the map on $(t=0)$ is non-chaotic. (Of course periodic points are then not necessarily dense. But we always get a semiconjugacy $\Phi : \mathcal{F} \to A$.)

LEMMA 2.9 *Let (x_j) be an arbitrary repelling periodic orbit for the dynamics in the space of lines and let $(p_j) = \Phi(x_j)$ be the corresponding periodic orbit in A. Then (p_j) is a saddle orbit. The unstable curve of any of the p_j and with respect to this periodic history is contained in A and is dense in A. Also the stable curve of p_j intersects A in a dense set.*

Proof: We prove that the unstable curve W^u for a periodic orbit is dense in A. It is clear that W^u intersects all lines. Let $q \in A$. Let $q_{-\ell} \in A$ be such that $f^\ell(q_{-\ell}) = q$. Now W^u intersects the line through $q_{-\ell}$ at Q. Then $f^\ell(Q)$ belongs to W^u and must be close to $f^\ell(q_{-\ell}) = q$ if $\ell >> 1$ since f^ℓ is contracting along the lines through 0 in B_A.

Also the stable manifold must contain all the intersections of the trapping region with the lines in the preimages of the periodic lines. Suppose $q \in A$, and let $y \in W^u$ be in the same line as q, close to q. Let Σ be a small irreducible neighborhood in W^u of y (in some branch of W^u). Then this Σ must intersect all nearby periodic lines. So W^s is dense in A.

■

LEMMA 2.10 *The set A is an attractor. Moreover $f|A$ is topologically transitive.*

Proof: The first statement was observed above. Recall that $f_{|A}$ is topologically transitive if given two relatively open sets U and V in A, there is an integer $n \geq 1$ so that $f^n(U) \cap V \neq \emptyset$. Since by Lemma 2.8 the periodic points are dense in A, we can assume that U contains a periodic saddle point p. The unstable curve associated to p is dense in A (Lemma 2.9) hence $\cup_n f^n(U)$ is dense in A.

■

2.3 Non-Algebraicity of A

In this section we will restrict to the case of degree $d = 2$ to simplify calculations.

Let z_f be the fixed point on A contained in the fixed line $(w = 0)$. Also let L be the unique other line in the preimage of $(w = 0)$ in the space of lines. We will make a condition on the coefficients of the polynomials P, Q, Q_1 that will imply that the image of the trapping region in L is disjoint from z_f. This will imply that the intersection of the attractor with the fixed line is infinite and hence non-algebraic.

We fix notation: Write:

$$\begin{aligned} P(z,w) &= z^2 + azw + bw^2 \\ Q(z,w) &= czw + dw^2 \\ Q_1(z,w) &= ez^2 + fzw + gw^2 \\ z_f &= [\alpha : 0 : 1] \end{aligned}$$

We assume that $e, \epsilon, c \neq 0$. We fix P, Q, Q_1 and let ϵ be small enough. The following is immediate from the form $z \to \frac{z^2}{1+\epsilon e z^2}$ of f on the fixed line.

LEMMA 2.11

$$\alpha = \frac{1 + \sqrt{1 - 4\epsilon e}}{2\epsilon e} \sim \frac{1}{\epsilon e}$$

Next we observe that the line $cz + dw = 0$ is the other preimage in the space of lines of the fixed line:

LEMMA 2.12 *The image of the point at infinity on the line* $cz + dw = 0$ *is the point* $[\frac{B}{\epsilon} : 0 : 1]$ *where*

$$B = \frac{d^2 - acd + bc^2}{(ed^2 - fcd + gc^2)}$$

The image under f *of the trapping region on the line* $cz + dw = 0$ *is contained in a disc around* $\frac{B}{\epsilon}$ *of radius about* C *for some fixed constant* C *independent of* ϵ. *In particular if* $B \neq \frac{\pm 1}{e}$ *then for all small enough* ϵ *the image under* f *of the attractor in* $cz + dw = 0$ *does not contain* z_f *nor it's preimage* $-z_f$ *on the fixed line. The second image is already closer to* z_f *than* C *(we can assume that* C *is the same constant as above). Since* f *is* $1 - 1$ *there, it follows that the intersection of the attractor with the fixed line is infinite and hence that the attractor is non-algebraic.*

COROLLARY 2.13 *The fixed point* z_f *has only one preimage in the attractor (itself). The same is then true for a whole neighborhood in* A. *The set of points with only one preimage is open. In particular, these attractors are not completely invariant.*

We don't need in the above arguments that the Julia set of $f_0 = [P : Q]$ is $\mathbf{P}^1$. It suffices to assume that f_0 has no attracting periodic points. Hence we have:

PROPOSITION 2.14 *Let* $f_0 = [P(z, w) : Q(z, w)]$ *be a rational map on* $\mathbf{P}^1$ *without attracting periodic points. Then for* $\epsilon \neq 0$ *small enough the map*

$$f_\epsilon = [P(z, w) : Q(z, w) : t^d + \epsilon Q_1(z, w)]$$

has a nontrivial attractor A. *If the Julia set of* f_0 *is different from* $\mathbf{P}^1$, *then* $f_{\epsilon|A}$ *is not topologically transitive and periodic points are not dense in* A. *When* f_0 *is of degree 2 and* $[1 : 0]$ *is a repelling fixed point for* f_0 *then* A *is non algebraic with the above conditions on the coefficients.*

Remark 2.15 *Assume* f_0 *has a Siegel disc* D *centered at the fixed point* $[z_0 : 1]$. *Then the sequence of iterates* $f^n_{\epsilon\,|D}$ *converges uniformly to a limit disc which is invariant and with the same rotation number as the original Siegel disc. This Siegel disc is contained in* A.

PROPOSITION 2.16 *Let $f \in \mathcal{H}_d$. If A is an attractor, $A \neq \mathrm{P}^k$, then A is disjoint from S_μ, the support of μ.*

Proof: Let U be a neighborhood of A such that $f(U) \subset\subset U$. Assume that $S_\mu \cap U$ is nonempty. Since by ([FS1]), $\cup_n f^n(S_\mu \cap U)$ covers P^k except for a pluripolar set, we get a contradiction by choosing U, $\overline{U} \neq \mathrm{P}^k$.

■

PROPOSITION 2.17 *Let $\delta > 0$. Then for all small enough ϵ each slice of the attractor for f_ϵ in a line through $[0 : 0 : 1]$ has Hausdorff dimension less than δ.*

Proof: We know from Lemma 2.7 that there exists a trapping region $W_{c\epsilon}$ where the derivative in the t direction is bounded by $C_1|\epsilon|^{d-1}$. It follows that the image of a disc of radius r in the t direction is contained in a disc of radius $C_1|\epsilon|^{d-1}r$, here $r << 1$.

For a given line ℓ_0 through 0, let $A_0 := A \cap \ell_0$. The closed set $A \cap f_\epsilon^{-n}(A_0)$ is contained in d^{2n} discs of radius $\epsilon_0 < 1$. Hence A_0 is contained in d^{2n} discs of radius $C_1^n|\epsilon|^{n(d-1)}\epsilon_0$. To estimate the Hausdorff measure of dimension δ, we calculate $d^{2n}(C_1^n|\epsilon|^{n(d-1)}\epsilon_0)^\delta$. This is finite if $d^2C_1|\epsilon|^{\delta(d-1)} < 1$. So it suffices to choose $\epsilon < \left(\frac{1}{d^2C_1}\right)^{\frac{1}{\delta(d-1)}}$.

■

THEOREM 2.18 *([Gu]) Every neighborhood of the support of a positive closed current of bidegree $(1,1)$ contains a compact complex curve. Hence every neighborhood of a nontrivial attractor contains a compact complex curve.*

2.4 Subhyperbolicity of attractors.

In this subsection we want to extend the subhyperbolicity of the critically finite maps $f_0 := [(z-2w)^2 : z^2]$ on P^1 to attractors for $f_\epsilon := [(z-2w)^2 : z^2 : t^2 + \epsilon Q_1(z,w)]$ and show that unstable curves are dense. This generalizes to other post-critically finite maps, but we use f_0 for computational simplicity.

The critical points of f_0 in $(w = 1)$ are $\{0, 2\}$ with orbits $2 \to 0 \to \infty \to 1 \hookleftarrow$. Then there is a metric $d\sigma$ on $\mathbf{P}^1$ ([CG]) which is smooth except at the points $0, \infty, 1$, for which the map is strictly expanding. The singularites are $d\sigma \sim |dz|/\sqrt{|z|}$ in local coordinates near 0 and $d\sigma \sim |dz|/|z|^{3/4}$ in local coordinates around ∞ and 1.

Next, we define a pseudometric $d\sigma^*$ on $\mathbf{P}^2$ near $(t = 0)$. Pick a smooth Hermitian metric ds, and pick a point (p, ξ) in the tangent bundle of $\mathbf{P}^2$, p near $(t = 0)$. We can decompose $\xi = \xi_1 + \xi_2$ where ξ_2 is along the line through 0 and p in $(t = 1)$ and ξ_1 is perpendicular to ξ_2 in the $ds-$ metric. There is a natural projection to $(t = 0)$ and this projects ξ_1 to a vector tangent to $(t = 0)$ which we also denote by ξ_1. We set

$$d\sigma^*(\xi) := \sqrt{(d\sigma)^2(\xi_1) + (ds)^2(\xi_2)}.$$

With this singular metric on the attractor, the map f is contracting in one direction and expanding along the line at infinity. This generalizes the concept of subhyperbolicity in one dimension.

Pick any p in the attractor A for the maps above. Let $\{p_{-n}\}$ be any history in A. For each n let Δ_n denote small discs of about the same radius around $[p_{-n}]$, the projection of p_{-n} to $(t = 0)$. Let $\Sigma_n := f^n(\Delta_n)$.

THEOREM 2.19 *The curves Σ_n converge to a nonconstant image of $\mathbb{C}$ contained and dense in A, and passing through p.*

Proof: First we give a brief outline of the proof. In the first part of the proof we show that the inverse maps in the space of lines are strongly contracting. Next we show that the inverse images of a small disc therefore will always stay far away from the critical orbit. This implies that the forward images in $\mathbf{P}^2$ of these small discs are all graphs, and hence have good convergence properties. This defines the local unstable manifolds and then the forward images of the discs Δ_n converge to the global unstable manifolds.

To proceed with the details of the proof, we will first assume that $[p]$ is not one of the points on the critical orbit. Fix a small $\delta-$ neighborhood, V of $[1]$.

Note that none of the points in the history are in the critical orbit. In local coordinates near $[1]$, the map is $z \to -4z$. We can assume that $[p]$ is further than δ from the critical orbit. To prove the theorem we will first construct the local unstable manifold through p with the given history. Our first step is to estimate the expansion of the map along the history and close to it.

If $[p_{-n}] \to [1]$, the growth of $|(f^n)'| \sim 4^n$. We assume next that $[p_{-n}]$ does not converge to $[1]$. Let $I = [p_{-k}, \dots, p_{-k+\ell}]$ be a maximal interval in V. Then the derivative of $f^{-\ell}$ at $p_{-k+\ell}$ is about $4^{-\ell}$ and the distance from $[p_{-k}]$ to $[1]$ is about $\delta/4^\ell$. This implies that p_{-k-1} has distance about $\delta/4^\ell$ from ∞ and that the derivative of $f^{-\ell-1}$ is about $4^{-\ell}$. Therefore p_{-k-2} is at distance about $\sqrt{\frac{\delta}{4^\ell}}$ from 0 and the derivative of $f^{-\ell-2}$ is about

$$\frac{1}{2\sqrt{\frac{\delta}{4^\ell}}}4^{-\ell} \sim \frac{1}{2\sqrt{\delta}}\frac{1}{\sqrt{4^\ell}}.$$

Next $[p_{-k-3}]$ is at distance about $\left(\frac{\delta}{4^\ell}\right)^{\frac{1}{4}}$ from $[2]$ and the derivative of $f^{-\ell-3}$ is about

$$\frac{1}{2\left(\frac{\delta}{4^\ell}\right)^{\frac{1}{4}}}\frac{1}{2\sqrt{\delta}}\frac{1}{\sqrt{4^\ell}} \sim \frac{1}{4\delta^{\frac{3}{4}}}\frac{1}{4^{\frac{\ell}{4}}}.$$

Next $[p_{-k-4}]$ is away from the critical orbit so the derivative follows the general expansion of the mapping measured in a given subhyperbolic metric (which is smooth there). Let next $\Delta_n' \subset \Delta_n$ denote the local preimages in $(t = 0)$ of Δ_0 containing $[p_{-n}]$. Since these estimates hold uniformly, we get that the diameters of the discs Δ_n' shrink exponentially and since they are always contained in sectors of small angles when they pass near critical points, we also get that $f^{-n} : \Delta_0 \to \Delta_n$ have well defined branches and are biholomorphic.

Next we consider the forward images in P^2 of Δ_n'. Set $f^k(\Delta_n') := D_{n,k}$, $0 \le k \le n$.

Suppose Γ is a graph over Δ_n' (or a subset), Γ contained in the trapping region. Since f, as a map on the space of lines is $1-1$ on Δ_n' and maps Δ_n' to Δ_{n-1}', $f(\Gamma)$ is a graph over Δ_{n-1}'. In particular, the sets $D_{n,k}$ are graphs over Δ_{n-k}.

In particular, we then get the graphs $f^n(\Delta_n')$, Γ_n over Δ_0. Moreover, these Γ_n converge uniformly over Δ_0 as graphs to a limit graph Γ_0 because of the uniform contraction in the t direction. Moreover, Γ_0 is contained in the attractor and is the local unstable manifold of p with the given history. Since $\{f^{-n}\}$ are strongly contracting in the space of lines, we can replace Δ'_n by minimal larger round discs $\Delta"_n \supset\supset (1+\delta)f^{-1}(\Delta''_{n-1})$ for a fixed $\delta > 0$. (Here the factor $(1+\delta)$ refers to an expansion of the radius.) For small enough $\delta > 0$ the discs $\Delta"_n$ are still shrinking exponentially. The inductive limit of $\{\Delta"_n\}$ is a $\mathbb{C}$. In particular, as map on the space of lines, for any n, $f^k(\Delta''_{n+k})$ cover P^1 in the space of lines for all $k \geq k(n)$. The forward images $f^n(\Delta"_n)$ converges to the global unstable set, which is dense in the attractor.

To complete the proof, we consider the case when the point $[p]$ is on the critical orbit. But in this case the points $[p_{-4}]$ are not in the critical orbit and we can forward iterate the unstable variety of p_{-4}.

■

3 When the compact set of points with bounded orbit is disjoint from the critical set

3.1 $\mathcal{J} = \mathrm{P}^2$

We investigate examples where $\mathcal{J} = \mathrm{P}^2$, the line $(t = 0)$ is preserved, and $\mathrm{Supp}(\mu)$ is a Cantor set.

Fix in this section $[P(z,w) : Q(z,w)]$ a map with Julia set P^1. Let $L_1, \ldots, L_r$ be those complex lines on which $\det(P,Q)' = 0$ and let $\tilde{L}_j := (P,Q)(L_j)$. Pick any point $(a,b) \notin \cup \tilde{L}_j$.

THEOREM 3.1 *If the complex number c is sufficiently large and*

$$f_c = [P(z,w) - cat^d : Q(z,w) - cbt^d : t^d]$$

then $\mathcal{J}_{f_c} = \mathrm{P}^2$.

Proof: We can assume that $a \neq 0$. There exist strictly positive constants A, B so that

$$A\|(z,w)\|^d \leq \|(P(z,w), Q(z,w))\| \leq B\|(z,w)\|^d.$$

Set $R = 2\left(\frac{\|(ac,bc)\|}{A}\right)^{\frac{1}{d}}$ and $U = \{\max(|z|,|w|) < R\}$.

LEMMA 3.2 *If $(z,w) \notin U$ and $|c|$ is sufficiently large, then $f_c^n(z,w) \to \infty$.*

Proof of the Lemma: Suppose that $(z,w) \notin U$. Then we get:

$$\begin{aligned}
\|f_c(z,w)\| &= \|(P(z,w) - ca, Q(z,w) - cb)\| \\
&\geq \|(P(z,w), Q(z,w))\| - |c|\|(a,b)\| \\
&\geq A\|(z,w)\|^d - \frac{\|(z,w)\|}{R}|c|\|(a,b)\| \\
&\geq \|(z,w)\|\left[A\|(z,w)\|^{d-1} - \frac{|c|\|(a,b)\|}{R}\right] \\
&\geq \|(z,w)\|\left[AR^{d-1} - \frac{|c|\|(a,b)\|}{R}\right] \\
&\geq \|(z,w)\|\left[A2^{d-1}\left(\frac{\|(ac,bc)\|}{A}\right)^{\frac{d-1}{d}} - \frac{|c|\|(a,b)\|}{2\left(\frac{\|(ac,bc)\|}{A}\right)^{\frac{1}{d}}}\right] \\
&\geq \|(z,w)\|A^{\frac{1}{d}}\|(ac,bc)\|^{\frac{d-1}{d}}\left[2^{d-1} - \frac{1}{2}\right] \\
\text{If } |c| \text{ is large enough:} &\geq 2\|(z,w)\|
\end{aligned}$$

So $|c| \geq c_0 := \frac{2^{\frac{d}{d-1}}}{A^{\frac{1}{d-1}}} \frac{1}{\left[2^{d-1}-\frac{1}{2}\right]^{\frac{d}{d-1}}} \frac{1}{\|(a,b)\|}$ will suffice.

■

Continuation of the Proof of the Theorem: It follows from the previous Lemma that the set K of points whose orbits are bounded is contained in

$$U_1 := \{(z,w) \in U; |P - ca| < R \wedge |Q - cb| < R\}.$$

LEMMA 3.3 *If $|c|$ is sufficiently large, the map f_c is uniformly expanding on U_1. More precisely, for some constant $\delta_0 > 0$, $|f_c'(z,w)(\xi)| \geq \delta_0|c|^{1-\frac{1}{d}}|\xi|$ on U_1.*

Proof of the Lemma: If $(z,w) \in U_1$, then $|P(z,w)| \geq |ca| - R$. Hence

$$\begin{aligned} B\|(z,w)\|^d &\geq |P(z,w)| \\ &\geq |ca| - R, \\ \|(z,w)\|^d &\geq \left|\frac{ca}{B}\right| - \frac{R}{B} \\ \|(z,w)\|^d &\geq \left|\frac{ca}{B}\right| - \frac{2\left(\frac{\|(ac,bc)\|}{A}\right)^{\frac{1}{d}}}{B} \\ \|(z,w)\| &\geq t|c|^{\frac{1}{d}} \text{ (for some fixed } t > 0,\ |c| \text{ large enough)} \end{aligned}$$

Hence, if we set $s = 4\left(\frac{\|(a,b)\|}{A}\right)^{\frac{1}{d}}$,

$$U_1 \subset \{t|c|^{\frac{1}{d}} \leq \|(z,w)\| \leq 2R \leq s|c|^{\frac{1}{d}}\}.$$

By homogeneity, there exists a continuous, nonnegative function $\lambda(\rho)$ on the space of lines, $\lambda(\rho) = 0 \Leftrightarrow L \in \{L_j\}$, and increasing near these lines, so that

$$|(P,Q)'(z,w)(\xi)| \geq \lambda([z:w])\|(z,w)\|^{d-1}|\xi|$$

for any tangent vector ξ.

Suppose that $(z,w) \in U_1$.

Then

$$\begin{aligned} |P(z,w) - ca| &\leq s|c|^{\frac{1}{d}} \\ |Q(z,w) - cb| &\leq s|c|^{\frac{1}{d}} \end{aligned}$$

Hence for large c the spherical distance

$$d([P(z,w):Q(z,w)],[a:b]) \lesssim |c|^{\frac{1}{d}-1}.$$

It follows that the spherical distances

$$d([P(z,w):Q(z,w)],\{\tilde{L}_j\}) \geq \frac{1}{2}d([a:b],\{\tilde{L}_j\})$$

for large enough $|c|$. Hence, for some $\tau > 0$ independent of c, $d([z:w],\{L_j\}) \geq \tau$ and

$$\begin{aligned}|(P,Q)'(z,w)(\xi)| &\geq \lambda([z:w])\|(z,w)\|^{d-1}|\xi| \\ &\geq \lambda([z:w])(t|c|^{\frac{1}{d}})^{d-1}|\xi| \\ &\geq 2|\xi| \text{ for large enough } |c|\end{aligned}$$

■

Remark 3.4 *We have in particular shown that there are no critical points in U_1 for large $|c|$.*

End of the Proof of the Theorem: The set K of points with bounded orbits is contained in U_1 and f_c is expanding there, hence these points are in the Julia set. However, points outside K are in the basin of the line $(t=0)$. Since the map is chaotic on $(t=0)$ it follows that $\mathcal{J}_{f_c} = \mathrm{P}^2$.

■

PROPOSITION 3.5 *If $|c|$ is large enough, K is a Cantor set, Supp(μ)=K and repelling periodic points are dense in K.*

Proof: The map $f_c : U_1 \to U$ is proper and has no critical value by the above remark. So we have d^2 well defined inverse branches on U, $g_1, \dots, g_{d^2}$. Each g_j is uniformly contracting and $g_j(U) \cap g_i(U) = \emptyset$ for $i \neq j$. Consequently every connected component of K is a point, and by symbolic dynamics, K has no isolated points, i.e. K is a Cantor set.

Using symbolic dynamics again, it is clear that periodic orbits are dense in K, and they are all repelling since f is expanding. Suppose $p \in \mathrm{P}^2 \setminus K$. Then there exists an $r > 0$ so that $\overline{\cup f^n(B(p,r))} \cap K = \emptyset$. By ([FS1]), see below, Supp $\mu \subset K$. To show the other inclusion, let $p \in K$ and $r > 0$. It is again clear from symbolic dynamics that $\cup f^n(B(p,r))$ contains K. As Supp μ is completely invariant and closed, this implies that p, hence $K \subset$ Supp μ.

Here we are using the following result from ([FS1]).

■

PROPOSITION 3.6 *Let $f \in \mathcal{H}_d$. A point q is in S_μ if and only if $\cup_{n\geq 0} f^n(B(q,r)) = \mathbb{P}^k \setminus E$ where E is pluripolar.*

Proof: If $q \in S_\mu$ then $\cup_{n\geq 0} f^n(B(q,r))$ is pluripolar, see ([FS1]). If $q \notin S_\mu$ then since S_μ is totally invariant and nonpluripolar ([FS1]), then $\cup_{n\geq 0} f^n(B(q,r))$ omits S_μ for r small enough.

■

PROPOSITION 3.7 *The Hausdorff dimension $\alpha(c)$ of K satisfies*

$$\alpha(c) \leq \frac{2\log d}{\log \delta_0 + \left(1 - \frac{1}{d}\right)\log|c|}$$

for all large enough $|c|$.

Proof: Recall from Lemma 3.3 that the map f_c is expanding on U_1 by $\delta_0 |c|^{1-\frac{1}{d}}$ where δ_0 is a fixed constant.

The d^2 components of $f^{-1}(U)$ have diameter at most $\operatorname{diam}(U)\frac{|c|^{\frac{1}{d}-1}}{\delta_0}$ so at each step the diameter of a component is multiplied by at most the (small) constant $\frac{|c|^{\frac{1}{d}-1}}{\delta_0}$. At the nth step K is covered by d^{2n} open sets each of diameter at most $\operatorname{diam}(U)\frac{|c|^{n(\frac{1}{d}-1)}}{\delta_0^n}$. It follows that $\alpha \leq \frac{2\log d}{\log\delta_0 + \left(1-\frac{1}{d}\right)\log|c|}$.

■

3.2 Support of μ.

We now consider the following situation. Let f be a polynomial map of algebraic degree d, $f : \mathbb{C}^k \to \mathbb{C}^k$. We assume that f extends as a holomorphic map into $\mathbf{P}^k$. So the hyperplane at infinity given by $(t = 0)$ is an attracting set for f. We define

$$K = \{z; (f^n)(z) \text{ is bounded}\}.$$

The assumption on f implies that K is a compact polynomially convex set in $\mathbb{C}^k$. Indeed, there is R such that for $\|z\| \geq R$, $(f^n(z))$ converge to infinity. So we have

$$K = \{z; |f^n(z)| \leq R \text{ for } n \geq 0\}.$$

Clearly the set K is totally invariant.

THEOREM 3.8 *Let f be a polynomial map as above. Let C be the critical set for f. Assume $K \cap C = \emptyset$. Then*
i) The map f is strictly expanding on K.
ii) Repelling periodic points are dense in K.
iii) $K =$ Support μ.

Proof: From the assumption, the critical set C is in the basin of attraction of the hyperplane at infinity. Let B denote the open ball of center 0 and radius R. Let $B_n := f^{-n}(B)$. We have $B_{n+1} \subset\subset B_n$ for every n. The map $f : B_{n+1} \to B_n$ is proper. We can choose n large enough so that B_n does not intersect the critical set C. Let K_ℓ denote the infinitesimal Kobayashi metric for B_ℓ.

Since f is a covering map from B_{n+1} to B_n we have

$$K_n(f(z), f'(z)\xi) = K_{n+1}(z, \xi)$$

for any vector ξ. Since $B_{n+1} \subset\subset B_n$, there is a constant $c > 0$ such that $K_{n+1}(z, \xi) \geq (1 + c)K_n(z, \xi)$, for z in B_{n+1}. We then have

$$K_n(f(z), f'(z)\xi) \geq (1 + c)K_n(z, \xi).$$

This proves that f is strictly expanding on K.

ii) Let $p \in K$. Assume that for a sequence n_j, $p_j := f^{n_j}(p)$ converges to q. We are going to show that q is a limit of periodic points (they are necessarily contained in repelling orbits). Fix $r > 0$ small. Then for j

large enough, there is an Ω_j, $p_j \in \Omega_j \subset\subset B(p_j, r/2) \subset B(p_{j+1}, r)$ for which $f^{n_{j+1}-n_j} : \Omega_j \to B(p_{j+1}, r)$ is a biholomorphism. Hence there is a fixed point q_j for $f^{n_{j+1}-n_j}$ in $B(p_j, r)$. So q is a limit of periodic orbits.

To prove that p is a limit of periodic points, it is enough to show that preimages of repelling periodic points are approximable by periodic orbits. We can assume that q_j as above has a preimage in $B(p, r)$. Indeed for n large enough $f^n(B(p, r))$ contains $B(p_j, r)$. Since q_j is on a periodic orbit we will then also have a repelling periodic orbit passing through $B(p, r)$. Assume $f^m(q) = q \in K$ and $(f^j(q))$ is repelling. Let $f^\ell(p_0) = q$. There is a curve γ from p_0 to q and a neighborhood U of γ, such that $U \cap [\cup_{n=0}^\infty f^n(C)] = \emptyset$. This is because C is in the domain of attraction of the hyperplane at infinity. It is then possible to define in U, inverse branches g_n of f^n such that $g_n(q) = q$. It follows from a Theorem by Ueda, ([U]) that the family g_n is equicontinuous. Hence (g_n) converges in U to the periodic orbit $(f^j(q))$. Let $B(p_0, r)$ be a small neighborhood of p_0 and consequently, since g_{nm} converges to q, there is a set $\Omega \subset\subset B(p_0, r)$ for which $f^{nm+\ell}(\Omega) = B(p, r)$ is biholomorphic. Therefore there is a periodic orbit passing through $B(p, r)$. This is the classical Julia's argument to construct homoclinic orbits, see Milnor ([Mi]) and Jonsson ([J2]).

iii) We show first that $S_\mu :=$ Support $\mu \subset K$. But if $p \notin K$, there exist small $r > 0$ for which $\overline{\cup f^n(B(p, r))} \cap K = \emptyset$. So by ([FS1]), $p \notin S_\mu$.

We want to show next that $K \subset S\mu$. We have proved that every point in K is non wandering. Since S_μ is totally invariant there is a neighborhood V of S_μ such that $f^{-1}(V) \subset\subset V$. So points in $V \setminus f^{-n}(V)$ are wandering. Hence $K \cap V = S_\mu$. Let $K_1 := K \setminus S_\mu$. We just proved that K_1 is closed and totally invariant. Since f is expanding on K_1 there is a neighborhood $V_1 \supset K$ such that $f^{-1}(V_1) \subset\subset V_1$. Given any point $q \in S_\mu$, if $B(q, r) \cap V_1 = \emptyset$ then $\cup f^n(B(q, r))$ is disjoint from V_1. This contradicts that $q \in S_\mu$. Hence K_1 is empty and $K = S_\mu$.

■

We now consider again in $\mathbb{C}^2$ the family of polynomial maps

$$f_c = (P(z, w) - ca, Q(z, w) - cb)$$

introduced in this paragraph. We assume that on $\mathbf{P}^1$ the map $[P : Q]$ is chaotic, i.e. it's Julia set is equal to $\mathbf{P}^1$.

The critical set for f_c is $(t=0) \cup_j L_j$. Define

$$\begin{aligned} K_c &= \{(z,w); (f_c^n(z,w)) \text{ is bounded}\} \\ \mathcal{H} &= \{(z,w,c); (f_c^n(z,w)) \text{ is bounded}\} \end{aligned}$$

We have proved in Lemma 3.2 that

$$\mathcal{H} \subset \{\|(z,w)\| < 4\frac{\|(a,b)c\|^{\frac{1}{d}}}{A}\}$$

if

$$|c| \geq c_0 := 2^{\frac{d}{d-1}} \frac{1}{\|(a,b)\|} \frac{1}{A^{\frac{1}{d-1}}} \frac{1}{\left[2^{d-1} - \frac{1}{2}\right]^{\frac{d}{d-1}}}.$$

So $\mathcal{H}$ is closed in $\mathbb{C}^3$ and each slice K_c is polynomially convex. By analogy with the Mandelbrot set for the quadratic family we define:

$$M := \{c; (\cup_j L_j) \cap K_c \neq \emptyset\}.$$

PROPOSITION 3.9 *With the previous notation the set M is closed and bounded. When c is in the unbounded component C_∞ of $\mathbb{C} \setminus M$, then K_c is a Cantor set.*

Proof: Lemma 3.3 implies that M is bounded. That M is closed is obvious. The result that K_c is a Cantor set was proved for large $|c|$, in Proposition 3.5. The previous result implies that for $c \notin M$, the map f_c is strictly expanding on K_c. It follows that in the connected component C_∞, we have a smooth family of strictly expanding map on K_c. A theorem in Ruelle ([Ru]) implies that for any $c_0 \in C_\infty$ there is a neighborhood $\Delta(c_0, \delta)$ and for every $c \in \Delta(c_0, \delta)$ a homeomorphism $\phi_c : K_{c_0} \to K_c$ such that $f_c \circ \phi_c = \phi_c \circ f_{c_0}$. It follows that K_c is a Cantor set for every $c \in C_\infty$.

■

PROPOSITION 3.10 *Theorem 3.8 remains valid for holomorphic maps on* $\mathbb{P}^2$ *when the critical set is in the basin of an attractor.*

PROPOSITION 3.11 *Let* f *be a polynomial map on* $\mathbb{C}^2$ *which extends as a holomorphic map on* $\mathbb{P}^2$. *Assume* $K \cap C = \emptyset$ *and that the restriction to the hyperplane at infinity is hyperbolic on its Julia set. Then* f *is* $s-$*hyperbolic.*

Proof: The notion of $s-$hyperbolicity was introduced in ([FS2]); It follows from Theorem 3.8 that the set S_2 is totally invariant. It is also clear that S_1 is totally invariant which is more strict than the notion of $s-$hyperbolicity.

■

4 Isolated repelling points

4.1 Isolated repelling orbits

In this section we investigate small perturbations of polynomial automorphisms of $\mathbb{C}^2$, in order to construct infinitely many isolated repelling points in the nonwandering set of an endomorphism of $\mathbf{P}^2$.

Let $f_0(z,w) = (P(z,w), Q(z,w))$ be a biholomorphism of $\mathbb{C}^2$ of degree d. Assume that the indeterminacy set of f_0 as a rational map on $\mathbf{P}^2$ is $I_+ = [0:1:0]$ and the indeterminacy set of f_0^{-1} is $I_- = [1:0:0]$. Let $\tilde{P}, \tilde{Q}$ denote the homogeneous polynomials of degree d, such that $\tilde{P}(z,w,1) = P(z,w), \tilde{Q}(z,w,1) = Q(z,w)$. Notice that $f((t=0)\setminus I_+) = I_-$. Hence $\tilde{Q}(z,w,0) = 0$. Since I_+ consists of only $[0:1:0]$ it follows that $\tilde{P}(z,w,0) = az^d$. We may assume that $a = 1$.

Using a result of Jung, see ([FM]), we can assume that f_0 is a finite composition of Hénon maps. We consider the endomorphisms f_ϵ of $\mathbf{P}^2$ defined by

$$f_\epsilon[z:w:t] = [\tilde{P}(z,w,t) : \tilde{Q}(z,w,t) + \epsilon w^d : t^d].$$

The restriction of f_ϵ to $(t=0)$ is given by $[z^d : \epsilon w^d : 0]$, whose Julia set is the circle $|z| = \epsilon^{\frac{1}{d-1}}$ in $w = 1$. Observe that $(t = 0)$ is an attracting set.

Moreover $f_\epsilon^{-1}(I_+) = I_+$ and $f_\epsilon^{-1}(I_-) = I_-$. So I_+ and I_- are superattractive fixed points for f_ϵ.

Let B_+, B_- denote the basins of attraction of I_+ and I_- respectively, they are clearly totally invariant and $\overline{B}_+ \cup \overline{B}_-$ contains a neighborhood in P^2 of $(t = 0)$, the common boundary near $(t = 0)$ is made of the stable manifolds corresponding to the Julia set $S_1 := \{|z| = \epsilon^{\frac{1}{d-1}}\}$ in $(t = 0)$. (The unstable manifold is just $(t = 0), z, w \neq 0$ for every such point.) We will consider also B_A, the basin of attraction of $(t = 0)$, it consists of $B_+ \cup B_-$ and the stable set of S_1.

For $R > 0$, define

$$\begin{aligned} V_R &= \{|z| \leq R, |w| \leq R\} \\ V_R^+ &= \{|z| \geq R, |w| \leq |z|\} \\ V_R^- &= \{|w| \geq R, |w| \geq |z|\} \end{aligned}$$

This decomposition of $\mathbb{C}^2$ was introduced by Friedland-Milnor ([FM]). They showed that for the automorphism f_0, there is R such that $f_0(V_R \cup V_R^+) \subset V_R \cup V_R^+$.

It is easy to check that for f_0, I_- is a superattractive fixed point.

LEMMA 4.1 *There exist $\epsilon_0 > 0$ and $R > 0$ such that for all $0 < \epsilon \leq \epsilon_0$, $f_\epsilon(V_R \cup V_R^+) \subset V_R \cup V_R^+$. Moreover every point in V_R has at most one preimage in V_R.*

Proof: In $\mathbb{C}^2$ we have

$$f_\epsilon(z, w) = (P(z, w), Q(z, w) + \epsilon w^d) = (z_1, w_1).$$

We can choose R and $\epsilon_1 > 0$ so that in $V_{\frac{R}{2}}^+$

$$|P(z, w)| \geq \frac{|z|^d}{2}$$

and

$$|w_1| \le \frac{|z_1|}{2} \text{ if } |\epsilon| \le \epsilon_1.$$

If $(z, w) \in V_R$ it is clear that, after possibly increasing R and decreasing ϵ_1, for $|\epsilon| \le \epsilon_1$ we have that $f_\epsilon(V_R) \subset V_R \cup V_R^+$.

Since for f_0 there is only at most one preimage of $(z, w) \in V_R$, the same holds for f_ϵ if ϵ is small enough.

■

From now on we omit the subscript R and we assume that the Jacobian of f_0, is larger than one in modulus.

LEMMA 4.2 *Assume p is a repelling periodic point for f_ϵ, $f_\epsilon^m(p) = p$, and $p \in V$. Then p is isolated in the nonwandering set Ω_ϵ of f_ϵ.*

Proof: Assume first that $f_\epsilon(p) = p$. Let U_0 be a neighborhood of p such that $U_0 \cap f^{-1}(U_0) \subset\subset U_0$, and $U_0 \subset V$.

Define by induction $U_j := f^{-1}(U_{j-1}) \cap U_0$, $j \ge 1$. We want to show that every point in $U_1 \setminus \{p\}$ is wandering. Suppose $q \in \Delta_j := U_{j-1} \setminus U_j$. By modifying U_0 slightly we can assume that q is in the interior of Δ_j. If q is nonwandering, there is a first n such that $f^n(\Delta_j) \cap U_{j-1} \ne \emptyset$.

So there is $y \in \Delta_j$ with $f^n(y) = x \in U_{j-1}$, but $f^{n-1}(y) \notin U_{j-1}$. On the other hand, since $F(V^+) \subset V^+$, $f^{n-1}(y) \in f^{-1}(U_{j-1}) \cap U_0 = U_j \subset U_{j-1}$, a contradiction. So q is isolated in Ω_ϵ.

If q is periodic, $f_\epsilon^m(q) = q$, the whole orbit of q is in V. We can just apply the above argument to f_ϵ^m.

■

THEOREM 4.3 *There is an endomorphism f_ϵ of $\mathbb{P}^2$ such that the nonwandering set Ω_ϵ contains a countable discrete set of repelling periodic orbits.*

Proof: We can start with a one parameter family g_t of volume decreasing complex Hénon maps with real coefficients so that g_0 has a generic homoclinic tangency for some saddle point in $V \cap \mathbb{R}^2$. It is known, see ([Ga]) that for generic perturbations the perturbed maps have infinitely many sinks. Define $f_t = g_t^{-1}$. For ϵ small enough the family $((f_t)_{\epsilon|V})^{-1})$ has a generic homoclinic tangency and hence for some t, $((f_t)_{\epsilon|V})^{-1})$ has infinitely many sinks. Consequently $(f_t)_\epsilon$ has infinitely many repelling orbits in V. It follows from Lemma 4.2 that they are isolated in Ω_ϵ.

■

Remark 4.4 *Support of μ_ϵ is disjoint from V. Indeed for q in V, $\cup_{n\geq 0} f^n(V)$ omits V^- so by Proposition 3.6, the point q cannot belong to Support(μ_ϵ). The existence of a repelling periodic point not in the support of μ_ϵ is already stated in ([HP], p. 345). We thank M. Jonsson for mentioning that.*

Remark 4.5 *It is possible to construct hyperbolic sets of dimension 1 and unstable dimension 3, for endomorphisms of $\mathbf{P}^3$, that are disjoint from Support of μ. It is enough to consider the map $g_\epsilon([z : w : t : u]) = [P(z,w,t) : Q(z,w,t) + \epsilon w^d : t^d : u^d]$. So the first three components are the components of f_ϵ. A repelling periodic orbit in V for f_ϵ gives rise to a circle of unstable dimension 3.*

We now restrict attention to the family

$$f_\epsilon(z,w) = (z^2 + aw, z + \epsilon w^2), \;\; |a| >> 1.$$

The critical set for f_ϵ is defined by the equation $4\epsilon zw - a = 0$.

PROPOSITION 4.6 *Suppose $|a| > 1$. Then for $0 < |\epsilon| < \epsilon_0(a)$, the only Fatou components for f_ϵ are B_+ and B_-. The compact set $K := \mathbb{C}^2 \setminus (B_+ \cup B_- \cup W^s(S_1)) = \mathbb{C}^2 \setminus B_A$ is the set of points in $\mathbb{C}^2$ with bounded orbit.*

Proof: For a given a, define V, V^+ and V^- as:

$$\begin{aligned} R &= 10|a|+3 \\ V &= \{(z,w); |z| \leq R,\ |w| \leq 8R\} \\ V^+ &= \{(z,w,); |z| > R,\ |w| \leq 8|z|\} \\ V^- &= \{(z,w); |w| > 8R,\ |z| < |w|/8\} \end{aligned}$$

If $(z,w) \in V^+$, $(z_1,w_1) = f_\epsilon(z,w)$, then

$$\begin{aligned} |z_1| &= |z^2+aw| \\ &\geq |z|^2 - 8|a||z| \\ &\geq |z|(|z|-8|a|) \\ &\geq 3|z| \\ |w_1| &= |z+\epsilon w^2| \\ &\leq |z| + 64|\epsilon||z|^2. \end{aligned}$$

We get that $|w_1| \leq 8|z_1|$ provided $|z|(1+64|\epsilon||z|) \leq 8|z|(|z|-8|a|)$ which follows if $|\epsilon| < \frac{15}{640}$. Next, f_ϵ maps V^+ to itself and multiplies the z variable by at least 3. If $(z,w) \in V$, then $w_1 = z+\epsilon w^2$, so $|w_1| \leq R + 64|\epsilon|R^2$. Then $|w_1| \leq 8R$ provided that $|\epsilon| \leq \frac{7}{64(10|a|+3)}$. In that case $f_\epsilon(V) \subset V \cup V^+$.

It follows that V^+ is in the basin of the line at infinity. Next we observe that points in V^- with $|w| \geq 2/\epsilon$ are also in the basin of the line at infinity. Also note that points in V^- with $|w| \leq 3/(4\epsilon)$ are iterated forward until they reach $V \cup V^+$.

We get on V that the Jacobian determinant $|4\epsilon zw - a| \geq |a| - 32|\epsilon|R^2 \geq \frac{|a|+1}{2}$ provided that $|\epsilon| \leq \frac{|a|-1}{64(10|a|+3)^2}$. Hence the only Fatou component intersecting V will be the one containing V^+, i.e. the basin of attraction of $[1:0:0]$.

Hence if there is a Fatou component B other than the basins of attraction of $[1:0:0], [0:1:0]$, then it must be contained in $U := \{(z,w, \in V^-; 3/(4|\epsilon|) < |w| < 2/|\epsilon|\}$. Since the Jacobian determinant is $4\epsilon zw - a$, B

must intersect the set of points in U for which $|4\epsilon zw - a| \leq 1$. At such a point

$$\begin{aligned} |z| &= \frac{4}{3}|z|\frac{3}{4} \\ &\leq \frac{4}{3}|z\epsilon w| \\ &\leq \frac{1}{3}\left[|4z\epsilon w - a| + |a|\right] \\ &\leq |a| + 1. \end{aligned}$$

However f_ϵ maps such points to V^+.

■

4.2 The compact set K of points with bounded orbit.

It is clear that the compact set $K := \mathbb{C}^2 \setminus B_A$ is the set of points defined by

$$K = \{(z,w); |f^n(z,w)| \leq \frac{4}{\epsilon} \text{ for every } n.\}$$

We define $K_1 := K \cap V$ and

$$K_2 := \{(z,w); f^n(z,w) \in K \cap \{(z,w); |z| < \frac{|w|}{8}, \frac{3}{4\epsilon} < |w| < \frac{2}{\epsilon}\} \,\forall\, n\}.$$

Observe that K_2 is totally invariant.

LEMMA 4.7 *f_ϵ is expanding on K_2 and Supp$(\mu_\epsilon) = K_2$.*

Proof: Let (u,v) be an arbitrary tangent vector at $(z,w) \in K_2$. Then

$$(u',v') := f'_\epsilon(u,v) = (2zu + av, u + 2\epsilon wv).$$

$$\begin{aligned}
\text{if } |u| &\leq |v|/4 : \\
\text{then } |v'| &\geq \frac{3}{2}|v| - \frac{|v|}{4} \\
&\geq \frac{5}{4}|v| \\
&\Rightarrow \\
\max\{|u'|,|v'|\} &\geq \frac{5}{4}\max\{|u|,|v|\} \\
\text{if } |u| &\geq |v|/4 : \\
\text{then } |u'| &\geq 2|z||u| - 4|a||u| \\
(i)\ |z| &\geq 2|a| + 3 \\
|u'| &\geq 6|u| \\
&\geq \frac{3}{2}\max\{|u|,|v|\}. \\
(ii)\ |z| &\leq 2|a| + 3 \\
&\Rightarrow \\
f_\epsilon(z,w) &\in V^+ \\
&\Rightarrow \\
(z,w) &\notin K_2.
\end{aligned}$$

Since $\mathrm{Supp}(\mu_\epsilon)$ is totally invariant and contained in the non-wandering set, it is necessarily contained in K_2. (Because of Proposition 3.6, the support of the measure cannot intersect the line at infinity.)
That $\mathrm{Supp}(\mu_\epsilon)$ is equal to K_2 follows from the argument in Theorem 3.8 of ([FS2]).

■

Let Ω denote the non wandering set for f_ϵ in $\mathbb{C}^2$. Define

$$\begin{aligned}
\Omega_1 &:= \Omega \cap V, \\
U &:= V^- \cap \{\frac{3}{4|\epsilon|} < |w| < \frac{2}{|\epsilon|}\}.
\end{aligned}$$

PROPOSITION 4.8 *The non wandering set Ω is the union of the two disjoint closed sets Ω_1, Supp μ_ϵ. The set Ω does not intersect the critical set C. The map f_ϵ is bijective on Ω_1.*

Proof: It is clear that Ω_1 is closed. Since f_ϵ is expanding on $\Omega \cap U$, and since Supp μ_ϵ is totally invariant, there is a neighborhood $U_1 \supset$ Supp μ_ϵ such that $f_\epsilon^{-\ell}(U_1) \subset\subset U_1$ for large ℓ. If a point in U is non wandering it has to be in K_2 hence in Supp(μ_ϵ).

The critical set C is disjoint from V so $C \cap \Omega = \emptyset$. We also know that any point in V has at most one preimage in V.

■

PROPOSITION 4.9 *The set K has no isolated point. The local Hausdorff dimension of any point in $K \setminus$ Supp μ_ϵ is at least 2.*

Proof: Given any $q \in$ Supp μ_ϵ, we know that Supp(μ_ϵ) is not pluripolar near q ([FS2]), so q is not isolated in K. We next show that the local Hausdorff dimension of any point in $K \setminus$Supp μ_ϵ is at least 2. Let $q \in K \setminus$Supp μ_ϵ. Fix $r > 0$ such that the ball $B(q,r) \cap$Supp $\mu_\epsilon = \emptyset$. So on $B(q,r)$ we have that $(dd^cG^+)^2 = 0$ and $K \cap B(q,r) = \{(z,w); (z,w) \in B(q,r), G^+(z,w) = 0\}$. The function G^+ is the solution in $B(q,r)$ of the Dirichlet problem for the Monge Ampère equation $(dd^cu)^2 = 0$ with $u_{|\partial B(q,r)} = G^+$. The solution u is equal to the maximum of plurisubharmonic functions on $B(q,r)$ which are smaller than G^+ on $\partial B(q,r)$ ([BT]). Hence q is in the polynomially convex hull of $K \cap \partial B(q,r)$. Hence, for every $r > 0$ the Hausdorff dimension of $K \cap \partial B(q,r)$ is at least one. Consequently the Hausdorff dimension of K at q is at least 2, see ([S2]).

■

We want to show next that for $|a| >> 1$ the maps f_ϵ satisfy strong hyperbolicity conditions.

Let $f = (z^2 + aw, az)$ be a hyperbolic Hénon map. Define

$$\begin{aligned}
V &= \{|z| \leq 2|a|^2+2, |w| \leq 4|a|^3+4|a|\}, \\
V^+ &= \{|z| > 2|a|^2+2, |w| \leq 2|a||z|\}, \\
V^- &= \{|w| > 4|a|^3+4|a|, 2|a||z| < |w|\}, \\
K^+ &= \{(z,w)\} \text{ with bounded orbits}, \\
K^- &= \{(z,w)\} \text{ with bounded inverse orbits}, \\
K &= K^+ \cap K^-.
\end{aligned}$$

Then $K \subset \operatorname{int}(V)$. Let next $S \subset K$ consist of the nonwandering points of f. We assume that f is hyperbolic on S. So $S = S_0 \cup S_1 \cup S_2$, disjoint compact sets, S_j has stable dimension j. (Of course S_0 or S_2 must be empty and they are anyhow finite sets.) By ([BS1]) periodic points are dense in S, i.e. S satisfies Axiom A.

THEOREM 4.10 *For all $\epsilon \neq 0$, $|\epsilon|$ small enough, the maps $f_\epsilon := [z^2+awt : azt+\epsilon w^2 : t^2]$ satisfy Axiom A on the non wandering set.*

Proof: Let $\Omega = \Omega_\epsilon$ denote the nonwandering set. Then $\Omega_0^\epsilon := \Omega \cap (t=0)$ contains two attracting fixed points, $[1:0:0]$ and $[0:1:0]$ in S_0 and a saddle set $\{|z| = |\epsilon w|\} \cap (t=0) \subset S_1$. Moreover, since $(t=0)$ is an attracting set, Ω_0^ϵ is isolated in the non wandering set. Also, f_ϵ satisfies Axiom A there.

The set V^+ is contained in the basin of attraction of $[1:0:0]$. The set V is mapped to $V \cup V^+$. If $(z,w) \in V^-, |w| \geq \frac{3}{2|\epsilon|}$, then $|az+\epsilon w^2| > \epsilon|w|^2 - \frac{|w|}{2} \geq |w|$ so (z,w) is in the basin of attraction of $(t=0)$. Also if $(z,w) \in V^-, |w| \leq \frac{1}{2|\epsilon|}$, then $|az+\epsilon w^2| < \frac{|w|}{2} + \frac{|w|}{2} = |w|$. So the orbit lands eventually in $V \cup V^+$. Hence the set $\Omega_\epsilon = \Omega_0^\epsilon \cup \Omega_1^\epsilon \cup \Omega_2^\epsilon$ where $\Omega_1^\epsilon \subset V$ and $\Omega_2^\epsilon \subset \tilde{V}^- := \{(z,w) \in V^-; \frac{1}{2|\epsilon|} < |w| < \frac{3}{2|\epsilon|}\}$.

Next consider the nonwandering points Ω_2^ϵ. Since $\Omega_2^\epsilon \subset K_2$, hence f_ϵ is expanding on Ω_2^ϵ. We prove as in Theorem 3.8 that periodic points are dense in Ω_2^ϵ. So f_ϵ satisfies Axiom A on Ω_2^ϵ.

Finally, it follows from Ruelle ([Ru]) that f_ϵ also is hyperbolic on Ω_1^ϵ and satisfies Axiom A there.

■

5 Examples of endomorphisms such that Supp $\mu =$ Julia set.

In general the Julia set for an endomorphism of P^2 doesn't coincide with the support of the equilibrium measure μ. The simplest example is

$$f([z:w:t]) = [z^d : w^d : t^d].$$

In this paragraph we give examples where the Julia set for f coincides with the support of μ. The Briend-Duval theorem ([BD]) will then guarantee that repelling periodic orbits are dense in the Julia set.

We first give a general criterion.

THEOREM 5.1 *Let $(f_\lambda)_{\lambda\in\Delta}$ be a holomorphic family of holomorphic endomorphisms of algebraic degree $d \geq 2$ in P^2. Here Δ denotes the unit disc in $\mathbb{C}$. Assume*
(a) i) There is an algebraic curve V_0 in P^2 such that $f_\lambda(V_0) = V_0,\ \lambda \in \Delta$ and the Julia set for the restriction $(f_\lambda)_{|V_0}$ is equal to V_0.
ii) There is an integer ℓ, such that $\mathrm{P}^2 \setminus \cup_{j=0}^{\ell} f_\lambda^{-j}(V_0)$ is Kobayashi hyperbolic and hyperbolically embedded for $\lambda \in \Delta$.
(b) There exists $p \in V_0 \cap Supp(\mu_0)$, a repelling fixed point for f_0.
Then there exists a $\delta > 0$ such that Supp $\mu_\lambda = \mathcal{J}_\lambda$ for $\lambda \in \Delta(0,\delta)$.

Remark 5.2 *If f_λ has in addition an attracting fixed point, we also have that Supp $\mu_\lambda = \mathcal{J}_\lambda \neq \mathrm{P}^2$. This is the case in examples below.*

LEMMA 5.3 *Let $(f_\lambda)_{\lambda\in\Delta}$ be a holomorphic family of endomorphisms in P^2 of algebraic degree $d \geq 2$. Assume that $f_\lambda(p) = p,\ \lambda \in \Delta$ and that $p \in$ Supp μ_0 and is repelling for f_0. Then there is a $\delta > 0$ such that $p \in$ Supp μ_λ for $\lambda \in \Delta(0,\delta)$.*

Proof: Recall that the support of $\mu_\lambda =: S_\lambda$ is a totally invariant closed set. A point q belongs to S_λ if and only if for every $r > 0$, $\cup_n f_\lambda^n(B(q,r))$ omits at most a pluripolar set. To simplify notation, assume p is the origin of coordinates in $\mathbb{C}^2$. Since it is a fixed repelling point, there is an $\ell > 1$, and $\delta_0 > 0, \rho > 0, c > 0$ such that

$$|f_\lambda^\ell(z)| \geq (1+c)|z|,\ |z| \leq \rho \text{ and } \lambda \in \Delta(0,\delta_0).$$

Hence for all $\lambda \in \Delta(0, \delta_0)$, given $r_\lambda > 0$ we have

$$\cup_{n \geq 1} f_\lambda^n(B(0, r_\lambda)) \supset B(0, \rho/2).$$

The total invariance of S_λ implies that if $0 \notin S_\lambda$ then there is $r_\lambda > 0$ so that $\mu_\lambda(\cup_{n\geq 1} f_\lambda^n(B(0, r_\lambda)) = 0$. As a consequence $\mu_\lambda(B(0, \rho/2) = 0$. On the other hand we know ([FS1]) that $\lambda \to \mu_\lambda$ is weakly continuous. Since we have assumed that $0 \in S_0$, $\mu_\lambda(B(0, \rho/2)) > 0$. It follows that we cannot have $0 \notin S_{\lambda_k}$ for a sequence λ_k converging to 0.

■

LEMMA 5.4 *Let $f_0 \in \mathcal{H}_d$. Assume there is an algebraic curve V_0 such that $f_0(V_0) = V_0$ and $(f_0)_{|V_0}$ is chaotic. Assume that there is a repelling fixed point $p_0 \in V_0 \cap S_{\mu_0}$. If there is an integer ℓ such that $\mathbf{P}^2 \setminus \cup_{j=0}^{\ell} f_0^{-j}(V_0)$ is Kobayashi hyperbolic and hyperbolically embedded, then Supp $\mu_0 = \mathcal{J}_0$, the Julia set of f_0.*

Proof: The total invariance of S_0 = Support μ_0 implies that $V_0 \subset S_0$. The assumption on Kobayashi hyperbolicity implies that $\mathbf{P}^2 \setminus S_0$ is Kobayashi hyperbolic and hyperbolically embedded. Since $\mathbf{P}^2 \setminus S_0$ is invariant under f_0, the family of maps $(f_0^n)_n$ is equicontinuous on $\mathbf{P}^2 \setminus S_0$, hence the Julia set $\mathcal{J}_0$ of f_0 satisfies $\mathcal{J}_0 \subset S_0$. The other inclusion is obvious.

■

Proof of the Theorem: Lemma 5.3 implies that the point p is repelling and in Supp μ_λ for $\lambda \in \Delta(0, \delta)$. One can then apply Lemma 5.4 to the maps f_λ for $\lambda \in \Delta(0, \delta)$.

■

We next give an example of the situation of the Theorem. Let $f = [P(z, w) : Q(z, w)]$ be a map of degree $d \geq 3$ on $\mathbf{P}^1$. Assume that f has no exceptional points. Also, let $S(z, w)$ be a homogeneous polynomial of degree $d - 1$. Let (p_α) be those points in $\mathbf{P}^1$ on which S vanishes.

We assume next that we have the following nondegeneracy condition: Let $p \in \mathbf{P}^1$. Then $f^{-1}(p)$ contains at least one point not in $\{p_\alpha\}$.

We extend f to $\mathbf{P}^2$ as a two parameter family:

$$f_{\lambda,c} := [P + ct^d : Q + ct^d : t^d + \lambda t S].$$

The next result shows that points in the Julia set of f belongs to S_μ.

THEOREM 5.5 *(i) There exists a $\lambda_0 > 0$ and for all $\lambda, |\lambda| \geq \lambda_0$ there is a $c_0(\lambda)$ so that if $c \in \mathbb{C}, |c| < c_0(\lambda)$, then $[p : q : 0]$ belongs to Support $\mu_{\lambda,c}$ whenever $[p : q]$ belongs to the Julia set of f. Moreover the maps $f_{\lambda,c}$ have an attracting fixed point.*
(ii) There exists $\lambda_0 > 0$ so that if $|\lambda| \geq \lambda_0$ and $|c| > 1$, then $[p : q : 0]$ belongs to Supp$(\mu_{\lambda,c})$ whenever $[p : q]$ belongs to the Julia set of f.

Proof: We show that if U is any neighborhood of $[p : q : 0]$, then $\cup_n f^n_{\lambda,c}(U)$ contains the complement of a pluripolar set (actually an attracting fixed point) provided we impose the indicated conditions on λ, c. It will follow then from ([FS1]) that $[p : q : 0]$ belongs to Supp $\mu_{\lambda,c}$.

Note that after finitely many iterations, $f^n_{\lambda,c}(U)$ contains a neighborhood of the line at infinity.

Let $R = R_{\lambda,c,U} \geq 0$ denote the smallest number so that $\|(z,w)\| > R \Rightarrow [z : w : 1] \in V := \cup_n f^n_{\lambda,c}(U) \setminus (t = 0)$. We want to show that $R = 0$.

We fix (small) discs $D_\alpha \subset (t = 0)$ around the p_α. If the discs are small enough, the nondegeneracy condition on f implies that $f(\mathbf{P}^1 \setminus \cup_\alpha \overline{D}_\alpha) = \mathbf{P}^1$.

Notice that if $\lambda, c = 0$ then the line at infinity $(t = 0)$ is an attracting set, which is why we need some assumptions on the constants. We will first consider the case when λ is large, but $c = 0$, and then consider the case $c \neq 0$ as a small perturbation.

There exist constants $C_1, C_2 > 0$ so that if $[z : w] \notin \cup D_\alpha$ then $|S(z,w)| \geq C_1\|(z,w)\|^{d-1}$, while $|S(z,w)| \leq C_2\|(z,w)\|^{d-1}$ everywhere. Choose constants $A, B > 0$ so that

$$A\|(z,w)\|^d \leq \|(P(z,w), Q(z,w))\| \leq B\|(z,w)\|^d.$$

Suppose that $\lambda_0 \geq \frac{4B}{C_1} + 2$ and that $\|(z,w)\|^{d-1}C_1 \geq 1$. If $[z:w] \notin \cup D_\alpha$ and $|\lambda| \geq \lambda_0$, then

$$\begin{aligned}
\|f_{\lambda,0}(z,w)\| &\leq \frac{\|(P(z,w),Q(z,w)\|}{|1+\lambda S(z,w)|} \\
|1+\lambda S(z,w)| &\geq |\lambda||S(z,w)|-1 \\
&\geq \left(\frac{4B}{C_1}+2\right)C_1\|(z,w)\|^{d-1}-1 \\
&\geq 1 > 0. \\
\|f_{\lambda,0}(z,w)\| &\leq \frac{B\|(z,w)\|^d}{|\lambda||S(z,w)|-1} \\
&\leq \frac{B\|(z,w)\|^d}{\frac{4B}{C_1}C_1\|(z,w)\|^{d-1}} \\
&\leq \frac{\|(z,w)\|}{4}
\end{aligned}$$

Since the maps $f_{\lambda,0}$ maps lines to lines, it follows that the forward images of U cover at least the complement of the ball of radius $\left(\frac{1}{C_1}\right)^{\frac{1}{d-1}}$.

Next, in case (i) assume that $\|(z,w)\| = \left(\frac{1}{C_1}\right)^{\frac{1}{d-1}}$.

$$\begin{aligned}
\|f_{\lambda,0}\|(z,w) &\leq \frac{\|(P,Q)\|}{\frac{|\lambda|}{2}C_1\|(z,w)\|^{d-1}} \\
&\leq \frac{B}{\frac{|\lambda|}{2}C_1}\|(z,w)\| \\
&\leq \frac{\tilde{K}}{|\lambda|}
\end{aligned}$$

for some constant $\tilde{K}$. Hence it follows that the forward images of U cover the complement of the ball of radius $\frac{\tilde{K}}{|\lambda|}$.

Next, we estimate the image of points in the ball of radius $\frac{\tilde{K}}{|\lambda|}$. We get, using the condition $d \geq 3$:

$$\begin{aligned}
\|f_{\lambda,0}(z,w)\| &\leq \frac{B\|(z,w)\|^d}{1-|\lambda|C_2\left(\frac{\tilde{K}}{|\lambda|}\right)^{d-1}} \\
&\leq 2B\left(\frac{\tilde{K}}{|\lambda|}\right)^{d-1}\|(z,w)\| \\
&\leq \frac{\|(z,w)\|}{2}
\end{aligned}$$

if λ_0 is chosen larger if necessary.

This shows that the forward images of U cover all of $\mathrm{P}^2 \setminus (0)$ when $c = 0$. If $c \neq 0$, but small, depending on λ, $f_{\lambda,c}$ has an attracting fixed point P near the origin and by a continuity argument, $f^n_{\lambda,c}(U)$ still covers the complement of the basin of attraction of P, hence $\mathrm{P}^2 \setminus (P)$.

Next, in case (ii), we estimate the images when $c \neq 0$. This is done by a perturbation argument. First of all notice that there are small discs Δ_z around each point z in $(t=0) \setminus \cup \overline{D}'_\alpha$ where D'_α are slight extensions of D_α of approximately the same size so that the image of the boundary of Δ_z does not contain the image of z. Also notice that for $\|(z,w)\|$ sufficiently large, dependent on c, we still have $\|f_{\lambda,c}\|(z,w) \leq \|(z,w)\|/2$. More precisely, there exists then an $L > 0$ such that if $\|(z,w)\| \geq L|c|^{\frac{1}{d}}$ in addition to the above conditions, we still get that the forward iterates of U cover the complement of the ball of radius $\max L|c|^{\frac{1}{d}}, \left(\frac{1}{C}\right)^{\frac{1}{d-1}}$.

Since we may assume that $|c| \geq 1$, we can, by increasing L, assume that $L|c|^{\frac{1}{d}} \geq \left(\frac{1}{C}\right)^{\frac{1}{d-1}}$ and hence that the forward iterates of U cover the complement of the ball of radius $L|c|^{\frac{1}{d}}$.

Next, assume that $\|(z,w)\| = L|c|^{\frac{1}{d}}$.

$$\begin{aligned}
\|f_{\lambda,c}\|(z,w) &\leq \frac{\|(P+c,Q+c)\|}{\frac{|\lambda|}{2}C\|(z,w)\|^{d-1}} \\
&\leq \frac{B}{\frac{|\lambda|}{2}C}\|(z,w)\| + \frac{|c|}{\frac{|\lambda|}{2}C\|(z,w)\|^{d-1}}
\end{aligned}$$

$$\leq \frac{\tilde{K}}{|\lambda|}|c|^{\frac{1}{d}}$$

Hence it follows that the forward images of U cover the complement of the ball of radius $\frac{\tilde{K}}{|\lambda|}|c|^{\frac{1}{d}}$.

Next, we estimate the image of the ball of radius $\frac{\tilde{K}}{|\lambda|}|c|^{\frac{1}{d}}$. We get

$$\begin{aligned}
\|f_{\lambda,c}(z,w)\| &\geq \frac{|c| - \|(P,Q)\|}{|1+\lambda S|} \\
&\geq \frac{|c| - \|(P,Q)\|}{1+|\lambda|\frac{C'}{|\lambda|^{d-1}}|c|^{\frac{d-1}{d}}} \text{ for some constant } C' \\
&\geq \frac{\tilde{K}}{|\lambda|}|c|^{\frac{1}{d}} \text{ if} \\
|\lambda| &\geq \text{some}\lambda_0 >> 1.
\end{aligned}$$

But this implies that the ball of radius $\frac{\tilde{K}}{|\lambda|}|c|^{\frac{1}{d}}$ is in the image of it's complement. This shows that the forward images of U cover all of $\mathbf{P}^2$.

∎

We next make some further hypothesis.

(1) $[P:Q] = [\nu(z-2w)^d : z^d]$, $\left(\frac{\nu-2}{\nu}\right)^d = 1$, which is critically finite and chaotic with repelling fixed point $[\nu:1]$. Then S_μ contains $(t=0)$ and all its preimages.

(2) $S(z,w) = z^{d-1}$, $d \geq 3$.

Then the preimages of $(t=0)$ contains the lines in $(t=1) : z = \left(-\frac{1}{\lambda}\right)^{\frac{1}{d-1}} =: c_{\lambda,j}, j = 1,\ldots,d-1$. The preimages of these lines are given by

$$\Sigma_j = \{\frac{\nu(z-2w)^d + c}{1+\lambda z^{d-1}} = c_{\lambda,j}\}.$$

Note that for each value of $z \neq c_{\lambda,j}$, there is at least one value for w for which $(z, w) \in \Sigma_j$, and as j varies these must be different. Also note that these are $(d-1)$ branched covers over the z axis. Hence Kobayashi hyperbolicity of the complement of S_μ follows. To show that $\mathrm{P}^2 \setminus \cup f^{-j}(V_0)$ is hyperbolically embedded, observe that for $c \neq 0$, each Σ_j intersects the line $\{z = c_{\lambda,i}\}$ at d distinct points.

If we apply Lemma 5.4 to each $f_{\lambda,c}$ and the repelling fixed point $[\nu : 1 : 0]$, we find that $\mathrm{Supp}(\mu_{\lambda,c}) = \mathcal{J}_{\lambda,c}$. We have thus obtained:

THEOREM 5.6 *There exists a $\lambda_0 > 0$ such that if $|\lambda| \geq \lambda_0$ and $|c| < c_0(\lambda)$, then Supp$(\mu_{\lambda,c}) = \mathcal{J}_{\lambda,c} \neq \mathrm{P}^2$ for the maps*

$$f_{\lambda,c} = [\nu(z-2w)^d + ct^d : z^d + ct^d : t^d + \lambda t z^{d-1}].$$

When $|c| > 1$ we only know that $\mathrm{Supp}(\mu_{\lambda,c}) = \mathcal{J}_{\lambda,c}$.

References

[BS1] Bedford, E., Smillie, J; *Polynomial diffeomorphisms of* $\mathbb{C}^2$, Invent. Math. 103 (1991), 69-99.

[BT] Bedford, E. Taylor, B. A; *The Dirichlet problem for a complex Monge-Ampére equation,* Invent. Math. 37 (1976), 1-44.

[BD] Briend, J.Y, Duval, J; *Exposants de Lyapunov et points periodiques d'endomorphismes de* P^k, preprint (1998).

[CG] Carleson, L., Gamelin, T.W; *Complex Dynamics,* Springer, (1993)

[F] Fornæss, J.E; *Dynamics in Several Complex Variables,* CBMS 87.

[FS1] Fornæss, J.E, Sibony, N; *Complex Dynamics in higher dimension II,* Ann. Math. Studies 137 (1995), 135-187.

[FS2] Fornæss, J.E, Sibony, N; *Hyperbolic maps on* P^2, Math. Ann. 311, 305-333 (1998).

[FS3] Fornæss, J.E, Sibony, N; *Complex Hénon mappings in* $\mathbb{C}^2$ *and Fatou-Bieberbach domains,* Duke Math. J. 65 (1992), 345-380.

[FS4] Fornæss, J.E, Sibony, N; *Oka's inequality for currents and applications,* Math. Ann. 301 (1995), 399-419.

[FW] Fornæss, J.E, Weickert, B; *Attractors on* $\mathbf{P}^2$, to appear.

[FM] Friedland, S, Milnor, J; *Dynamical Properties of plane polynomial automorphisms,* Ergodic Theory and Dynamical Systems. 9 (1989), 67-99.

[Ga] Gavosto. E; *Attracting basins in* $\mathbf{P}^2$, J. Geometric Analysis, to appear.

[Gu] Guedj, V; *Approximation of currents on complex manifolds,* Math. Ann. 313, 437-474 (1999).

[HP] Hubbard, J, Papadopol; *Superattractive fixed points in* $\mathbb{C}^n$, Indiana Univ. Math. J. 43 (1994), 321-365.

[J1] Jonsson, M; *A nonalgebraic attractor in* $\mathbf{P}^2$, preprint (1998).

[J2] Jonsson, M; Critically finite paper Michigan Math Journal

[JW] Jonsson, M, Weickert, B; preprint 1998.

[Mi] Milnor, J; *Dynamics in one complex variable: Introductory lectures,* Preprint SUNY Stony Brook (1990).

[Ru] Ruelle, D; *Elements of differentiable dynamics and bifurcation theory,* Academic Press (1989).

[S1] Sibony, N; *Dynamique des applications rationelles de* $\mathbf{P}^k$, Panoramas et Synthèses, to appear.

[S2] Sibony, N; *Valeurs au bord de fonctions holomorphes et ensembles polynomialement convexes,* Springer-Verlag, Lecture Notes in Math. 578 (1977), 300-312.

[T] Tsuji; *Potential theory in modern function theory,* Tokyo Mazuren, 1959.

[U] Ueda, T; *Critical orbits of holomorphic maps on projective spaces,* J. Geom. Anal. (to appear).

John Erik Fornæss
Mathematics Department
The University of Michigan
East Hall, Ann Arbor, Mi 48109
USA

Nessim Sibony
CNRS UMR8628
Mathematics Department
Université Paris-Sud
Batiment 425
Orsay Cedex
France

Contemporary Mathematics
Volume **269**, 2001

3 Lectures on Foliations and Laminations on 3-Manifolds

David Gabai*

§0
Introduction

The study of foliations on manifolds has a long history and has undergone a rapid development since the 1940's. In particular, the geometric topological theory of foliations and laminations in 3-manifolds has experienced a huge development thanks in part to the works of Reeb [R], Haefliger [H], Novikov [No] and Thurston [Th1].

This manuscript is a survey of how foliations and laminations are used to derive topological information about 3-manifolds. The material chosen here includes many of the main results of the field as well as several central open problems in the subject. We have attempted to explain the key ideas behind a number of the results mentioned. We do not attempt to state results in their greatest generality.

The paper is organized as follows. Many classical and some recent results on codimension-2 and codimension-1 foliations are presented in Chapter 1. Chapter 2 is devoted to the theory of taut foliations. Chapter 3 gives an exposition of the theory of essential laminations.

Standard texts for the theory of 3-manifolds are [He] and [J]. For expository books on foliations see [Go], [La], [HH], [CN] and [CaCo] and for the theory of essential laminations in 3-manifolds see [GO] and [G6].

Notation and Terminology. For a topological space E, $|E|$ denotes the number of connected components, $\overset{\circ}{E}$ denotes the interior of E, and $N(E)$ the regular neighborhood of E in its ambient space. The compact 3-manifold M is *atoroidal* if every $\mathbb{Z}\oplus\mathbb{Z}\subset\pi_1(M)$ arises from a boundary torus. A *Haken* manifold is a compact, orientable, irreducible 3-manifold which contains an incompressible surface. *Irreducible* means that every smooth S^2 bounds a 3-cell and here an *incompressible* surface is a properly embedded surface, without 2-sphere components, which is π_1-injective.

Acknowledgement. This paper is based on lectures given at the May 1998 Foliations conference at SUNY Stony Brook and lectures given in August 1998 at the Morningside Center of Mathematics, Chinese Academy of Science, Beijing.

2000 *Mathematics Subject Classification.* Primary 57M50, 57R30.

*Partially supported by NSF Grant DMS-9505253

The author is deeply indebted to Fengchun Lei of Jilin University who provided a detailed and augmented set of lecture notes complete with references. He is grateful to Sergio Fenley, Will Kazez and Rachel Roberts for their many comments.

§1
Some Classical And Recent Results On Foliations

DEFINITION 1.1. A codimension-1 *foliation* $\mathcal{F}$ of a 3-manifold M is a decomposition of M into connected surfaces called *leaves* such that M is covered by a collection of *foliation charts* of the form $\mathbb{R}^2 \times \mathbb{R}$ where the leaves pass through a given chart in slices of the form $\mathbb{R}^2 \times x, x \in \mathbb{R}$. The foliation $\mathcal{F}$ is of class C^r, possibly analytic, if all the various coordinate transformations of the foliation charts are class C^r. We assume that M has an underlying smooth structure. The foliation is *tangentially smooth* if all leaves are smoothly immersed surfaces in M. (There exist tangentially smooth non C^1 foliations of the torus.) In this paper we will assume that all foliations are tangentially smooth. (See 5.3 [G6].) A *transversal* to $\mathcal{F}$ is a smooth curve transverse to the leaves of $\mathcal{F}$. Foliations $\mathcal{F}$ of M and $\mathcal{G}$ of N are *topologically equivalent* means that there is a homeomorphism $f : M \to N$ which takes leaves of $\mathcal{F}$ to leaves of $\mathcal{G}$.

In a similar manner one can define codimension-k foliations of n-manifolds. Here the foliations charts take the form $\mathbb{R}^{n-k} \times \mathbb{R}^k$ and the leaves are immersed manifolds of dimension $n - k$ which pass through the charts in slices of the form $\mathbb{R}^{n-k} \times x, x \in \mathbb{R}^k$. Although this paper is primarily about codimension-1 foliations and laminations on 3-manifolds, we begin with a brief discussion of codimension-2 foliations on 3-manifolds.

Codimension-2 Foliations Of 3-Manifolds. The simplest type of 1-dimensional foliation is a *circle bundle* over a closed surface. By definition each circle fibre has a neighborhood of the form $D^2 \times S^1$ such that the foliation restricts to the product foliation by circles. The next simplest foliation is the *Seifert fibering* of a compact 3-manifold. Again, the manifold is foliated by circle leaves and all but finitely many leaves have neighborhoods as above. There may be finitely many leaves, called *singular fibres* which have $D^2 \times S^1$ neighborhoods which are foliated as follows. Start with the product foliation of $D^2 \times I$, by intervals. Now identify $D^2 \times 0$ with $D^2 \times 1$ by a $2\pi\frac{p}{q}$ rotation about the origin, where p and q are relatively prime integers and give the resulting $D^2 \times S^1$ the induced foliation. The core circle of the $D^2 \times S^1$ is a leaf and all the other leaves wrap q times about the solid torus.

A detailed analysis of the rich structure of Seifert fibred spaces was carried out in Seifert's 1933 paper [Se], now in English translation. In particular Seifert classified these spaces up to fibre preserving homeomorphism. Barring a few exceptions [OVZ] Seifert fiberings are unique. Seifert fibred spaces play a key role in Thurston's [Th2] geometrization Conjecture. Seifert fibred spaces correspond to six of the eight geometries conjectured by Thurston [Th2] to comprise the building blocks of all closed 3-manifolds.

THEOREM 1.2 (EPSTEIN [E1-2]). *If M^3 is a closed orientable foliation by circles, then M is a Seifert fibred space.*

If M has a Seifert fibering and $|\pi_1(M)| = \infty$, then a generic fibre generates infinite cyclic normal subgroup of $\pi_1(M)$. Conversely we have

THEOREM 1.3 (... SCOTT, MESS, CASSON-JUNGREIS [CJ], GABAI [G5]). *If M is a closed orientable irreducible 3-manifold and $\pi_1(M)$ has an infinite cyclic normal subgroup, then M is a Seifert fibred space.*

Anosov and pseudo-Anosov flows give rise to important codimension-2 foliations of 3-manifolds. See for example [Fe] and [Mo]. The stable and unstable laminations of these flows give rise to "Reebless" foliations and/or essential laminations of which more will be said later.

Codimension-1 Foliations of 3-Manifolds. For the remainder of this paper, unless stated otherwise, all foliations and laminations will be of codimension one.

As far as I know, the first application of foliations to derive a topological theorem is

THEOREM 1.4 (ALEXANDER [A2]). *Every PL embedded $S^2 \subset \mathbb{R}^3$ bounds a PL 3-ball.*

IDEA OF PROOF. View $\mathbb{R}^3$ as $\mathbb{R}^2 \times \mathbb{R}$ with the product foliation. After a small isotopy we can assume that the S^2 is transverse to the horizontal planes except at finitely many local maxima, minimal and saddle points. One readily checks that if the 2-sphere has exactly one maximum and one minimum, then it bounds a 3-ball. Now proceed by induction on the number of critical points. If it has more than 2 such critical points, then use Morse theory to isotope the S^2 to reduce that number. (Morse theory had yet to be discovered!) Figure 1.1 shows the the most elementary simplifying move (drawn in the smooth category). Usually one pushes down on a local max until it encounters a saddle, then one does an isotopy of one sort or another to cancel the max and saddle, though one must take care to keep the S^2 embedded. One should read Alexander in the original for it is but three pages, very well written and efficiently organized.

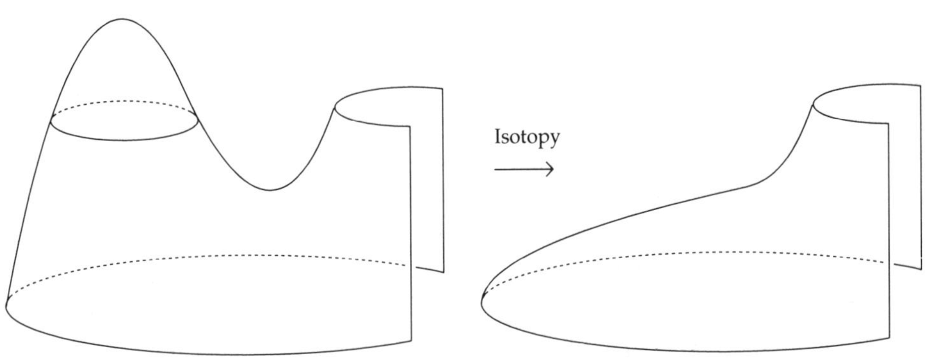

Figure 1.1

Alexander's proof is the model for many arguments of increasing complexity. For detailed accounts of such arguments see [Rou], [G3] or [G7].

Constructions of Foliations. The birth of the theory of foliations on 3-manifolds was heralded by Reeb's 1948 discovery [Re] of a foliation on the 3-sphere. A $D^2 \times S^1$ (resp. $I \times S^1$) together with the foliation described on it by Figure 1.2b (resp. 1.2a) is called a Reeb (resp. 2-dimensional Reeb) component. Reeb foliated S^3 by appropriately gluing together two Reeb components along their common torus bounday. A more analytic way of constructing a Reeb component is by starting with $\bar{\mathbb{H}}^3 = \mathbb{R}^2 \times [0, \infty)$ with the product foliation by planes. Now delete $(0,0,0)$ and act on $\bar{\mathbb{H}}^3$ by the $\mathbb{Z}$-action generated by $(x, y, z) \to 2(x, y, z)$. The quotient manifold $(\bar{\mathbb{H}}^3 - (0,0,0))/\mathbb{Z}$ is a solid torus and the induced foliation is the Reeb foliation.

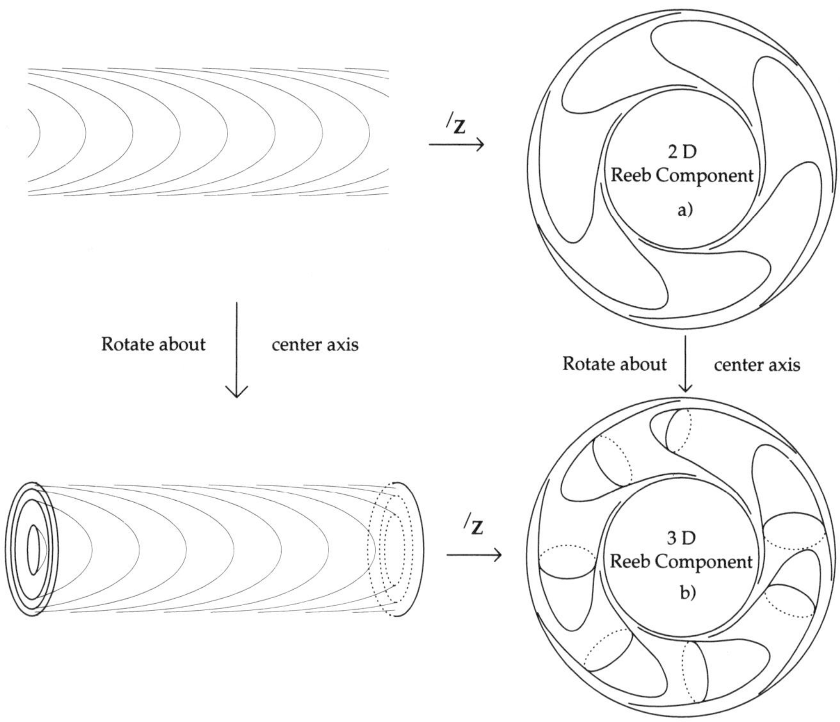

Figure 1.2

THEOREM 1.5 (LICKORISH [LICK], ZIESCHANG [NO]). *Every closed orientable 3-manifold M has a codimension-1 foliation.*

IDEA OF PROOF. By [Wl] and [A1], M is obtained by surgery on a braid L in S^3. Make this braid transverse to the Reeb foliation of S^3 by putting it near the core of a Reeb component. By a process called *turbularization* perturb the foliation, to obtain a foliation of $S^3 - \overset{\circ}{N}(L)$ with $\partial N(L)$ leaves. M is obtained by attaching solid tori to $\partial N(L)$. Extend the foliation on $S^3 - \overset{\circ}{N}(L)$ to all of M by filling the attached solid tori with Reeb foliations. This is schematically shown in Figure 1.3

REMARK. All of these constructions require the use of Reeb components. We say a foliation is *Reebless* if it contains no Reeb component.

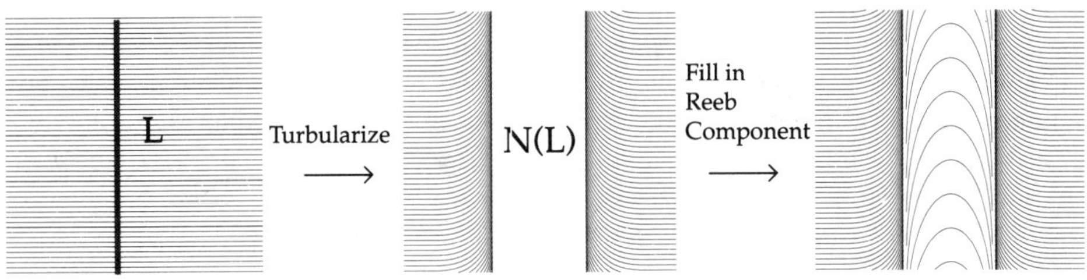

Figure 1.3

Holonomy 1.6. If L is a leaf of the foliation $\mathcal{F}$ and $x \in L$, then define a map $\Phi : \pi_1(L, x) \to \{$germs of homeomorphisms of $(-1,1)$ fixing $0\}$ as follows. If γ is a closed loop based at x, then γ is covered by a chain of foliation charts $U_1, \cdots, U_n$. Let $(-1,1)$ parametrize a transversal of $\mathcal{F}$ with 0 identified with x. If $y \in (-1,1)$ is sufficiently close to 0, then using the foliation charts, the path γ starting at x, can be shadowed by a path starting at y. Define $\Phi(\gamma)(y)$ to be the endpoint of this path. An argument similar to one in the proof of the Seifert - Van Kampen theorem shows that two paths that are homotopic rel endpoints give rise to the same germ. If the foliation is of class C^r, then so is each $\Phi(\gamma)$. In a similar manner one can define holonomy for codimension-k foliations. If $\Phi(\gamma) = id$, then γ is said to have *trivial holonomy.* If each curve in L (resp. each leaf of $\mathcal{F}$) has trivial holonomy, then we say L (resp. $\mathcal{F}$) has trivial holonomy. Given Φ and a compact subset K of a leaf, one can recover the structure of $\mathcal{F}$ near K.

REEB STABILITY THEOREM 1.7 [R]. *Let $\mathcal{F}$ be a codimension-1 foliation on M. If K is a compact subset of a leaf L, and each closed curve in K has trivial holonomy, then there exists a product neighborhood U homeomorphic to $K \times (-1,1)$ such that $\mathcal{F}|U$ is the product foliation.*

THEOREM 1.8 (REEB[R]). *If $\mathcal{F}$ is a codimension-1 foliation on the closed orientable 3-manifold M and $\mathcal{F}$ has an S^2 leaf, then either $M = S^2 \times S^1$ or $P^3 \# P^3$. In the first case $\mathcal{F}$ is the product foliation and in the second case M is covered by $S^2 \times S^1$ and $\mathcal{F}$ lifts to the product foliation.*

Using Foliations To Study The Fundamental Group.

THEOREM 1.9 (HAEFLIGER [H]). *If $\mathcal{F}$ is an analytic foliation on M, then*
i) $|\pi_1(M)| = \infty$ and
ii) If γ is a closed transversal to $\mathcal{F}$, then $1 \neq [\gamma]^n \in \pi_1(M)$ for any integer $n \neq 0$.

IDEA OF THE PROOF OF II). If γ is homotopically trivial, then let $D \subset M$ be an immersed disc which spans γ. By perturbing the map slightly we can assume that D is transverse to $\mathcal{F}$ except at finitely many generic tangencies. Since $\chi(D^2) = 1$, and ∂D is transverse to $\mathcal{F}$, the Poincare - Hopf index formula applied to the induced foliation on D implies that there exists a singularity of index 1 on $\mathcal{F}|D$ at some point $x \in \mathring{D}$. This implies that a foliation chart containing x can be parametrized so that under projection to the $\mathbb{R}$ coordinate of the $\mathbb{R}^2 \times \mathbb{R}$, D has a local maximum at x. This in turn implies that near $x, \mathcal{F}|D$ is a collection of concentric circles converging to x. Thus all leaves of $\mathcal{F}|D$ sufficiently close to index 1 singularities have trivial holonomy in $\mathcal{F}|D$. By taking a limit of trivial holonomy curves on $\mathcal{F}|D$ one finds a curve η whose holonomy germ $\phi : (-\epsilon, \epsilon) \to (-1, 1)$ satisfies $\phi|(-\epsilon, 0] = \text{id}$, but the germ of $\phi|[0, \epsilon)$ is not equivalent to the identity. (For example, the holonomy is trivial on one side of η, but is conjugate to the contraction $t \to \frac{1}{2}t$ on the other side.) This contradicts the fact that ϕ is a real analytic map. Since one readily finds a closed transversal to any foliation of a closed manifold, ii) implies i).

THEOREM 1.10 (NOVIKOV[NO]). *Let M^3 be a closed orientable 3-manifold with Reebless foliation $\mathcal{F}$, then*

i) If L is a leaf of $\mathcal{F}$, then the homomorphsim $\pi_1(L) \to \pi_1(M)$ induced by inclusion is injective.

ii) If γ is a closed transversal to $\mathcal{F}$, then $1 \neq [\gamma]^n \in \pi_1(M)$ for any integer $n \neq 0$. (Again this implies that $\pi_1(M)$ is infinite.)

iii) Either $\pi_2(M) = 0$ or $\mathcal{F}$ is the product foliation on $M = S^2 \times S^1$

REMARKS 1.11. Using Novikov's theorem and Alexander's argument, Rosenberg [Ro] improved iii) to the statement, "If $M \neq S^2 \times S^1$, then M is irreducible" (i.e. every smoothly embedded $S^2 \subset M$ bounds a 3-ball), thereby ruling out the possibility that M contains fake 3-cells. Thus if M supports a Reebless foliation, then $\tilde{M}$ contains no submanifold which gives rise to a counter example to the Poincare conjecture. This result was further extended [Pa] to show that in the non $S^2 \times S^1$ case, $\tilde{M} = \mathbb{R}^3$. See 1.18 for the precise statement.

IDEA OF THE PROOF OF II). Let γ be a homotopically trivial closed transversal. Let D be an immersed disc that γ bounds. To capture the idea of the proof, make the simplififying assumptions that D is embedded, $\mathcal{F}|D$ appears as in Figure 1.4 a) and D is transverse to $\mathcal{F}$ except at the single point x where D has a local maximum in some foliation chart. From this 2-dimensional picture we can derive the 3-dimensional picture shown in Figure 1.4 b). Each circle of $\mathcal{F}|D$ near x bounds a disc. Use the Reeb stability theorem and Haefliger's theorem [H] (that the limit of compact leaves is compact) to span more and more of these circles by discs. Since some closed leaf of $\mathcal{F}|D$ has holonomy, not every closed leaf of $\mathcal{F}|D$ bounds a disc in its leaf. By Reeb the set of such leaves is closed and hence there exists an innermost closed leaf ψ of $\mathcal{F}|D$ which does not bound a disc in its leaf. Leaves interior to ψ bound discs of arbitrarily large area, else ψ would bound a disc which is a limit of such discs. Since M has finite volume, the only possibility is that the discs sweep back over previously defined ones as in Figure 1.4 c). Passing to the limit one finds the Reeb component. See [CN] for a detailed proof.

Other Important Foliations/Topological Theorems.

THEOREM 1.12 (SACKSTEDER [SA]). *If $\mathcal{F}$ is a C^2-foliation without holonomy on the compact 3-manifold M, then $\mathcal{F}$ is topologically equivalent to a foliation defined by a closed nonsingular 1-form.*

THEOREM 1.13 (ROSENBERG-SONDOW [RO]). *If the closed 3-manifold M has a C^2-foliation by planes, then $M = T^3$.*

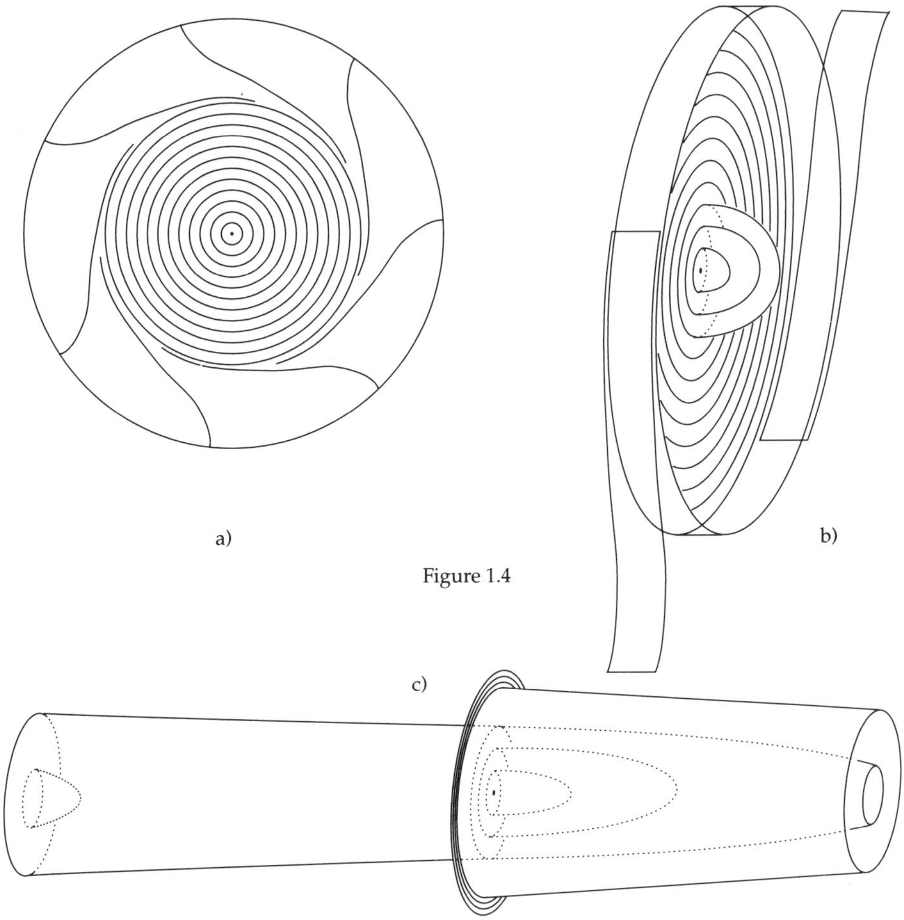

a) b) c)

Figure 1.4

THEOREM 1.14 (IMANISHI [I], [G4]). *If the closed 3-manifold M has a C^0-foliation by planes, then $M = T^3$.*

THEOREM 1.15 (TISCHLER [TI]). *If the foliation $\mathcal{F}$ on the compact 3-manifold M is defined by a closed 1-form, then M fibres over S^1.*

The Space Of Leaves. The *space of leaves* of a foliation $\mathcal{F}$ on M is the quotient space of M obtained by contracting each leaf of $\mathcal{F}$ to a point. This in general is a rather intractible object; however if M is a simply connected n-manifold and each leaf of $\mathcal{F}$ is a plane of dimension n-1, then the space of leaves of $\mathcal{F}$ is a possibly non Hausdorff simply connected 1-manifold. Let $\mathcal{T}(\mathcal{F})$ denote the space of leaves of the lift $\tilde{\mathcal{F}}$ of $\mathcal{F}$ to the universal covering space $\tilde{M}$ of M. The following result was motivated by Kaplan [Ka1-2].

THEOREM 1.16 (HAEFLIGER-REEB [HR]). *There exists a 1-1 correspondence between the sets* {*second countable, connected and simply connected 1-manifolds*} *and* {*Foliations on* $\mathbb{R}^2$}/ $\sim$.

Very roughly speaking the equivalence relation on the foliations corresponds to turning Reeb strips upside down. See Figure 1.5b.

THEOREM 1.17 (PALMEIRA [PA]). *If $\mathcal{F}_1$ and $\mathcal{F}_2$ are C^2 foliations by (n-1)-planes on the simply connected n manifold $M, n \geq 3$, then $\mathcal{F}_1$ is topologically conjugate to $\mathcal{F}_2$ if and only if $\mathcal{T}(F_1) = \mathcal{T}(F_2)$.*

COROLLARY 1.18 [PA]. *If $\mathcal{F}$ is a C^2 Reebless foliation on the closed 3-manifold M, then $\tilde{\mathcal{F}}$ is topologically conjugate to a product foliation of the form (foliation on $\mathbb{R}^2$) $\times \mathbb{R}$. In particular $\tilde{M} = \mathbb{R}^3$.*

EXERCISE 1.19. Show that $\lambda_1 \neq \lambda_2$ but $\lambda_1 \times \mathbb{R} = \lambda_2 \times \mathbb{R}$, where λ_1 and λ_2 are the foliations of $\mathbb{R}^2$ by lines shown in Figure 1.5.

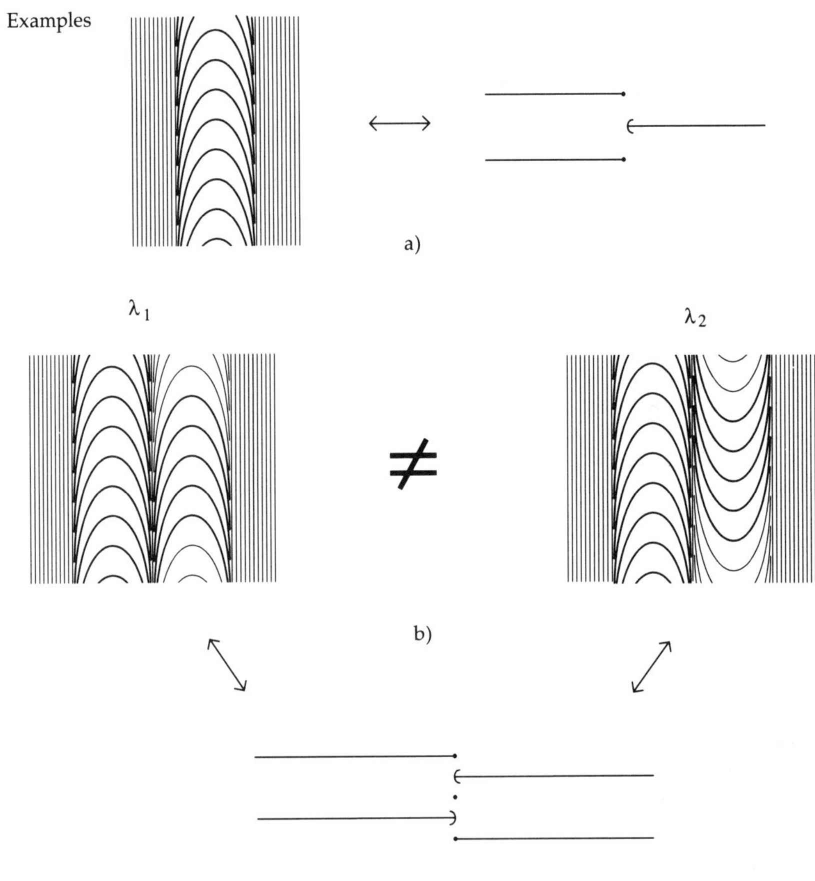

Figure 1.5

Given a path in a leaf L connecting two points x, y one similary defines a holonomy map, which is a germ of a homeomorphism of $(-1,1)$ fixing 0. The germ is obtained by parallel transporting a short transversal through L at x to a transversal through y. The following amazing theorem says that almost all the time one can define a holonomy map over infinite paths.

THEOREM 1.20 (THURSTON [TH4]). *Let $\mathcal{F}$ be a codimension-1 foliation of a closed n-manifold M such that each leaf of $\mathcal{F}$ is dense in M and $\mathcal{F}$ has no holonomy invariant transverse measure.*

1) For each $x \in M$ and for almost every ray $\rho : [0, \infty) \to L$ based at x and geodesic in its leaf L, there exists a transversal $[-\epsilon, \epsilon]$ of $\mathcal{F}$ with 0 identified with x such that for each subpath $\rho|[0,t]$, the transversal $[-\epsilon, \epsilon]$ is contained in the domain of an associated holonomy map h_t. Furthermore $lim_{t\to\infty}$ length($h_t([-\epsilon, \epsilon])$) = 0.

2) If dimension $M = 3$ and each leaf of $\mathcal{F}$ is negatively curved, then there exists an action of $\pi_1(M)$ on $\mathcal{U}$, where $\mathcal{U}$ is a circle called the universal circle. $\mathcal{U}$ has the property that if $\tilde{L}$ is a leaf of $\tilde{\mathcal{F}}$, then $S^1_\infty(\tilde{L})$, the circle at infinity to $\tilde{L}$, is obtained as a monotone quotient of $\mathcal{U}$. This quotient is compatible with the action of $\pi_1(M)$ on the disjoint union $\{S^1_\infty(\tilde{L})\}_{\tilde{L} \text{ leaf of } \tilde{\mathcal{F}}}$.

CONJECTURE 1.21 (THURSTON [TH4]). *If $\mathcal{F}$ is a taut foliation on the atoroidal 3-manifold such that every leaf of $\mathcal{F}$ is dense in M, then the action of $\pi_1(M)$ on $\mathcal{U}$ gives rise a 2-dimensional genuine lamination transverse to $\mathcal{F}$.*

§2
Taut foliations

A foliation $\mathcal{F}$ is *transversely orientable* if there exists a vector field transverse to the leaves. Any foliation on M lifts to a transversely orientable one by passing to an appropriate 2-fold covering space. A *dead end component* [No] of a transversely orientable foliation $\mathcal{F}$ in a compact 3-manifold M is a codimension-0 manifold N such that for some transverse orientation on $\mathcal{F}$, the vectors point into N along the points of ∂N. An elementary Euler characteristic argument [Good] shows that ∂N is a union of tori.

DEFINITION. A transversely orientable foliation $\mathcal{F}$ on a compact orientable 3-manifold is taut if there exists a closed transversal which intersect each leaf of $\mathcal{F}$.

THEOREM 2.1 (GOODMAN [GOOD]). *A transversely orientable foliation $\mathcal{F}$ is taut if and only if $\mathcal{F}$ has no dead end components. In particular a taut foliation is Reebless.*

IDEA OF PROOF. It suffices to show that each leaf meets a closed transversal. This is readily verified for non compact leaves [H]. Since the union of the closed leaves is a closed set in M, there are finitely many compact leaves $L_1, \cdots, L_n$ such that any other closed leaf L cobounds with some L_i a foliated $L \times I$ with $\mathcal{F}|L \times I$ transverse to the I-fibres. Thus it suffices to show that there exists a closed transversal through each of these L_i's. Partially order the L_i's by declaring that $L_i \leq L_j$ if there is an oriented transversal from L_i to L_j and show that if $1 \leq i, j \leq n$ then $L_i \leq L_j$. To do this use the following fact. Let N be a compact foliated submanifold of M with ∂N consisting of leaves of $\mathcal{F}$. If $I_1, \cdots, I_p$ denote the components of ∂N with transverse orientations pointing in, and $O_1, \cdots, O_m$ denote the components with transverse orientation pointing out, then $\sum_i \chi(I_i) = \sum_j \chi(O_j)$.

THEOREM 2.2 (SULLIVAN [SU1]). *A C^2 foliation $\mathcal{F}$ is taut if and only if there exists a volume preserving flow transverse to the leaves of $\mathcal{F}$.*

MINIMAL SURFACE THEOREM 2.3 (SULLIVAN [SU2], HARVEY - LAWSON [HL]). *If $\mathcal{F}$ is a C^2 taut foliation on the 3-manifold M, then there exists a Riemannian metric on M such that leaves of $\mathcal{F}$ are homologically area minimizing. I.e., if K*

is a compact surface lying in a leaf L *of* $\mathcal{F}$, T *is an immersed surface in* M *with* $\partial T = \partial K$ *and* $[K] = [T] \in H_2(M, \partial K)$, *then area*$(K) \leq$ *area*(T).

The groundbreaking work of Sullivan (inspired by Gluck and using work of Rummler) showed that the leaves are locally area minimizing. Subsequently Harvey and Lawson showed that the leaves are in fact calibrated and hence homologically area minimizing.

IDEA OF THURSTON'S PROOF OF SULLIVAN'S THEOREM. Fix a transverse orientation on $\mathcal{F}$ and find a finite set $\gamma_1, \cdots, \gamma_n$ of oriented closed transversals which go through every foliation chart. Let $\Gamma = \cup \gamma_i$. By considering algebraic intersection number it follows that if K is a compact subsurface of the leaf L of $\mathcal{F}$, T is an immersed surface in M with $\partial T = \partial K$ and $[K] = [T] \in H_2(M, \partial K)$, then $|\Gamma \cap K| \leq |\Gamma \cap T|$. Diffuse this "discrete metric", given by geometric intersection number with Γ, to obtain the desired Riemannian metric.

TRANSVERSE HOMOTOPY THEOREM 2.4 [G7]. *Let* S *be a compact immersed incompressible surface in the compact 3-manifold* M *with a* C^0 *taut foliation, such that each component of* ∂S *is either a leaf of* $\mathcal{F}|\partial M$ *or transverse to* $\mathcal{F}|\partial M$. *Then* S *can be homotoped rel boundary to be either a leaf of* $\mathcal{F}$ *or a surface transverse to* $\mathcal{F}$ *except at isolated saddles.*

Early results of this kind can be found in [A2] and [Ro]. A very important precursor is the Roussarie [Rou] - Thurston [Th0] theorem independently proven in 1971. They established the above theorem for the case when S is embedded. [G1] 1981 established the above theorem for $\mathcal{F}$ finite depth. Hass [Hs] 1983 established the above theorem for C^2 foliations by combining Sullivan's theorem with the following theorem of Schoen - Yau [SY]. "If S is a compact immersed incompressible surface in a compact Riemannian 3-manifold M with convex boundary, then S is homotopic to a least area properly embedded surface."

Thurston Norm 2.5 [Th1]. Let M be a compact, orientable, irreducible 3-manifold such that ∂M is a union of incompressible tori and every incompressible torus in M is boundary parallel. Define $\chi_- : H_2(M, \partial M) \to \mathbb{Z}$ by

$$\begin{aligned}\chi_-(\alpha) = \min\{|\chi(S)| \mid (S, \partial S) &\text{ is properly embedded in } (M, \partial M),\\ [S] = \alpha \in H_2(M, \partial M), &\text{ and } S \text{ has no 2-sphere components}\}\end{aligned}$$

A surface which minimizes χ_- is called *Thurston norm minimizing.*

EXAMPLE 2.6. Let $M = S^3 - \overset{\circ}{N}(k)$ where k is a knot in S^3 and $\overset{\circ}{N}(k)$ is an open regular neighborhood of k. A Thurston norm minimizing surface representing a generator of $H_2(M, \partial M)$ corresponds to a minimal genus Seifert surface for k.

THEOREM 2.7 (THURSTON [T1]). χ_- *extends to a norm* $x : H_2(M, \partial M; \mathbb{R}) \to \mathbb{R}$ *which is*

i) Linear on rays: $x(t\alpha) = tx(\alpha)$ *for* $t \in R$

ii) Symmetric about 0: $x(-\alpha) = x(\alpha)$

iii) Convex: $x(\alpha + \beta) \leq x(\alpha) + x(\beta)$

iv) The unit ball $B_x = \{\alpha \in H_2(M, \partial M; \mathbb{R}) | x(\alpha) = 1\}$ *is a finite sided polyhedron.*

v) If $\mathcal{F}$ is a Reebless foliation on M such that each component of ∂M is either transverse to $\mathcal{F}$ or is a leaf of $\mathcal{F}$ and S is a compact leaf of $\mathcal{F}$, then S is norm minimizing.

IDEA OF THE PROOF OF V). Let T be an incompressible surface such that $[T] = [S]$. For simplicity assume that T is connected, M is closed, and $\mathcal{F}$ is taut. We may assume by the Roussarie - Thurston theorem that T is either a leaf of $\mathcal{F}$ or transverse to $\mathcal{F}$ except at isolated saddle tangencies. Let $E \in H^2(M, \partial M)$ be the Euler class of the bundle of 2-planes tangent to $\mathcal{F}$. Since this bundle restricts to the tangent bundle to S we obtain $E(S) = \chi(S)$. On the other hand, a direct calculation shows

$$|E(T)| = \begin{cases} |\chi(T)|, & \text{if } T \text{ is a leaf or} \\ \sum \sigma(x_i) & \text{otherwise,} \end{cases}$$

where, in the latter case, the summation is taken over all $x_i \in T$ which are saddle tangencies with $\mathcal{F}$ and $\sigma(x_i) = 1$ or -1 according to whether or not the transverse orientation to $\mathcal{F}$ at x_i agrees with the oriented normal to T at x_i. Thus

$$|\chi(S)| = |E([S])| = |E([T])| = |\sum \sigma(x_i)| \leq \sum |\sigma(x_i)| = |\chi(T)|$$

where the last equality follows by the Poincare - Hopf index formula.

Using the transverse homotopy theorem we obtain

THEOREM 2.8 [G1]. *Let S be a compact leaf of the Reebless foliation $\mathcal{F}$ on M. If T is immersed in M with no 2-sphere components and $[T] = [S] \in H_2(M, \partial M)$, then $|\chi(S)| \leq |\chi(T)|$.*

REMARK 2.9. i) Actually the proof of Theorem 2.8 shows that compact subsurfaces of leaves are homologically norm minimizing. More precisely "If K is a compact surface lying in a leaf L of $\mathcal{F}$, T is an immersed surface in M with $\partial T = \partial K$ and $[K] = [T] \in H_2(M, \partial K)$, then $|\chi(K)| \leq |\chi(T)|$."

ii) Theorem 2.4 was not available in 1981, but the transverse homotopy theorem for finite depth foliations together with Theorem 2.11 provided sufficient technology to establish Theorem 2.8.

The C^2 versions of the previous results on taut foliations can be summarized by the following

THEOREM 2.10. *If $\mathcal{F}$ is a C^2 transversely orientable foliation on a compact orientable 3-manifold, then the following are equivalent.*

1) $\mathcal{F}$ is taut, i.e. there exists a closed transversal intersecting each leaf of $\mathcal{F}$.

2) $\mathcal{F}$ has no dead end components.

3) The leaves of $\mathcal{F}$ are homologically norm minimizing (among immersed surfaces).

4) There exists a Riemannian metric such that the leaves of $\mathcal{F}$ are homologically area minimizing (among immersed surfaces).

5) There exists a volume preserving flow transverse to $\mathcal{F}$.

REMARK. 1) if and only if 2) if and only if 3) are true for C^0 foliations.

Constructions Of Taut Foliations And Applications. Let M be a compact oriented 3-manifold with a codimension-1 foliation $\mathcal{F}$. We say a leaf L is of *depth* 0 if L is compact. Having defined *depth* $j \leq k$ *leaves* we say that L is a depth $k+1$ leaf if $\bar{L} - L$ is a union of depth $j \leq k$ leaves and contains a depth k leaf. $\mathcal{F}$ is of *depth* k if depth(L) is defined for every leaf of $\mathcal{F}$ and $k = \max\{\text{depth}(L) | L \text{ a leaf of } \mathcal{F}\}$. Every leaf of a finite depth foliation is proper, i.e. its topology as a leaf is equal to its topology as a subspace of M.

THEOREM 2.11 [G1]. *Let M be a compact, orientable, irreducible 3-manifold such that ∂M is a (possibly empty) union of tori. If S is a norm minimizing surface in M, then there exists a taut foliation $\mathcal{F}$ of M such that $\mathcal{F}$ is transverse to ∂M, S is a leaf of $\mathcal{F}$, $\mathcal{F}$ is of finite depth and $\mathcal{F}|\partial M$ is a suspension of homeomorphisms of S^1, i.e. $\mathcal{F}|\partial M$ has no Reeb components.*

REMARK. There exists a C^∞ foliation of M with all of the above conclusions, except for the finite depth property.

MOTIVATING IDEA OF PROOF. Construct the foliation $\mathcal{F}$ leaf by leaf, the first leaf S_0 being S. Suppose that M is closed and $H_2(M) = \mathbb{Z}\oplus\mathbb{Z}$. Let $0 \neq z \in H_2(M)$ be disjoint from the ray containing $[S]$. The fact that the unit ball is a finite sided polyhedron implies that there is a positive integer n such that if T_n is a norm minimizing surface representing $z + n[S]$ then $T_n + S$ is a norm minimizing surface where $+$ denotes oriented cut and paste. The second leaf S_2 is T_n spun infinitely many times about S_0. This is schematically shown in Figure 2.1. To deal with the case $H_2(M) = \mathbb{Z}$ and/or prove the theorem in general, we need to develop a relative version of these ideas. The sutured manifold theory was created for precisely that reason.

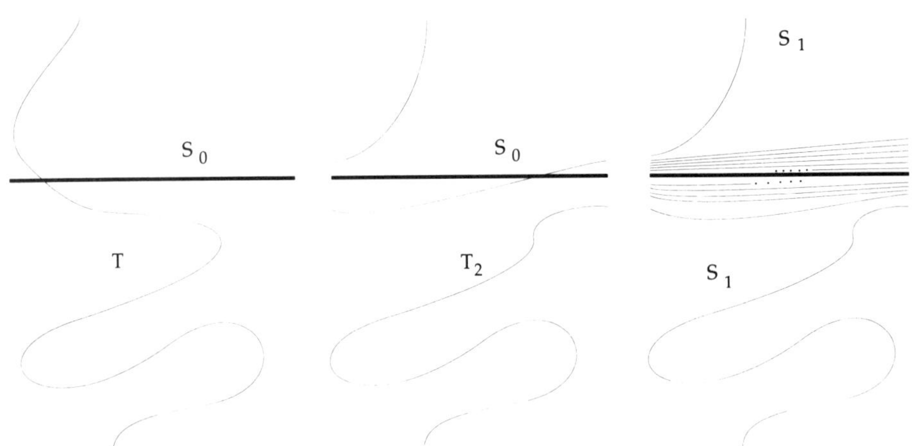

Figure 2.1

COROLLARY 2.12 [G1]. *Let M be a compact orientable irreducible 3-manifold. If $H_2(M, \partial M) \neq 0$, then M has a taut foliation.*

COROLLARY 2.13 [G1]. *If S is a minimal genus Seifert surface for a knot $k \subset M$, where M is a homology 3-sphere, then there exists a taut finite depth foliation of $M - \overset{\circ}{N}(k)$, transverse to $\partial N(k)$, such that S is a compact leaf.*

Define the *immersed genus* of the knot $k \subset S^3$ to be the smallest g such that k bounds an immersed surface S of genus g which is embedded near the boundary, so if $f : S \to S^3$, then $f^{-1}(k) = \partial S$.

THEOREM 2.14 [G1]. *If k is a knot in S^3, then the immersed genus of k is equal to the embedded genus of k.*

PROOF. Construct a finite depth, taut foliation $\mathcal{F}$ on $S^3 - \overset{\circ}{N}(k)$ with S as a compact leaf and genus(S) = genus(k). By Theorem 2.8 genus(S) = genus(T) if T is any minimal genus immersed surface in $S^3 - \overset{\circ}{N}(k)$ representing the same homology class as S.

Theorem 2.14 is a special case of a general theorem which more or less says that homology classes (relative or not) can be represented as efficiently by embedded surfaces as they can by immersed surfaces. Indeed if χ_s denotes the norm on $H_2(M, \partial M)$ based on singular surfaces, and g denotes the Gromov norm on $H_2(M, \partial M)$ then we have

THEOREM 2.15 [G1]. $\chi_s = \chi = \frac{1}{2}g$.

THEOREM 2.16 [G3]. *If S is a minimal genus Seifert surface for the knot $k \subset S^3$, then there exists a taut, finite depth foliation $\mathcal{F}$ of $S^3 - \overset{\circ}{N}(k)$ such that S is a leaf of $\mathcal{F}$ and $\mathcal{F}|\partial N(k)$ is a foliation by circles. If genus(k) $\neq 1$, then $\mathcal{F}$ can be constructed to be C^∞, though not of finite depth.*

It took 5 years to go from Corollary 2.12 to Theorem 2.16. The former is about homology spheres while the latter is about S^3. The key point is that $S^3 - \{2 \text{ points}\}$ has a foliation $\mathcal{G}$ by 2-spheres which in turn severely constrains the foliations which can exist therein. For example, if $\mathcal{F}$ is a taut foliation of $S^3 - \overset{\circ}{N}(k)$, then using the idea of *thin position* $\mathcal{F}$ and $N(k)$ can be isotoped so that some S^2 leaf of $\mathcal{G}$ is transverse to $\mathcal{F}$ except at saddle tangencies. Since $\chi(S^2) = 2$, the number of such tangencies is $|k \cap S^2| - 2$. The existence of such a 2-sphere is crucial to the entire investigation.

If k is a knot in S^3, then denote by $k(0)$ the manifold obtained by 0-frame surgery on k.

COROLLARY 2.17 [G3]. *If S is a minimal genus Seifert surface for a knot $k \subset S^3$, then $k(0)$ has a taut, finite depth foliation $\mathcal{F}$ such that the capped off surface S is a leaf.*

Can surgery on a knot in S^3 produce a $S^1 \times S^2$? The "Property R conjecture" asserts that the answer is no for all but 0-surgery on the unknot. Can surgery on a knot in S^3 produce a $S^1 \times S^2 \# M$, where $\#$ denotes connected sum? The stronger Poenaru conjecture asserts that the answer is again no for all but 0-surgery on the unknot, in which case $M = S^3$.

THEOREM 2.18 [G3]. *The Property R and Poenaru Conjectures are true.*

PROOF. For homological reasons the only way to obtain a homology $S^2 \times S^1$ by surgery on a knot is via 0-surgery. By Corollary 2.17 and [Ro] (See Remarks 2.11) the only way to obtain a reducible manifold by 0-surgery on a knot is via 0-surgery on the unknot.

Uniformization Of Taut Foliations. The following is a special case of a general theorem about abstract laminations of all dimensions.

THEOREM 2.19 (CANDEL [C]). *If $\mathcal{F}$ is a taut foliation of the closed, atoroidal, aspherical 3-manifold M, then there exists a Riemannian metric on M whose restriction to the leaves of $\mathcal{F}$ is a metric of constant curvature -1. The metric varies smoothly on the leaves but varies continuously in the transverse direction.*

§3
Essential Laminations

During the last 40 years, researchers have proved a wide array of spectacular results about compact aspherical manifolds containing incompressible surface. For example Haken's recognition of the unknot and the algorithmic computation of knot genus [Ha], Waldhausen's topological rigidity of Haken manifolds [Wa] and Thurston's geometrization of Haken manifolds [Th2]. More recently the essential lamination, which is a simultaneous generalization of both the incompressible surface and the taut foliation, has been introduced. Such manifolds appear to be far more widespread than Haken manifolds or manifolds with taut foliations.

DEFINITION 3.1. A codimension-1 *lamination* λ on a 3-manifold M is a codimension-1 foliation on a closed subset of M, i.e. it is defined by charts of the form $\mathbb{R}^2 \times \mathbb{R}$ such that $\lambda | \mathbb{R}^2 \times \mathbb{R} =$ product lamination on $\mathbb{R}^2 \times K$, where K is closed in $\mathbb{R}$. Unless stated otherwise all laminations in this chapter are codimension-1 in a 3-manifold.

DEFINITION 3.2 [GO]. A *closed complementary region* V of a lamination λ is the completion of a component $\mathring{V}$ of $M - \lambda$ with respect to the path metric on $\mathring{V}$. If λ has no isolated leaves, then V is the union of $\mathring{V}$ and its *boundary leaves* and is topologized by the path metric. The *closed complement* is the disjoint union of the closed complementary regions. Let $x \in \partial D^2$. An *end-compression* of a manifold V is a properly embedded $D^2 - x \subset V$ such that $D^2 - x$ is not properly homotopic in V rel $\partial D^2 - x$ to a mappping into ∂V.

A codimension-1 lamination λ on the compact, orientable 3-manifold M is *essential* if

0) λ has no S^2 leaves

1) No torus leaf of λ bounds a solid torus

2) If V is a closed complementary region, then V is irreducible, end-incompressible and ∂V is π_1-injective in V.

M is *laminar* if it contains an essential lamination.

EXAMPLE 3.3 [GO]. Let k be a knot in S^3 and suppose that $M_k = S^3 - \mathring{N}(k)$ fibres over the circle with fibre S and pseudo-Anosov monodromy $f : S \to S$. The suspension λ of the stable lamination of f is a lamination on M_k. One readily checks that 0)-2) hold and hence λ is an essential lamination of M_k disjoint from ∂M_k.

The cusps of the complementary region of λ containing ∂M_k are parallel to $n > 0$ copies of an essential simple closed curve $\beta \subset \partial M_k$. Let $M_k(\alpha)$ denote the manifold obtained by Dehn filling M_k along a slope α and let $\lambda(\alpha)$ denote the lamination λ viewed inside $M_k(\alpha)$. Again one readily checks that $\lambda(\alpha)$ is essential if and only if $| < \alpha, n\beta > | > 1$. Thus if k is a hyperbolic knot in S^3, then except for possibly a line in Dehn surgery space, λ remains essential in all manifolds obtained by surgery on k. A more elaborate construction due to Gabai - Mosher [Mo] yields similar results for all hyperbolic knots in S^3 and more generally all compact hyperbolic manifolds with single torus boundaries.

See [G6] for a discussion of some of the many other beautiful and very different constructions of laminations in 3-manifolds due to Christy, Delman, Fried, Gabai, Hatcher, Naimi, Oertel, Roberts, Thurston, and Wu. For example, Delman - Roberts [DR] show that if k is a non-torus alternating knot in S^3, than every manifold obtained by nontrivial surgery on k is laminar.

DEFINITION 3.4 [FO]. A (generic) *branched surface* B in a 3-manifold M is a smooth 2-complex locally modeled on the objects seen in Figure 3.1a. It has a *fibred neighbourhood* $N(B)$, with a fibering by intervals denoted by $\mathcal{V}$. The frontier of $N(B)$ is divided into $\partial_h N(B)$ (the horizontal boundary) and $\partial_v N(B)$ (the vertical boundary). The horizontal boundary is transverse to $\mathcal{V}$ and the vertical boundary is tangent to $\mathcal{V}$. One must look at Figure 3.1.b carefully to see $\partial_v N(B)$.

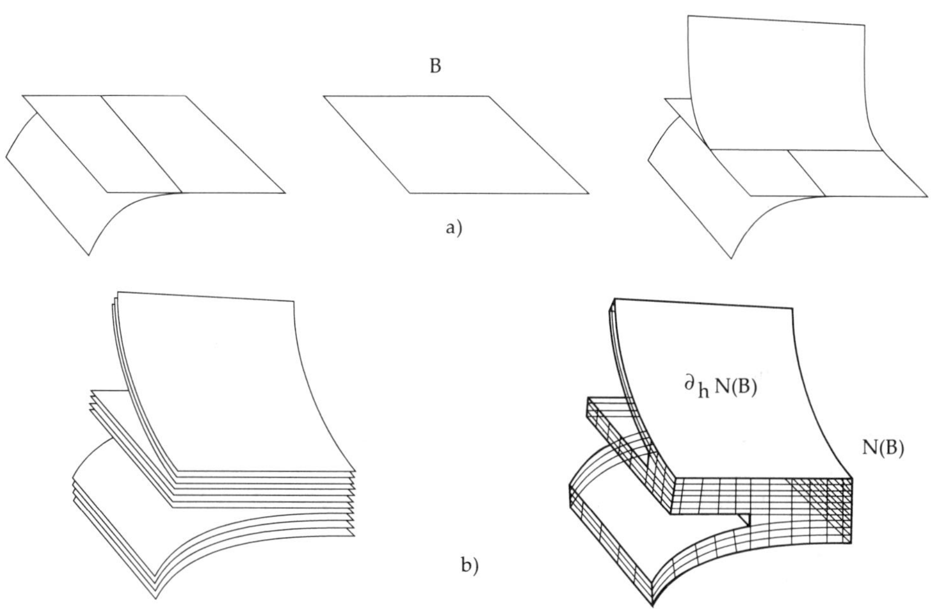

Figure 3.1

A lamination λ is *carried* by B if $\lambda \subset N(B)$ and is transverse to $\mathcal{V}$, and λ is *fully carried* by $N(B)$ if it also intersects every fibre of $\mathcal{V}$.

It is shown in [GO] that every nowhere dense essential lamination is carried by an essential branched surface.

We say that a curve γ transverse to λ is *efficient* if there is no compact subarc $\beta \subset \gamma$ such that $\beta \cap \lambda = \partial\beta$ and β is homotopic rel endpoints to an arc lying in a leaf of λ.

THEOREM 3.5 [GO]. *Let λ be an essential lamination in M. Then*

i) If L is a leaf of λ, then the homomorphism $\pi_1(L) \to \pi_1(M)$, induced by inclusion, is injective.

ii) If γ is a transverse efficient closed curve, then $1 \neq [\gamma]^n \in \pi_1(M)$ for any integer $n \neq 0$

iii) The universal cover $\tilde{M}$ of M is $\mathbb{R}^3$, in particular M is irreducible. □

THEOREM 3.6 (BRITTENHAM [BR1], CLAUS [CL]).

1) Let λ be an essential lamination in a Seifert fibred space M. Then λ has a sublamination which is either horizontal or vertical, i.e. is a union of Seifert fibres.

2) If M is non Haken, then λ can be isotoped to be horizontal (i.e. is transverse to the Seifert fibres).

REMARK. As a consequence of this theorem, Brittenham and Claus show that there exist nonlaminar aspherical 3-manifolds. Indeed a horizontal lamination can be extended to a horizontal foliation and by Milnor - Wood [Mi], [Wo], there exist many aspherical Seifert fibred spaces without horizontal foliations. A remarkable theorem of Milnor, Wood, Eisenbud-Hirsch-Neumann [EHN], Jankins-Neumann [JN1-2], and Naimi [Na], completely classifies which Seifert fibred spaces have horizontal foliations.

Problems.

3.7.1 What 3-manifolds are laminar? Are all closed hyperbolic 3-manifolds laminar? Is the manifold obtained by 35/2 surgery on the (-2,3,7) pretzel knot laminar? Are all manifolds obtained by nontrivial surgery on 10_{139} laminar?

3.7.2 Are all closed hyperbolic 3-manifolds finitely covered by laminar manifolds?

3.7.3 Is there an algorithm to decide if a closed 3-manifold is laminar?

3.7.4. Is there an algorithm to decide if a branched surface in a 3-manifold carries a lamination?

3.7.5 Are there useful branched surfaces which carry nothing?

Warning 3.8 (Mosher [G6]) The analogue of Problem 3.7.4 is undecidable for branched surfaces that abstractly carry laminations. On the other hand Christy [Ch] has shown that many branched surfaces do not embed in 3-manifolds.

For some very recent developments in the theory of laminations and branched surfaces and their applications to 3-manifold theory see Tao Li's papers [Li1] and [Li2].

Perhaps the following weaker and perhaps more plentiful object is of some interest.

DEFINITION 3.9. A lamination λ on the compact, orientable, irreducible 3-manifold M is called *loosesse* and M is called *loose laminar* if λ satisfies:

0) λ has no S^2 leaves and

1) for any leaf L of λ, the homomorphism $\pi_1(L) \to \pi_1(M)$ induced by inclusion is injective, and for any closed complementary region V, the homomorphism $\pi_1(V) \to \pi_1(M)$ induced by inclusion is injective.

CONJECTURE 3.10. *If $\mathcal{L}$ is a loosesse lamination on the closed orientable 3-manifold M and $\tilde{\mathcal{L}}$ is the lift to the universal cover of M, then $\tilde{\mathcal{L}}$ is topologically equivalent to a product lamination of the form (lamination on $\mathbb{R}^2$)$\times\mathbb{R}$. In particular $\tilde{M} = \mathbb{R}^3$.*

REMARK. The conjecture is true if each leaf of $\tilde{L}$ is properly embedded in $\tilde{M}$. The argument is similar to that of Theorem 4.4 [GK1].

CONJECTURE 3.11. *If $M = S^3 - \mathring{N}(k)$, where k is a fibred hyperbolic knot, then all but finitely many of the laminations $\lambda(\alpha)$ are loosesse. More generally, If M is a compact hyperbolic 3-manifold with single torus boundary, then all but finitely many fillings on M are loosesse laminar.*

REMARK. The terminology, loosesse, suggests that leaves of a non essential loosesse lamination are floppy as opposed to the non floppiness of leaves of taut foliations.

Let M be a closed, orientable, atoroidal Haken manifold.

THEOREM 3.12 (WALDHAUSEN [WA]). *If M is a closed Haken manifold, then*

i) $\tilde{M} = R^3$

ii) if N is irreducible and $f : N \to M$ is a homotopy equivalence, then f is homotopic to a homeomorphism

iii) if the two homeomorphisms $f, g : M \to M$ are homotopic, then f is isotopic to g.

THEOREM 3.13 (JOHANNSON [JO]). *If M is a closed orientable atoroidal Haken manifold, then $|\pi_0(Diff(M))| < \infty$.*

THEOREM 3.14 (THURSTON [TH2]). *If M is a closed orientable atoroidal Haken manifold, then M has a hyperbolic structure.*

One of the main goals of the subject is to extend the above results to laminar 3-manifolds. While there has been surprisingly little progress to date for 3-manifolds with taut foliations there has been some progress for the genuinely laminar manifolds.

DEFINITION 3.15. A *genuine* lamination λ [GK1] is an essential lamination whose closed complement is not an I-bundle.

A non genuine lamination λ readily extends to a Reebless foliation. Indeed, if λ has no isolated leaves, then it also arises from a Reebless foliation by splitting open leaves. Thus an essential lamination is either genuine or more or less a Reebless foliation.

The closed complement of an essential lamination can be uniquely decomposed (up to isotopy) into a union of $\mathcal{I}(\lambda)$ and $\mathcal{G}(\lambda)$. The *interstitial bundle* $\mathcal{I}(\lambda)$ is a maximal union of maximal connected non compact I-bundle's or I-bundles over closed surfaces or I-bundles over surfaces of negative Euler characteristic. The *gut* $\mathcal{G}(\lambda)$ is a compact manifold such that $\mathcal{I}(\lambda) \cap \mathcal{G}(\lambda)$ is a union of properly embedded essential annuli. The *gut number* $\mathcal{GN}(\lambda)$ is the number of components of the gut of λ.

Our definition [G8] of gut is different from that of [GK1] for it allows for components of the interstitial bundle which are I-bundles over surfaces of negative Euler characteristic. The same argument as [GK1] shows that the gut is unique up to isotopy. (Uniqueness is lost if we allow I-bundles over the annulus or Mobius band.)

It follows by Gabai - Mosher [Mo], that if M is a compact 3-manifold whose interior is a complete hyperbolic 3-manifold with a single cusp, then away from at most two lines in Dehn filling space all manifolds obtained by filling M are genuinely laminar.

THEOREM 3.16 [GK3],[GK4],[GK1]. *If M is atoroidal and genuinely laminar, then*

i) $\pi_1(M)$ is negatively curved (i.e. Gromov-hyperbolic),

ii) $|\pi_0(Diff(M))| < \infty$, and

iii) homotopic homeomorphisms of M are isotopic, if some non I-bundle complementary region is a solid torus.

REMARK. The above results are obtained by exploiting the negative curvature concentrated in the gut.

Order Trees. M contains an incompressible surface S if and only if $\pi_1(M)$ acts nontrivally on a simplicial tree T. Given S, then T is the tree dual to the lifts of S to $\tilde{M}$, the universal covering of M. More generally, Morgan and Shalen [MS] showed that M contains an incompressible measured lamination if and only if $\pi_1(M)$ acts nontrivally on an $\mathbb{R}$-tree (see 3.17 for a definition). If M contains a taut foliation $\mathcal{F}$, then $\pi_1(M)$ acts nontrivially on a (possibly non Hausdorff) simply connected 1-manifold $\mathcal{T}(\mathcal{F})$. This $\mathcal{T}(\mathcal{F})$ is the space of leaves of $\tilde{\mathcal{F}}$, the lift of $\mathcal{F}$ to $\tilde{M}$. The analogous dual object to an essential lamination λ without isolated leaves is $T(\lambda) = \tilde{M}/\sim$ where $\sim$ is the equivalence relation which corresponds to collapsing each (connected) closed complementary region and each (connected) non boundary leaf to a point. It is shown in [GO] that $\mathcal{T}(\lambda)$ is an *order tree* (defined below).

DEFINITION 3.17 [GO]. An *order tree* is a set $\mathcal{T}$ together with a collection $\mathcal{P}$ of linearly ordered subsets called *segments*, each with distinct least and greatest elements called *initial* and *final* ends. If σ is a segment, $-\sigma$ denotes the same subset with the reverse order, and is called the *inverse* of σ. The following conditions are satisfied:

(1) $\sigma \in \mathcal{P} \Rightarrow -\sigma \in \mathcal{P}$.

(2) Any closed subinterval of a segment is a segment if it has more than one element.

(3) Any two elements of $\mathcal{T}$ can be joined by a sequence $\sigma_1, \sigma_2, \cdots, \sigma_k$ with the final end of σ_i equal to the initial end of σ_{i+1} $(i = 1, 2, \ldots, k-1)$.

(4) given a cyclic word $\sigma_0\sigma_1 \cdots \sigma_{k-1}$ (subscripts mod$k$) with the final end of σ_i equal to the initial end of σ_{i+1}, there is a subdivision of the σ_i's yielding a cyclic word $\rho_0\rho_1 \ldots \rho_{n-1}$ which becomes the trivial word when adjacent inverse segments are cancelled.

(5) If σ_1 and σ_2 are segments whose intersection is a single element which is the final element of σ_1 and the initial element of σ_2, then $\sigma_1 \cup \sigma_2$ is a segment.

Topologize an order tree by declaring U open if for every segment σ of $\mathcal{T}$, $U \cap \sigma$ is open in σ.

An $\mathbb{R}$-order tree is an order tree which is a countable union of segments which are order isomorphic to compact intervals of $\mathbb{R}$. An $\mathbb{R}$-tree is a metric space (X, d) such that X is an $\mathbb{R}$-order tree and each segment in X is isometric to a compact interval of $\mathbb{R}$.

The notion of order tree is simultaneously a generalization of the notions of $\mathbb{R}$-tree and simply connected 1-manifold. Frequently if λ is essential, then $\mathcal{T}(\lambda)$ is a topological $\mathbb{R}$-tree but $\pi_1(M)$ does not act on $\mathcal{T}(\lambda)$ via isometries for any $\mathbb{R}$-tree metric structure on $\mathcal{T}(\lambda)$.

In what follows all order trees under discussion are assumed to be $\mathbb{R}$-order trees.

THEOREM 3.18 [GK2].

i) There exists a 1-1 correspondence between the set of all $\mathbb{R}$-order trees and the set of the spaces $\mathcal{T}(\tilde{\lambda}) = \mathbb{R}^2/\sim$ where $\tilde{\lambda}$ is a lamination of $\mathbb{R}^2$ by lines, $\tilde{\lambda}$ has no isolated leaves and $\mathbb{R}^2$ is covered by lamination charts, each of which meets every leaf at most once. (This is stronger than saying that $\tilde{\lambda}$ is a lamination by properly embedded lines.)

ii) If λ is essential in M and $\tilde{\lambda}$ denotes the induced lamination in the universal cover, then $\tilde{\lambda} =$ (lamination on $\mathbb{R}^2$ by lines) $\times\mathbb{R}$. Furthermore if γ is a non-trivial simple closed curve on a leaf and $\tilde{\gamma}$ the lift of γ to $\tilde{M}$, then this equality can be achieved so that $\tilde{\gamma} =$ (discrete points on $\mathbb{R}^2$) $\times\ \mathbb{R}$.

Problems. Develop a theory of $\pi_1(M^3)$ groups acting on order trees.

3.19.1 Which 3-manifold groups have such actions?

3.19.2 If M is closed, oriented and irreducible, and $\pi_1(M)$ has such an action, is M laminar? Can anything else be said about M, e.g. is $\tilde{M} = \mathbb{R}^3$?

For more related problems and information, see [G6].

THEOREM 3.20 (ROBERTS - STEIN [RS]). *If M is a non Haken, Seifert fibred space, then every nontrivial minimal action of $\pi_1(M)$ on an order tree $\mathcal{T}$ is dual to a foliation transverse to the Seifert fibres. Thus if $\pi_1(M)$ acts nontrivially and minimally on $\mathcal{T}$, then $\mathcal{T} = \mathbb{R}$.*

Essential laminations and Kneser normal form.

DEFINITION 3.21. Let M be a compact orientable 3-manifold with a fixed triangulation τ. A surface F in M is *normal* [Ha] with respect to τ, if F intersects each tetrahedron $\sigma \in \tau$ in a collection of disks all of whose boundary curves intersect each edge of σ at most once, i.e., in normal triangles and quadrilaterals. Generalizing the definition of normal surface, we say a lamination $\mathcal{L}$ in M is *normal* with respect to τ if $\mathcal{L}$ is in general position with respect to τ, and for every tetrahedron σ of τ, $\mathcal{L} \cap \sigma$ is a lamination consisting of compact disks, each of which meets the 1-skeleton of σ, and such that each disk D of $\mathcal{L} \cap \sigma$ meets each edge of σ at most once.

THEOREM 3.22 (KNESER [K], HAKEN [HA]). *An incompressible surface in an irreducible triangulated 3-manifold is isotopic to a normal surface.*

IDEA OF PROOF. If the surface S is not normal, then do an obvious isotopy to reduce the pair (n, m) lexicographically ordered, where $n = |S \cap \tau^1|$ and $m = |S \cap \tau^2|$.

THEOREM 3.23 (BRITTENHAM [BR2]). *Let M be 3-manifold with a triangulation τ. If λ is an essential lamination in M, then there exists a normal essential lamination $\mathcal{L}$ in M.*

IDEA OF PROOF. Via isotopy, make λ normal with respect to one 3-simplex at a time. Of course cleaning up λ near one 3-simplex messes up λ within others. Repeatedly cycle through the 3-simplices. In the limit, one is left with a lamination $\mathcal{L}'$. Delete the Reeb laminations of $\mathcal{L}'$ to obtain $\mathcal{L}$.

By carefully examining the isotopy process we obtain,

THEOREM 3.24 [G8]. *Let λ be a nowhere dense essential lamination in the closed orientable atoroidal 3-manifold M with triangulation τ. Then at least one of the following occurs.*

1) After possibly splitting λ open along a finite number of leaves, λ is isotopic to a normal lamination.

2) There exists a normal essential lamination $\mathcal{L}$ in M such that $\mathcal{GN}(\mathcal{L}) > \mathcal{GN}(\lambda)$, and $\mathcal{L}$ is obtained from λ by evacuating a taut sutured manifold (N, γ).

Applications.

THEOREM 3.25 [GK4]. *If M is genuinely laminar, then $\frac{\text{Homeo}(M)}{\text{Homeo}_0(M)}$ is finite.*

THEOREM 3.26 [G8]. *Assume that M is closed, orientable and atoroidal.*

1) If M has a triangulation with t 3-simplices and λ is essential in M, then $\mathcal{GN}(\lambda) < 2^t$.

2) Up to isotopy, every nowhere dense essential lamination is carried by one of finitely many branched surfaces.

3) There exists a triangulation τ of M such that any essential lamination or taut foliation can be (finitely) isotoped to be normal to τ.

By a foliation being *normal* to $\mathcal{T}$ we mean that for each 3-simplex σ of $\mathcal{T}$ there exists a foliation chart of the form $\mathbb{R}^2 \times \mathbb{R}$ such that σ appears as a linear simplex with vertices at distinct heights.

COROLLARY 3.27 [G8]. *If M is a closed atoroidal 3-manifold, there exists finitely many taut foliations $\mathcal{F}_1, \ldots, \mathcal{F}_n$ such that for any $\epsilon > 0$ any taut foliation $\mathcal{F}$ on M is ϵ-coarse isotopic to one of the $\mathcal{F}$'s*

This means that a given taut foliation $\mathcal{F}$ there exists an $\mathcal{F}_i$ such that $\mathcal{F}$ (resp. $\mathcal{F}_i$) can be isotoped to a foliation $\mathcal{G}$ (resp. $\mathcal{G}_i$) such that the tangent plane fields of $\mathcal{G}$ and $\mathcal{G}_i$ are pointwise ϵ-close to being tangent. (Actually the isotopy of $\mathcal{F}_i$ is "very" small.) As a consequence we obtain an elementary proof of

THEOREM 3.28 (KRONHEIMER - MROWKA [KM]). *There are only finitely many homotopy classes of plane fields of taut foliations.*

BIBLIOGRAPHY

[A1] J. W. Alexander, *A Lemma on Systems of Knotted Curves*, Proc. N. Acad. Sci. **9** (1923), 93–95.

[A2] ———, *On the Subdivision of 3-Space by a Polyhedron*, Proc. Nat. Acad. Sci. **10** (1924), 6-8.

[Br1] M. Brittenham, *Essential Laminations in Seifert Fibred Spaces*, Topology **32** (1993), 61-85.

[Br2] ———, *Essential Laminations and Haken Normal Form I*, Pacific J. **168** (1995), 217–234.

[Br3] ———, *Essential Laminations and Deformations of Homotopy Equivalences: From Essential Pullback to Homeomorphism*, Top. and App **53** (1994), 317–324.

[CN] C. Camacho and A. Neto, translated by S. Goodman, *Geometry of Foliations*, Birkhauser 1986.

[C] A. Candel, *Uniformization of Surface Laminations*, Ann. Ecole Norm. Sup. **26** (1993), 489-516.

[CaCo] A. Candel and L. Conlon, *Foliations I*, AMS Grad. Stud. in Math. **23** (2000).

[CJ] A. Casson and D. Jungreis, *Seifert Fibered Spaces and Convergence Groups*, Invent. Math. **114** (1994), 441–456.

[Ch] J. Christy, *Immersing Branched Surfaces in Dimension-3*, Proc. AMS **115** (1992), 853-861.

[Cl] E. Claus, 1990 U. Texas, Austin thesis.

[DR] C. Delman and R. Roberts, *Alternating Knots Satisfy Strong Property P*, . Comm. Math. Helv. **74** (1999), 376–397.

[EHN] D. Eisenbud, U. Hirsch and W. Neumann, *Transverse Foliations of Seifert Bundles and self Homeomorphisms of the Circle*, Comm. Math. Helv. vol 56 (1981), 638-660.

[E1] D. Epstein, *Periodic Flows on 3-Manifolds*, Ann. Math. **95** (1975), 66-82.

[E2] ———, *Pointwise Periodic Homeomorphisms*, Proc. London Math. Soc **42** (1981), 415-460.

[E] C. Ehresmann and W. Shih, *Sur les Espaces Feuilletees: Theorem de Stabillite*, C. R. A. S. Paris **243** (1956), 344–346.

[Fe] S. Fenley, *Anosov flows in 3-Manifolds*, Ann. of Math. **139** (1994), 79-115.

[FO] W. Floyd and U. Oertel, *Incompressible Surfaces via Branched Surfaces*, Topology **23**, 117-125.

[G1] D. Gabai, *Foliations and the Topology of 3-Manifolds*, J. Diff. Geom. **18** (1983), 479–536.

[G2] ———, *Foliations and the Topology of 3-Manifolds II*, J. Diff. Geom. **26** (1987), 461-478.

[G3] ———, *Foliations and the Topology of 3-Manifolds III*, J. Diff. Geom. **26** (1987), 479-536.

[G4] ———, *Foliations and 3-Manifolds*, Proc. ICM Kyoto-1990 **1** (1991), 609–619.

[G5] ———, *Convergence Groups are Fuchsian Groups*, Ann. Math. **136** (1992), 447-510.

[G6] ———, *Problems in Foliations and Laminations*, AMS/IP Stud. Adv. Math. **2.2** (1997), 1–34.

[G7] ———, *Taut Foliations and Combinatorial Volume Preserving Flows*, to appear, Comm. Math. Helv.

[G8] ———, *Essential Laminations and Kneser Normal Form*, to appear, J. Diff. Geom..

[GK1] D. Gabai and W. H. Kazez, *Homotopy, Isotopy and Genuine Laminations of 3-Manifolds*, AMS/IP Stud. Adv. Math. **2.1** (1997), 123–138.

[GK2] ———, *Order Trees and Laminations of the Plane*, Math. Res. Lett. **4** (1997), 603–616.

[GK3] D. Gabai and W. H. Kazez, *Group Negative Curvature for 3-Manifolds with Genuine Essential Laminations*, Geom. and Top. **2** (1998), 65–77.

[GK4] ———, *Finiteness of the Mapping Class Group for Atoroidal 3-Manifolds with Genuine Laminations*, J. Diff. Geom **50** (1998), 123–127.

[GO] D. Gabai and U. Oertel, *Essential Laminations in 3-Manifolds*, Ann. of Math. **(2) 130** (1989), 41–73.

[Gh] E. Ghys, *Topologie des Feuilles Generiques*, Ann. Math. **141** (1995), 387–422.

[Go] C. Godbillon, *Feuilletages*, Prog. in Math. **93** (1991).

[Good] S. Goodman, *Closed Leaves in Foliations of Codimension One*, Comment. Math. Helv. **50**, 383–388.

[H] A. Haefliger, *Variétés Feuilletées*, Ann. Scuola Norm. Sup. Pisa. **16** (1962), 367-397.

[HR] A. Haefliger and G. Reeb, *Varietes (non Separees) a une Dimension et Structures Feuilletees du Plan*, L'Ensign. Math. **3** (1957), 107-125.

[Ha] W. Haken, *Theorie der Normalflächen*, Acta Math. **105** (1961), 245–375.

[HL] Harvey and B. Lawson, *Calibrated Foliations (Foliations and Mass-Minimizing Currents*, Amer. J. Math. **104**, 607–633.

[Hs] J. Hass, *Minimal Surfaces in Foliated Manifolds*, Comment. Math. Helv. **61** (1986), 1-32.

[HH] G. Hector and U. Hirsch, *Introduction to the Geometry of Foliations A,B*, Aspects of Math. **E1,E3**.

[I] Imanishi, *On the Theorem of Denjoy-Sacksteder for Codimension One Foliations without Holonomy*, J. Math. Kyoto Univ. **14** (1974), 607–634.

[He] J. Hempel, *3-Manifolds*, Ann. of Math. Stud. **86** (1976).

[J] W. Jaco, *Lectures on 3-Manifold Topology*, CBMS Lecture notes **43** (1980).

[JN1] M. Jankins and W. Neumann, *Homomorphisms of Fuchsian Groups to PSL(2,R)*, Comm. Math. Helv. **60** (1985), 480-495.

[JN2] M. Jankins and W. Neumann, *Rotation Numbers of Products of Surface Homeomorphisms*, Math. Annalen **271** (1985), 381- 400.

[Jo] K. Johannson, *Homotopy Equivalence of 3-Manifolds with Boundary*, Springer LNM **761** (1979), 1–303.

[Ka1] W. Kaplan, *Regular Curves - Families Filling the Plane I*, Duke Math. J. **7** (1940), 154–155.

[Ka2] ______, *Regular Curves - Families Filling the Plane II*, Duke Math. J. **8** (1941), 11–46.

[K] H. Kneser, *Geschlossene Flachen in Dreidimensionalen Mannigfaltigkeiten*, Jahres. der Deut. Math. Verein. **38** (1929), 248–260.

[KM] P. Kronheimer and T. Mrowka, *Monopoles and Contact Structures*, Invent. Math. **130** (1997), 209–255.

[La] B. Lawson, *The Quantitative Theory of Foliations*, CBMS Lecture Notes **27** (1977).

[Li1] T. Li, *Boundary Circles of Surfaces with 4-Plane Property*, preprint.

[Li2] ______, *Laminar Branched Surfaces in 3-Manifolds*, preprint.

[Lick] R. Lickorish, *A Foliation for 3-Manifolds*, Ann. of Math. **82** (1965), 414–420.

[Mi] J. Milnor, *On the Existence of a Connection with Curvature Zero*, Comm. Math. Helv. **32** (1957), 215-223.

[MS] J. Morgan and P. Shalen, *Degenerations of Hyperbolic Structures III*, Ann. of Math. **127** (1988), 457-519.

[Mo] L. Mosher, *Laminations and Flows Transverse to Finite Depth Foliations I*, preliminary version preprint.

[Na] R. Naimi, *Foliations Transverse to Fibers of Seifert Manifolds*, Comment. Math. Helv. **69** (1994), 155–162.

[No] S. P. Novikov, *Topology of Foliations*, Trans. Moscow Math. Soc. (1965), 268-304.

[OVZ] P. Orlik, E. Vogt and H. Zieschang, *Zur Topologie Gefaserter Dreidimensionaler Mannigfaltigkeiten*, Topology **6** (1967), 49–64.

[Pa] C. F. B. Palmeira, *Open Manifolds Foliated by Planes*, Ann. Math. **107** (1978), 109-131.

[R] G. Reeb, *Sur certaines proprietes Topologiques des Variétés Feuilletées*, Actualites Sci. Indust., Hermann, Paris **1183** (1952), 91–158.

[RS] R. Roberts and M. Stein, *Group Actions on Order Trees*, in preparation.

[Ro] H. Rosenberg, *Foliations by Planes*, Topology **6** (1967), 131–138.

[Rou] R. Roussarie, *Plongements dans les Variétés Feuilletées et Classification de Feuilletages sans Holonomie*, Inst. Hautes Études Sci. Publ. Math. **43** (1973), 101-142.

[Sa] R. Sacksteder, *Foliations and Pseudogroups*, Am. J. Math. **87** (1965), 79–102.

[SY] R. Schoen and S. T. Yau, *Existence of Incompressible Minimal Surfaces and the Topology of Three Dimensional Manifolds with Non-Negative Scalar Curvature*, Ann. Math. **110** (1979), 127-142.

[Se] H. Seifert, *Topologie Dreidimensionales Gefaserter Raum*, Acta. Math. **60** (1933), 147-288; English transl. by Wolfgang Heil in Acad. Press Monographs in Pure and Ap. Math. **89** (1980), 360–418.

[Su1] D. Sullivan, *Cycles for the Dynamical Study of Foliated Manifolds and Complex Manifolds*, Invent. Math. **36** (1976), 225–255.

[Su2] ______, *A Homological Characterization of Foliations Consisting of Minimal Surfaces*, Comm. Math. Helv. **54** (1979), 218-223.

[Th0] W. P. Thurston, *Foliations of Three-Manifolds which are Circle Bundles*, UC Berkeley Ph.D. Thesis (1972).

[Th1] ———, *A Norm For the Homology of 3-Manifolds*, AMS Memoirs **339** (1986), 99-130.

[Th2] ———, *Three-Dimensional Manifolds, Kleinian groups and Hyperbolic geometry*, Proc. Symp. Pure Math. **39** (1983), 87-111.

[Th3] ———, *Three-Dimensional Geometry and Topology. Vol. 1. Edited by Silvio Levy*, Princeton Math. Series **35** (1997).

[Th4] ———, *Three Manifolds, Foliations and Circles II, the Transverse Asymptotic Geometry of Foliations*, in preparation.

[Ti] D. Tischer, *On fibering certain foliated manifolds over S^1*, Topology **9** (1970), 153–154.

[Wa] F. Waldhausen, *On Irreducible 3-Manifolds which are Sufficiently Large*, Annals of Math. **87** (1968), 56–88.

[Wl] A. Wallace, *Modifications and Cobounding Manifolds*, Can. J. Math. **12** (1960), 503–528.

[Wo] J. Wood, *Bundles with Totally Disconnected Structure Group*, Comment. Math. Helv. **46** (1971), 271-273.

California Institute of Technology, Pasadena, CA 91125-0001 USA
E-mail address: gabai@its.caltech.edu

Contemporary Mathematics
Volume **269**, 2001

Rational Laminations of Complex Polynomials

Jan Kiwi

ABSTRACT. We characterize the equivalence relations in $\mathbb{Q}/\mathbb{Z}$ which arise as the rational lamination of a complex polynomial with connected Julia set.

1. Introduction

External rays have played a key role in the study of the dynamics of complex polynomials ever since they were introduced by Douady and Hubbard [**DH1**]. The study of the pattern in which external rays approach the Julia set has been useful to describe the dynamics of a given polynomial and to understand how polynomials are organized in parameter space according to their dynamics. This pattern may be encoded in the form of "combinatorial objects". One of these combinatorial objects is the rational lamination.

The rational lamination $\lambda_{\mathbb{Q}}(f)$ of a monic polynomial f with connected Julia set $J(f)$ contains the information regarding the pattern in which rational external rays land at $J(f)$. More precisely, the *rational lamination* $\lambda_{\mathbb{Q}}(f)$ is the equivalence relation in $\mathbb{Q}/\mathbb{Z}$ which identifies two arguments t and t' if and only if the external rays with arguments t and t' land at a common point (compare with Section 6.4 of [**McM**]). Rational laminations are implicit in the work of Douady and Hubbard. Although rational laminations were not defined as above by Thurston, they may be regarded as part of his preprint [**Th**] where "invariant laminations" are studied. Moreover, many of the ideas and techniques throughout this paper stem from [**Th**].

This paper explores the properties that characterize the equivalence relations in $\mathbb{Q}/\mathbb{Z}$ which appear as the rational lamination of a polynomial. More precisely, we will show that an equivalence relation in $\mathbb{Q}/\mathbb{Z}$ for which five simple conditions hold is the rational lamination of some polynomial and vice versa.

THEOREM 1.1. *An equivalence relation $\lambda_{\mathbb{Q}}$ in $\mathbb{Q}/\mathbb{Z}$ is the rational lamination of a degree d monic polynomial f with connected Julia set if and only if the following hold:*

(Q1) $\lambda_{\mathbb{Q}}$ is closed in $\mathbb{Q}/\mathbb{Z} \times \mathbb{Q}/\mathbb{Z}$.
(Q2) Each $\lambda_{\mathbb{Q}}$-equivalence class A is a finite subset of $\mathbb{Q}/\mathbb{Z}$.
(Q3) If A is a $\lambda_{\mathbb{Q}}$-equivalence class, then $d \cdot A$ is a $\lambda_{\mathbb{Q}}$-equivalence class.
(Q4) If A is a $\lambda_{\mathbb{Q}}$-equivalence class, then $A \mapsto d \cdot A$ is consecutive preserving.
(Q5) $\lambda_{\mathbb{Q}}$-equivalence classes are pairwise unlinked.

2000 *Mathematics Subject Classification.* 37F10, 37F20.

Supported by "Proyecto Fondecyt #1990436", "Fundación Andes" and "Cátedra Presidencial en Geometría".

In Section 3, we discuss in detail the definitions involved in the statement of the theorem.

The combinatorics of polynomials with one critical point (in particular, quadratic polynomials) has already been studied in great detail and is fairly well understood (see [**Th, La1, Sch1, Sch2, Sch3, M2**]). When the results of this paper are restricted to the case of polynomials with one critical point either they become well known or easier proofs should follow from the results contained in the articles of the references above.

We next describe how the paper is organized, as well as give a rough outline of the proof of the Theorem 1.1.

Section 2 introduces some basic notions and results from holomorphic dynamics. In Section 3, we discuss the definitions involved in the statement of our main result and prove that the rational lamination of a polynomial has the five properties listed in Theorem 1.1. In Subsection 3.1, we deal with properties (Q2) through (Q5) which are straightforward and well known. In Subsection 3.2, we show that rational laminations of polynomials are closed in $\mathbb{Q}/\mathbb{Z} \times \mathbb{Q}/\mathbb{Z}$.

We temporarily say that an equivalence relation $\lambda_{\mathbb{Q}}$ in $\mathbb{Q}/\mathbb{Z}$ for which (Q1) through (Q5) hold is a *formal rational lamination.* The rest of the paper is devoted to find a polynomial f with rational lamination $\lambda_{\mathbb{Q}}$, where $\lambda_{\mathbb{Q}}$ is a formal rational lamination.

Section 4 gives a first step in this direction. Namely, we construct a topological branched covering $P : \mathbb{C} \to \mathbb{C}$ which has topological dynamics closely related to that of our desired polynomial f. We call the map P a *topological realization of* $\lambda_{\mathbb{Q}}$. From a topological viewpoint, the dynamics of P will be just as the one of a polynomial with locally connected Julia set. The map P is obtained via a "pinched disk construction" (see [**D**]). To perform the pinched disk construction we need to study the *extension* $\lambda_{\mathbb{R}}$ *of* $\lambda_{\mathbb{Q}}$ *to* $\mathbb{R}/\mathbb{Z}$. In Subsection 4.1, the extension $\lambda_{\mathbb{R}}$ is defined as the smallest equivalence relation in $\mathbb{R}/\mathbb{Z}$ which contains the closure of $\lambda_{\mathbb{Q}}$. In Subsection 4.2, we show that $\lambda_{\mathbb{R}}$ inherits the "good properties" of $\lambda_{\mathbb{Q}}$. These properties of $\lambda_{\mathbb{R}}$ allow us to construct a topological branched covering $P : \mathbb{C} \to \mathbb{C}$ which topologically realizes $\lambda_{\mathbb{R}}$ (see Subsection 4.3). More precisely, the boundary of the basin of infinity of P is homeomorphic to $(\mathbb{R}/\mathbb{Z})/\lambda_{\mathbb{R}}$. We say that this boundary is the *topological Julia set* $J^{top}(P)$ *of* P. The dynamics of P over $J^{top}(P)$ is topologically conjugate to the projection of multiplication by d acting on $(\mathbb{R}/\mathbb{Z})/\lambda_{\mathbb{R}}$.

It is worth pointing out that in Section 4 we discuss in greater generality when this type of constructions may be used. That is, we list five properties which an equivalence relation $\lambda_{\mathbb{R}}$ in $\mathbb{R}/\mathbb{Z}$ must have to assure the existence of a topological realization (see Proposition 4.3 and Definition 4.4). We call such equivalence relations $\mathbb{R}$*eal laminations.*

Since the object of this paper is to characterize rational laminations, we leave out, for a future occasion, a detailed discussion of the relation between the topology of polynomial Julia sets and that of the topological Julia sets built in Section 4. Nevertheless, it is worth to mention that every topological Julia set is always locally connected. Moreover, every locally connected Julia set of a polynomial can be identified with a corresponding topological Julia set. The interested reader may refer to [**D, Sch2**] for a closely related discussion.

Section 5 discusses some "separation lemmas" which are a technical aid for the proof of our main result.

Since our topological realization P has topological dynamics very similar to that of a polynomial, some of the notions and constructions associated to a polynomial may be associated to P. For example, we call *topological filled Julia set* the object corresponding to the filled Julia set, *topological external ray* the object corresponding to an external ray, etc. Section 6 generalizes some constructions and results of Poirier [**P**]. Namely, in Subsection 6.1, to a topological realization $P : \mathbb{C} \to \mathbb{C}$ we assign a *critical marking* Θ which abstractly captures the location of the critical points of P in its *topological filled Julia set.* In Subsection 6.2, we mimic the work of Poirier and show how, via symbolic dynamics in $\mathbb{R}/\mathbb{Z}$, we may obtain $\lambda_{\mathbb{Q}}$ back from the critical marking Θ.

In Section 7, we apply a realization theorem of Poirier to find a sequence of critical markings Θ_n of post-critically finite polynomials f_n such that Θ_n converges (in a certain sense) to Θ and f_n converges to a polynomial f. The polynomial f is our desired polynomial. The notion of convergence of the sequence Θ_n is chosen so that the rational laminations $\lambda_{\mathbb{Q}}(f_n)$ of f_n become similar to $\lambda_{\mathbb{Q}}$ as n increases. Then, based on the separation lemmas of Section 5, we are able to show that $\lambda_{\mathbb{Q}}(f) = \lambda_{\mathbb{Q}}$.

The paper concludes with an Appendix where we prove a more abstract version of a result contained in the author's thesis [**K2**] which bounds the number of rays landing at points with infinite forward orbit. We use this result in the extension of $\lambda_{\mathbb{Q}}$ to $\mathbb{R}/\mathbb{Z}$, as well as in the construction of the topological realization P.

Acknowledgments. Many of the ideas used in this paper stem from the author's thesis [**K2**]. The author is grateful to John Milnor for his enlightening comments and suggestions during his graduate studies. I would like to thank Alfredo Poirier for explaining his work to me, as well as to Genadi Levin for his questions and comments. Also, I am grateful to the referee for the suggestions made to improve the exposition.

2. Preliminaries

In this section, we introduce some definitions and recall some results from iteration of complex polynomials. We assume certain familiarity with the basic ideas from complex dynamics and refer the reader to [**M1, CG, Be, St**] for a more detailed exposition.

Consider a monic centered polynomial

$$\begin{array}{rccl} f: & \mathbb{C} & \to & \mathbb{C} \\ & z & \mapsto & z^d + a_{d-2}z^{d-2} + \cdots + a_0 \end{array}$$

of degree $d \geq 2$. Let $\Omega(f)$ denote the basin of infinity. Throughout, $K(f) = \mathbb{C} \setminus \Omega(f)$ will denote the filled Julia set, and $J(f) = \partial K(f) = \partial \Omega(f)$ will denote the Julia set of f.

In this article we are mainly concerned with polynomials f that have connected Julia set $J(f)$ or, equivalently, connected filled Julia set $K(f)$. A result of Fatou shows that $K(f)$ is connected if and only if every critical point of f is non-escaping.

Moreover, according to Douady and Hubbard [**DH0**], Branner and Hubbard [**BH**], and Lavaurs [**La2**], the set of all monic centered polynomials with connected Julia sets is a compact, cellular and connected subset of parameter space. More precisely:

THEOREM 2.1. *The set*

$$\mathcal{C}_d = \{(a_0, \dots, a_{d-2}) \in \mathbb{C}^{d-2} : J(z^d + a_{d-2}z^{d-2} + \dots + a_0) \textit{ is connected}\}$$

is compact, connected and cellular.

When the Julia set $J(f)$ is connected there exists a unique conformal isomorphism $\phi : \mathbb{C} \setminus \overline{\mathbb{D}} \to \mathbb{C} \setminus K(f) = \Omega(f)$ which is asymptotic to the identity at infinity and conjugates $w \mapsto w^d$ with $z \mapsto f(z)$ (i.e., $f \circ \phi(w) = \phi(w)^d$). This conformal isomorphism ϕ is called the Böttcher map.

The boundary behavior of the map ϕ is intimately related to topological features of $J(f)$ and the dynamics of f. A first understanding of the boundary behavior of ϕ is obtained via the study of radial limits. Following Douady and Hubbard [**DH1, M1**], the external ray R^t_f of argument $t \in \mathbb{R}/\mathbb{Z}$ is defined as the image of $(1, \infty)e^{2\pi i t}$ under the Böttcher map ϕ. When f is clear from the context, we drop the subscript and simply denote by R^t the external ray of argument t. We say that R^t lands at $z \in J(f)$ if the radial limit $\lim_{r \searrow 1} \phi(re^{2\pi i t}) = z$.

Note that f maps external rays onto external rays acting as the multiplication by d map m_d on the arguments of the external rays (i.e., $f(R^t) = R^{dt}$). This leads to a basic dichotomy according to whether $t \in \mathbb{Q}/\mathbb{Z}$ or $t \notin \mathbb{Q}/\mathbb{Z}$. In fact, an argument $t \in \mathbb{R}/\mathbb{Z}$ has finite forward orbit under m_d if and only if $t \in \mathbb{Q}/\mathbb{Z}$.

According to Douady and Hubbard, each external ray R^t with periodic or preperiodic argument t lands at a periodic or preperiodic point, respectively. In the former case, we say that R^t is a periodic ray. In the latter, we say that R^t is a preperiodic ray. Moreover, if z is the landing point of a periodic ray R^t, then all the other rays landing at z are periodic of the same period (see Section 18 of [**M1**]).

3. Rational Laminations

As discussed in Section 2, according to Douady and Hubbard, each rational external ray R^t of a polynomial f with connected Julia set has a well defined limit as it approaches the Julia set $J(f)$ (see Section 18 of [**M1**]). The landing pattern of rational external rays is organized in the rational lamination $\lambda_{\mathbb{Q}}(f)$ of f.

DEFINITION 3.1. *Consider a monic polynomial $f : \mathbb{C} \to \mathbb{C}$ with connected Julia set. We say that the* **rational lamination** *of f is the equivalence relation $\lambda_{\mathbb{Q}}(f)$ in $\mathbb{Q}/\mathbb{Z}$ given by:*

$t, s \in \mathbb{Q}/\mathbb{Z}$ are $\lambda_{\mathbb{Q}}(f)$-equivalent if and only if R^t_f and R^s_f land at the same point.

Example 1. Consider the cubic polynomial $f(z) = z^3 - 1.743318z + 0.50322$ with critical points $c_{\pm} = \pm 0.7623030893$ (these numbers are approximations). The second iterate of the critical point c_- is c_+. The forward orbit of c_+ is ordered in the real line just as the critical orbit of the real quadratic Feigenbaum map. In Figure 1, the fixed external rays with arguments 0 and $1/2$ are drawn. Also, the rays that map in at most two iterations to one of the periodic rays with arguments $1/4, 3/4, 1/10, 3/10, 9/10, 7/10$ are shown. Some equivalence classes of $\lambda_{\mathbb{Q}}(f)$ are: $\{0\}$, $\{1/10, 9/10\}$ and $\{1/4, 3/4\}$.

Example 2. The polynomial $g(z) = z^3 - iz^2 - 1.1z$ has two period two attracting cycles (the immediate basins are "the wings of the butterfly" in Figure 1). Here $\{1/4, 3/4, 5/8, 7/8\}$ is a $\lambda_{\mathbb{Q}}(g)$-class. The figure shows the rays that map in at most two iterations to one with argument in this class.

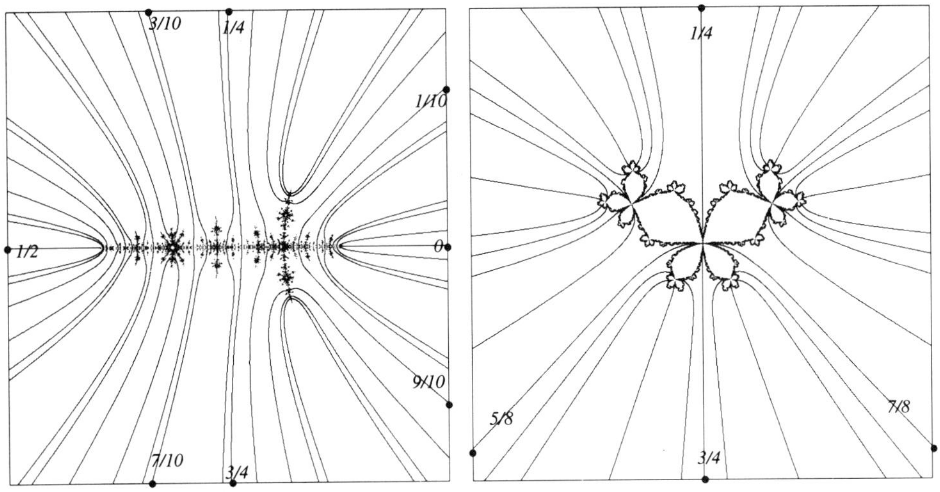

FIGURE 1. The Julia set of Example 1 (left) and of Example 2 (right).

The object of this paper is to characterize the equivalence relations in $\mathbb{Q}/\mathbb{Z}$ that arise in the study of iterations of complex polynomials. Our main result is that five properties characterize rational laminations. In this section we prove that the rational lamination of a polynomial has these five properties. More precisely:

PROPOSITION 3.2. *Let $\lambda_{\mathbb{Q}} = \lambda_{\mathbb{Q}}(f)$ be the rational lamination of a polynomial f with connected Julia set. Then:*

(Q1) $\lambda_{\mathbb{Q}}$ is closed in $\mathbb{Q}/\mathbb{Z} \times \mathbb{Q}/\mathbb{Z}$.

(Q2) Each $\lambda_{\mathbb{Q}}$-equivalence class A is a finite subset of $\mathbb{Q}/\mathbb{Z}$.

(Q3) If A is a $\lambda_{\mathbb{Q}}$-equivalence class, then $d \cdot A$ is a $\lambda_{\mathbb{Q}}$-equivalence class.

(Q4) If A is a $\lambda_{\mathbb{Q}}$-equivalence class, then multiplication by d maps A to $d \cdot A$ in a consecutive preserving fashion.

(Q5) $\lambda_{\mathbb{Q}}$-equivalence classes are pairwise unlinked.

Subsection 3.2 is devoted to the proof of (Q1), while Subsection 3.1 contains the proof of the more basic properties (Q2)–(Q5). Before proceeding to prove these basic properties, let us introduce the definitions involved in the previous proposition.

In general, we say that an equivalence relation λ in a topological space X is **closed** if $\lambda \subset X \times X$ is a closed set.

In property (Q3), as well as in (Q4), we use the notation $d \cdot A$ to denote the image of A under the multiplication by d map m_d acting on $\mathbb{R}/\mathbb{Z}$.

Loosely speaking, we say that multiplication by d maps A onto $d \cdot A$ in a "consecutive preserving" fashion if the following holds. With respect to circular order, if t is the element of A immediately after s in A, then dt is the element of $d \cdot A$ immediately after ds.

In order to make this precise, endow $\mathbb{R}/\mathbb{Z}$ with its standard orientation and use interval notation accordingly with the agreement that $(t,t) = \mathbb{R}/\mathbb{Z} \setminus \{t\}$.

DEFINITION 3.3. *We say that multiplication by d maps a finite subset A of* $\mathbb{R}/\mathbb{Z}$ *to* $d \cdot A$ *in a* **consecutive preserving fashion** *if whenever* (s,t) *is a connected component of* $\mathbb{R}/\mathbb{Z} \setminus A$*, we have that* (ds, dt) *is a connected component of* $\mathbb{R}/\mathbb{Z} \setminus d \cdot A$.

"Consecutive preserving" is a more general notion than "cyclic order preserving". Both notions coincide when multiplication by d maps A onto $d \cdot A$ injectively. However, the multiplication by d map m_d may fail to be injective on A and still map A onto $d \cdot A$ in a consecutive preserving fashion. As we show in Subsection 3.1, property (Q4) captures the fact that $f : \mathbb{C} \to \mathbb{C}$ is orientation preserving.

Remark. Suppose $A \subset \mathbb{Q}/\mathbb{Z}$ is a non-empty set formed by the arguments of the external rays landing at a critical point ω of a polynomial f. The multiplication by d map m_d is not cyclic order preserving on A. However, m_d is consecutive preserving on A. Hence we cannot replace "consecutive preserving" by "cyclic order preserving" in Proposition 3.2.

Two or more external rays landing at a common point z chop the complex plane into two or more connected components. All the external rays landing at a point w different from z must be contained in one of these components. This fact is expressed in the circle $\mathbb{R}/\mathbb{Z}$ by the "unlinked" property.

DEFINITION 3.4. *We say two finite sets A and B contained* $\mathbb{R}/\mathbb{Z}$ *are* **unlinked** *if A is contained in a connected component of* $\mathbb{R}/\mathbb{Z} \setminus B$.

Note that A and B are unlinked if and only if there exists disjoint intervals I_A and I_B such that $A \subset I_A$ and $B \subset I_B$.

REMARK 3.5. Consider three finite subsets A, B and B' of $\mathbb{R}/\mathbb{Z}$ such that A and B are unlinked, and A and B' are unlinked. Suppose $B \cap B' \neq \emptyset$. Then A and $B \cup B'$ are unlinked. In fact, both B and B' are contained in the same connected component J of $\mathbb{R}/\mathbb{Z} \setminus A$. Hence $B \cup B'$ is contained in J.

Remark. Alternative descriptions of the above definition are as follows. Regard A and B as subsets of $\partial\mathbb{D}$ via the standard identification of $\mathbb{R}/\mathbb{Z}$ with $\partial\mathbb{D}$, where $\mathbb{D}$ denotes the open unit disk contained in $\mathbb{C}$. Now $A \subset \partial\mathbb{D} \cong \mathbb{R}/\mathbb{Z}$ and $B \subset \partial\mathbb{D} \cong \mathbb{R}/\mathbb{Z}$ are unlinked if one of the following equivalent conditions hold (see Figure 2):

a.- The Euclidean convex hulls of A and B are disjoint subsets of $\overline{\mathbb{D}}$.

b.- There exist disjoint trees $\Gamma_A \subset \overline{\mathbb{D}}$ and $\Gamma_B \subset \overline{\mathbb{D}}$ each with one vertex in $\mathbb{D}$ and the rest of the vertices being $A \subset \partial\mathbb{D}$ and $B \subset \partial\mathbb{D}$, respectively.

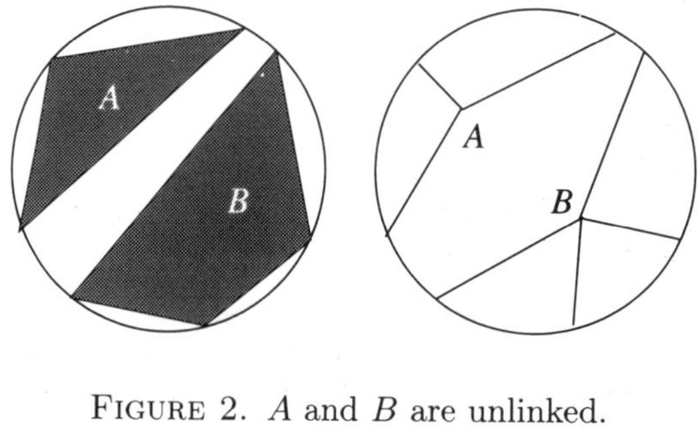

FIGURE 2. A and B are unlinked.

3.1. Basic Properties. Later, in Subsection 3.2, we show that rational laminations are closed in $\mathbb{Q}/\mathbb{Z} \times \mathbb{Q}/\mathbb{Z}$, i.e., (Q1) of Proposition 3.2. We now provide proofs of the fairly well known and straightforward properties (Q2)–(Q5) of Proposition 3.2.

PROOF OF PROPOSITION 3.2 [(Q2) through (Q5)]: Concerning (Q3), let z be the landing point of the external rays with arguments in a $\lambda_{\mathbb{Q}}(f)$-class A. Continuity of f and the identity $f(R^t) = R^{dt}$ implies that the external rays with arguments in $d \cdot A$ land at $f(z)$. We must show that these are all the rays which land at $f(z)$. In fact, if R^s lands at $f(z)$ then lift R^s to an external ray R^t joining z with ∞. It follows that $t \in A$ and $s = dt \in d \cdot A$.

Concerning (Q2), recall that there are at most finitely many rational external rays landing at a periodic point (compare with Lemma 18.3 in [**M1**]). The general case follows from property (Q3).

Now, given a $\lambda_{\mathbb{Q}}(f)$-class A, we let Γ_A denote the graph in $\mathbb{C}$ formed by the union of the external rays with arguments in A and the landing point z_A of these rays.

Note that there is a one-to-one correspondence between connected components U of $\mathbb{C} \setminus \Gamma_A$ and connected components of $\mathbb{R}/\mathbb{Z} \setminus A$ given by:

$$\pi_\infty U = \{t \in \mathbb{R}/\mathbb{Z} : R^t \subset U\} \subset \mathbb{R}/\mathbb{Z} \setminus A.$$

For property (Q5), consider two distinct $\lambda_{\mathbb{Q}}(f)$-classes A and A'. The corresponding graphs Γ_A and $\Gamma_{A'}$ are disjoint. Hence $\Gamma_{A'}$ lies in a connected component U of $\mathbb{C} \setminus \Gamma_A$ and A' is contained in a connected component of $\mathbb{R}/\mathbb{Z} \setminus A$.

It remains to prove the consecutive preserving property (Q4). Let (t, s) be a connected component of $\mathbb{R}/\mathbb{Z} \setminus A$, where A is a $\lambda_{\mathbb{Q}}(f)$-class. Note that there exists a unique connected component V of $\mathbb{C} \setminus f^{-1}(\Gamma_{d \cdot A})$ which contains the external rays with arguments $t + \epsilon$ and $s - \epsilon$, for $\epsilon > 0$ small enough. Under f, the component V maps onto a connected component U of $\mathbb{C} \setminus \Gamma_{d \cdot A}$ which contains the external rays with argument $dt + d\epsilon$ and $ds - d\epsilon$, for $\epsilon > 0$ small enough. It follows that $\pi_\infty U = (dt, ds)$ is a connected component of $\mathbb{R}/\mathbb{Z} \setminus d \cdot A$. □

3.2. Rational Laminations are Closed. In this subsection, we show that rational laminations are closed in $\mathbb{Q}/\mathbb{Z} \times \mathbb{Q}/\mathbb{Z}$. We begin by establishing some closely related basic facts.

Recall that we endow the circle $\mathbb{R}/\mathbb{Z}$ with the standard orientation and use interval notation accordingly with the agreement that $(t, t) = \mathbb{R}/\mathbb{Z} \setminus \{t\}$.

DEFINITION 3.6. *We say that a sequence $\{t_n\}$ converging to t is* **increasing** *if $t_{n+1} \in (t_n, t)$. Throughout, $t_n \nearrow t$ will denote that $\{t_n\}$ is an increasing sequence converging to t. Similarly, a sequence $\{s_n\}$ converging to s is* **decreasing** *if $s_{n+1} \in (s, s_n)$. Also, $s_n \searrow s$ will denote that s_n is a decreasing sequence converging to s.*

LEMMA 3.7. *For $n \geq 0$, suppose $t_n, s_n \in \mathbb{R}/\mathbb{Z}$ are the arguments of external rays landing at a common point w_n such that $w_n \neq w_m$, for $n \neq m$. Assume that $t_n \to t \in \mathbb{R}/\mathbb{Z}$ and that $s_n \to s \in \mathbb{R}/\mathbb{Z}$. Then:*

(i) There exist subsequences $\{t_{n(k)}\}$ and $\{s_{n(k)}\}$ such that either $t_{n(k)} \nearrow t$ and $s_{n(k)} \searrow s$, or $t_{n(k)} \searrow t$ and $s_{n(k)} \nearrow s$.

(ii) If $t \in \mathbb{Q}/\mathbb{Z}$, then $s \in \mathbb{Q}/\mathbb{Z}$. Moreover, if t is a period p argument, then s is a period p argument.

PROOF OF LEMMA 3.7. Without loss of generality, we assume that there exists a subsequence $\{t_{n(k)}\}$ such that $t_{n(k)} \nearrow t$. Then $\{s_{n(k)}\}$ must be a decreasing sequence converging to s (i.e., $s_n \searrow s$). This follows easily from the fact that the collection formed by the pairs $\{t_n, s_n\}$ is pairwise unlinked.

To prove that if t is a period p argument, then s is also a period p argument, we may pass to an iterate and assume that t is fixed. Remark that the rays with arguments $dt_{n(k)}$ and $ds_{n(k)}$ also land at a common point $f(w_{n(k)})$. If t is fixed and s is not, then $ds \in (s,t)$ or $ds \in (t,s)$. In the former case, we conclude that $\{ds, dt = t\}$ and $\{t_{n(k)}, s_{n(k)}\}$ are linked, for k large enough. Hence $\{ds_{n(\hat{k})}, dt_{n(\hat{k})}\}$ and $\{t_{n(k)}, s_{n(k)}\}$ are linked, for k large enough. Thus, for k large enough, $w_{n(k)}$ is a fixed point, which contradicts the hypothesis of the lemma. Similarly, it is not difficult to obtain a contradiction from the latter case. We conclude that s is also a fixed argument and the lemma follows. □

The fact that rational laminations are closed in $\mathbb{Q}/\mathbb{Z} \times \mathbb{Q}/\mathbb{Z}$ will be obtained as an immediate consequence of a result involving left and right prime end impressions.

DEFINITION 3.8 (Left and Right Prime end Impressions). *Let f be a polynomial with connected Julia set. We say that $z \in \mathrm{Imp}^{\pm}(t)$ if and only if there exists $w_n = r_n e^{2\pi i t_n} \in \mathbb{C} \setminus \overline{\mathbb{D}}$ converging to $e^{2\pi i t}$, with $t_n \to t^{\pm}$, and such that the image $\{\phi(w_n)\}$ of the sequence $\{w_n\}$ under the Böttcher map ϕ converges to $z \in J(f)$. Here we say that $t_n \to t^+$ (resp. $t_n \to t^-$) if t is the limit of the sequence $\{t_n\}$ and there exists $\epsilon > 0$ such that, for all $n \geq 0$, we have that $t_n \notin (t-\epsilon, t)$ (resp. $t_n \notin (t, t+\epsilon)$).*

Lemma 3.7 together with the next lemma immediately imply that the rational lamination of a polynomial is closed in $\mathbb{Q}/\mathbb{Z} \times \mathbb{Q}/\mathbb{Z}$.

LEMMA 3.9. *Let f be a polynomial with connected Julia set. Consider two sequences $\{t_n\}$ and $\{s_n\}$ in $\mathbb{R}/\mathbb{Z}$ such that:*
- *the external rays $R_f^{t_n}$ and $R_f^{s_n}$ land at the same point w_n.*
- *$t_n \nearrow t$ and $s_n \searrow s$ (see Definition 3.6).*
- *$t, s \in \mathbb{Q}/\mathbb{Z}$.*

Then the external rays R_f^t and R_f^s land at the same point z_0. Moreover, the left prime end impression $\mathrm{Imp}^-(t)$ and the right prime end impression $\mathrm{Imp}^+(s)$ are trivial, that is,

$$\mathrm{Imp}^-(t) = \mathrm{Imp}^+(s) = \{z_0\}.$$

PROOF OF LEMMA 3.9. We begin by establishing the lemma for s and t periodic under multiplication by d. By passing to an iterate of f, if necessary, we may assume that R^s and R^t are fixed rays.

Let N be large enough so that the following conditions hold:
- Under multiplication by d, the interval (t_N, t) is mapped in a one-to-one fashion onto (dt_N, dt). Similarly, (s, s_N) is also mapped in a one-to-one fashion onto (ds, ds_N).
- The intervals $(dt_N, dt = t)$ and $(s = ds, ds_N)$ are disjoint.

For $n > N$, let U_n be the simply connected domain bounded by the external rays R^{t_N}, R^{s_N}, R^{s_n} and R^{t_n} together with their landing points (see Figure 3). It follows that the union

$$U = \cup_{n>N}^{\infty} U_n$$

is a simply connected domain. One boundary component of U is formed by R^{t_N} and R^{s_N} together with their common landing point. The other boundary component of U is formed by R^s, R^t, and a connected set $X \subset J(f)$ which joins them.

We claim that $X = \mathrm{Imp}^-(t) \cup \mathrm{Imp}^+(s)$. Indeed, every point $x \in X$ is a limit point of $x_n \in R^{s_n} \cup R^{t_n}$. Hence $X \subset \mathrm{Imp}^-(t) \cup \mathrm{Imp}^+(s)$. It remains to prove the converse. The right impression $\mathrm{Imp}^+(s)$ is disjoint from U_n, for all $n > N$, and thus $\mathrm{Imp}^+(s)$ is disjoint from the union U of these sets U_n. Since, for $\epsilon > 0$ small enough, U contains the image of $\{z \in \mathbb{C} \setminus \overline{\mathbb{D}} : s < \arg(z) < s + \epsilon\}$ under the Böttcher map, we have that $\mathrm{Imp}^+(s) \subset \overline{U} \setminus U = \partial U$. Since $\mathrm{Imp}^+(s) \subset \partial U \cap J(f)$ and $\mathrm{Imp}^+(s)$ does not contain the landing point of R^{s_N}, we conclude that $\mathrm{Imp}^+(s) \subset X$. Similarly, $\mathrm{Imp}^-(t) \subset X$, and we have proven the claim.

To complete the proof of the lemma for periodic angles s and t, it suffices to show that X is a singleton.

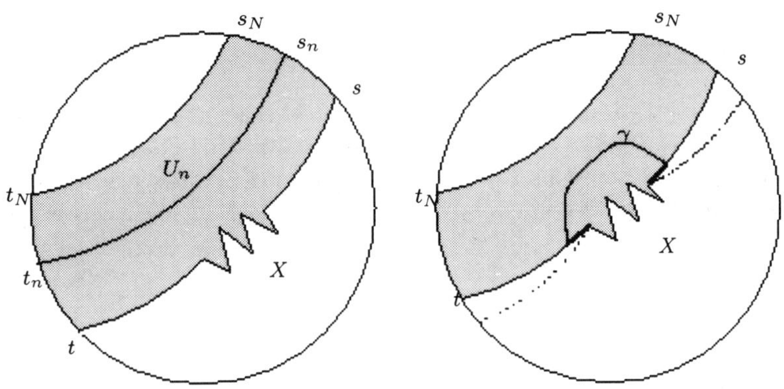

FIGURE 3. The domains U_n (left) and the arc γ (right) in the proof of Lemma 3.9. The shaded region corresponds to U.

The map $f : U \to f(U)$ is univalent, since U_n maps onto its image in a one-to-one fashion. Moreover $U \subset f(U)$. In fact, given n, for k large enough, $U_n \subset f(U_k)$. Furthermore, U extends to a simply connected domain V which contains R^t and R^s where f is again univalent and such that $V \subset f(V)$.

We next construct an arc $\gamma \subset V \cup \{z_0, w_0\}$ starting at the landing point z_0 of R^t which continues along R^t and then crosses over to R^s through the domain U and continues along R^s to end at the landing point w_0 of R^s (see Figure 3).

Now take successive preimages of γ to obtain a sequence $\{\gamma_n\} \subset \overline{U}$ of arcs that join the landing points z_0 and w_0 of R^t and R^s. We claim that the Euclidean diameter of $\gamma_n \to 0$, as $n \to \infty$. In fact, local fixed point theory implies that the diameter of $\gamma_n \cap R^t$ as well as of $\gamma_n \cap R^s$ become arbitrarily small as n increases (see sections 8 and 10 of [**M1**]). Moreover, the diameter of $\gamma_n \cap U$ in the hyperbolic metric of V is uniformly bounded by the diameter of $\gamma \cap U$, since $f^{-1} : f(V) \to V$ is a contraction in this metric. By the standard comparison between the Euclidean and the hyperbolic metric (see Theorem 4.3 of [**CG**]), it follows that the Euclidean diameter of $\gamma_n \cap U$ also becomes arbitrarily small. Thus the diameter of γ_n converges to zero.

We conclude that R^t and R^s land at a common point and that $\{\gamma_n\}$ are in fact closed loops. The domain bounded by γ_n must contain X. Hence X must be a singleton and we have proven the lemma for periodic arguments t and s.

To complete the proof of the lemma, consider the case in which s and t are preperiodic. That is, there exists $l \geq 1$ such that $d^l t$ and $d^l s$ are periodic. We proceed by induction. Remark that the hypothesis of the lemma is invariant under multiplication by d. Suppose $\mathrm{Imp}^-(dt) = \mathrm{Imp}^+(dt) = \{z\}$. Constructing a domain U as in the periodic case yields that $X = \mathrm{Imp}^-(t) \cup \mathrm{Imp}^+(s)$ is a compact and connected set. Since $f(\mathrm{Imp}^{\pm}(\theta)) = \mathrm{Imp}^{\pm}(d\theta)$, it follows that X must be a trivial compact and connected set (i.e., a singleton). □

4. Topological realization

In order to save ink and paper we temporarily introduce the following definition.

DEFINITION 4.1 (Formal Rational Laminations). *An equivalence relation $\lambda_{\mathbb{Q}}$ in $\mathbb{Q}/\mathbb{Z}$ such that (Q1) through (Q5) of Theorem 1.1 hold is called a* **formal rational lamination**.

Given a formal rational lamination $\lambda_{\mathbb{Q}}$, our goal is to find a polynomial f with rational lamination $\lambda_{\mathbb{Q}}(f)$ equal to $\lambda_{\mathbb{Q}}$. A first step in this direction is to build a topological branched covering $P : \mathbb{C} \to \mathbb{C}$ which roughly captures the dynamical features of our desired polynomial f. We call the map P a **topological realization** of $\lambda_{\mathbb{Q}}$.

The construction of $P : \mathbb{C} \to \mathbb{C}$ will be the outcome of the following three steps:
Step 1. In Subsection 4.1, we extend $\lambda_{\mathbb{Q}}$ to an equivalence relation $\lambda_{\mathbb{R}}$ in $\mathbb{R}/\mathbb{Z}$. More precisely, $\lambda_{\mathbb{R}}$ is introduced as the smallest equivalence relation in $\mathbb{R}/\mathbb{Z}$ that contains the closure of $\lambda_{\mathbb{Q}} \subset \mathbb{Q}/\mathbb{Z} \times \mathbb{Q}/\mathbb{Z}$ in $\mathbb{R}/\mathbb{Z} \times \mathbb{R}/\mathbb{Z}$. Also, Subsection 4.1 contains the statement of Proposition 4.3 which shows that $\lambda_{\mathbb{R}}$ inherits properties (R1)–(R5) that are similar to (Q1)–(Q5). Equivalence relations in $\mathbb{R}/\mathbb{Z}$ such that (R1)–(R5) hold are called $\mathbb{R}$eal laminations. Subsection 4.2 is devoted to the proof of Proposition 4.3.

Step 2. In Subsection 4.3, we use the standard identification of $\mathbb{R}/\mathbb{Z}$ with the boundary of the unit disk $\partial\mathbb{D} \subset \mathbb{C}$ and extend the $\mathbb{R}$eal lamination $\lambda_{\mathbb{R}}$ obtained in the previous step to an equivalence relation $\lambda_{\mathbb{C}}$ in $\mathbb{C}$. More precisely, each non-trivial equivalence class of $\lambda_{\mathbb{C}}$ is introduced as the Euclidean convex hull in $\overline{\mathbb{D}}$ of $A \subset \partial\mathbb{D}$, where $A \subset \mathbb{R}/\mathbb{Z} \cong \partial\mathbb{D}$ is a non-trivial equivalence class of $\lambda_{\mathbb{R}}$.

Step 3. In subsections 4.3 and 4.4, we construct a topological realization $P : \mathbb{C} \to \mathbb{C}$ as follows. A result of Moore will imply that the quotient topological space $\mathbb{C}/\lambda_{\mathbb{C}}$ is homeomorphic to $\mathbb{C}$. Invariance of $\lambda_{\mathbb{R}}$ under multiplication by d will allow us to project $z \mapsto z^d$ in $\mathbb{C} \setminus \mathbb{D}$ to a map $P : q(\mathbb{C} \setminus \mathbb{D}) \to q(\mathbb{C} \setminus \mathbb{D})$, where $q : \mathbb{C} \to \mathbb{C} \cong \mathbb{C}/\lambda_{\mathbb{C}}$ is the quotient map. We then extend P to the complex plane $\mathbb{C} = q(\mathbb{C})$ in such a way that the dynamics of P will be, from a topological viewpoint, "similar" to that of a polynomial with locally connected Julia set. In Subsection 4.3, we make precise the word "similar" of the previous sentence and, in Subsection 4.4, prove that such a P exists.

Given a polynomial g with connected Julia set, we may consider the topological realization P of its rational lamination $\lambda_{\mathbb{Q}}(g)$. Provided that g has no irrationally neutral cycle and that every critical point of g, which is in the boundary of a bounded Fatou component, has finite forward orbit, we would expect that the action of g on its Julia set $J(g)$ be semiconjugate to the action of P on $q(\partial\mathbb{D})$. The preimage of a point under the semiconjugacy should be a fiber in the sense of Schleicher [**Sch2**]. Since our goal is to characterize rational laminations, we leave this discussion for a future occasion.

4.1. Extension to $\mathbb{R}/\mathbb{Z}$. We extend formal rational laminations $\lambda_{\mathbb{Q}}$ to equivalence relations in $\mathbb{R}/\mathbb{Z}$ by considering the smallest equivalence relation $\lambda_{\mathbb{R}}$ in $\mathbb{R}/\mathbb{Z}$ which contains the closure of $\lambda_{\mathbb{Q}}$.

DEFINITION 4.2. *We say that $\lambda_{\mathbb{R}}$ is the* **extension to $\mathbb{R}/\mathbb{Z}$** *of the formal rational lamination $\lambda_{\mathbb{Q}}$ if $\lambda_{\mathbb{R}}$ is the smallest equivalence relation in $\mathbb{R}/\mathbb{Z}$ that contains the closed set $\overline{\lambda_{\mathbb{Q}}} \subset \mathbb{R}/\mathbb{Z} \times \mathbb{R}/\mathbb{Z}$.*

The relation $\overline{\lambda_{\mathbb{Q}}}$ may fail to be an equivalence relation only due to lack of transitivity. It follows that two arguments $s, t \in \mathbb{R}/\mathbb{Z}$ are $\lambda_{\mathbb{R}}$-equivalent if and only if there exists a chain $s_0 = s, s_1, \dots, s_n = t$ such that s_i is $\overline{\lambda_{\mathbb{Q}}}$-related to s_{i+1}.

Our aim is to show that extensions to $\mathbb{R}/\mathbb{Z}$ have properties similar to those that define formal rational laminations.

PROPOSITION 4.3. *Consider a formal rational lamination $\lambda_{\mathbb{Q}}$. Let $\lambda_{\mathbb{R}}$ be its extension to $\mathbb{R}/\mathbb{Z}$. Then:*

(R1) $\lambda_{\mathbb{R}}$ is closed in $\mathbb{R}/\mathbb{Z} \times \mathbb{R}/\mathbb{Z}$.

(R2) Each $\lambda_{\mathbb{R}}$-equivalence class A is a finite subset of $\mathbb{R}/\mathbb{Z}$.

(R3) If A is an $\lambda_{\mathbb{R}}$-equivalence class, then $d \cdot A$ is a $\lambda_{\mathbb{R}}$-equivalence class.

(R4) If A is an $\lambda_{\mathbb{R}}$-equivalence class, then $A \mapsto d \cdot A$ is consecutive preserving.

(R5) $\lambda_{\mathbb{R}}$-equivalence classes are pairwise unlinked.

The proof of this proposition is the main content of 4.2.

If $J(f)$ is locally connected, then we may consider the equivalence relation $\lambda_{\mathbb{R}}(f)$ in $\mathbb{R}/\mathbb{Z}$ which captures the landing pattern of the external rays of f (i.e., $t, s \in \mathbb{R}/\mathbb{Z}$ are $\lambda_{\mathbb{R}}(f)$-equivalent if and only if R^t and R^s land at a common point). We call this equivalence relation $\lambda_{\mathbb{R}}(f)$ the **landing relation** of f. It is fairly easy to show that (R1)–(R5) hold for $\lambda_{\mathbb{R}}(f)$, except (R2) which follows from Theorem A.1 of the Appendix.

DEFINITION 4.4 ($\mathbb{R}$eal Laminations). *An equivalence relation $\lambda_{\mathbb{R}}$ in $\mathbb{R}/\mathbb{Z}$ such that (R1) through (R5) of Proposition 4.3 hold is called a* **$\mathbb{R}$eal lamination**.

The author believes that $\mathbb{R}$eal laminations are in a one-to-one correspondence to Thurston's invariant laminations. Since we do not need this correspondence in this paper, we do not make it precise. Moreover, the previous definition and the construction of the topological realization that we describe here are inspired and motivated by Thurston's work [**Th**] (also, compare with [**D**] and [**BL**]).

We remark that there exist polynomials f with locally connected Julia set such that $\lambda_{\mathbb{R}}(f)$ is NOT the extension $\lambda_{\mathbb{R}}$ of the rational lamination $\lambda_{\mathbb{Q}}(f)$. Examples of this kind are a consequence of results by Faught [**Fa**] and Petersen [**Pe**]. More precisely, consider a quadratic polynomial f_c with a fixed Siegel disk of constant type rotation number. According to Petersen, such a polynomial has locally connected Julia set. The rational lamination $\lambda_{\mathbb{Q}}(f_c)$ is trivial (i.e., all classes are singletons). Hence, its extension $\lambda_{\mathbb{R}}$ to $\mathbb{R}/\mathbb{Z}$ is also trivial. But the critical point of f_c belongs to the Julia set $J(f_c)$. A critical point in a locally connected Julia set is automatically the landing point of at least two rays. Thus $\lambda_{\mathbb{R}}(f_c)$ is non-trivial.

A similar situation occurs for cubic polynomials with a fixed critical point c_1 when the other critical point c_2 is not preperiodic and lies in the boundary of the immediate basin of c_1. According to Faught, these polynomials have locally connected Julia sets and, again, they have trivial rational laminations.

Let us emphasize a conclusion of the previous discussion. There are $\mathbb{R}$eal laminations which are not the extension of formal rational laminations. Moreover, these $\mathbb{R}$eal laminations appear in polynomial dynamics. This justifies the fact that, in some results contained in subsections 4.3 and 4.4, we do not restrict ourselves to $\mathbb{R}$eal laminations which are the extension of formal rational laminations.

4.2. $\mathbb{R}$eal Laminations. Here we prove Proposition 4.3 which shows that the extension of rational laminations are $\mathbb{R}$eal laminations. The proof of Proposition 4.3 requires some definitions and lemmas after which the reader may find it at the end of this subsection.

DEFINITION 4.5. *Let λ be a formal rational lamination or a $\mathbb{R}$eal lamination. We say $t, s \in \mathbb{R}/\mathbb{Z} \setminus \mathbb{Q}/\mathbb{Z}$ are λ-unlinked if, for all non-trivial λ-equivalence classes A, we have that $\{t, s\}$ and A are disjoint and unlinked.*

Equivalently, $t, s \in \mathbb{R}/\mathbb{Z} \setminus \mathbb{Q}/\mathbb{Z}$ are λ-unlinked if no λ-equivalence class A "separates" s and t (i.e., $\{s, t\}$ is contained in a connected component of $\mathbb{R}/\mathbb{Z} \setminus A$).

The λ-unlinked relation is, in fact, an equivalence relation. To verify that it is transitive, we proceed as follows. If t, s are λ-unlinked, and if s, r are also λ-unlinked, then for any λ-class A, the sets $\{t, s, r\}$ and A are unlinked (see Remark 3.5). In particular, $\{t, r\}$ and A are unlinked. Therefore, t and r are λ-unlinked.

Though for the moment we are mainly concerned with the case in which λ is a formal rational lamination, later, in Subsection 4.4, we will employ the above definition and the next four lemmas in the case that λ is a $\mathbb{R}$eal lamination.

To illustrate the previous definition, consider an equivalence relation $\lambda_{\mathbb{Q}}$ which is the rational lamination $\lambda_{\mathbb{Q}}(f)$ of a polynomial f with locally connected Julia set $J(f)$. Irrational arguments t, s of rays which land at the same point in $J(f)$ are a typical example of $\lambda_{\mathbb{Q}}$-unlinked arguments. Also, irrational arguments t, s of rays that land in the boundary of the same bounded Fatou component are an example of $\lambda_{\mathbb{Q}}$-unlinked arguments.

LEMMA 4.6. *Let λ be a formal rational lamination or a $\mathbb{R}$eal lamination. If L is a λ-unlinked class, then $d \cdot L$ is a λ-unlinked class. Moreover, if $L \mapsto d \cdot L$ is one-to-one, then $L \mapsto d \cdot L$ is cyclic order preserving.*

PROOF OF LEMMA 4.6. Given an irrational argument $t \in \mathbb{R}/\mathbb{Z} \setminus \mathbb{Q}/\mathbb{Z}$ and a λ-class A we let $I_A(t)$ denote the component of $\mathbb{R}/\mathbb{Z} \setminus A$ which contains t. The proof of the lemma relies on the following claim:

Claim. Consider a finite collection $\mathcal{C}$ of λ-classes. Let $\tilde{\mathcal{C}}$ be the collection formed by all λ-classes that are mapped onto one which belongs to $\mathcal{C}$ in one step. Consider $t \in \mathbb{R}/\mathbb{Z} \setminus \mathbb{Q}/\mathbb{Z}$. Let

$$\tilde{J} = \cap_{C \in \tilde{\mathcal{C}}} I_C(t) \quad \text{and} \quad J = \cap_{C \in \mathcal{C}} I_C(dt).$$

Then $d \cdot \tilde{J} = J$. Moreover if multiplication by d maps $\tilde{J}$ bijectively onto J, then this map is cyclic order preserving.

Remark. The interval $\tilde{J}$ might not map bijectively onto J. In fact, consider the rational lamination of the basilica $z \mapsto z^2 - 1$. Clearly, if $\mathcal{C} = \{\{1/3, 2/3\}\}$ and $t = 2/9$, then $\tilde{J} = (1/6, 1/3) \cup (2/3, 5/6)$ and $J = (1/3, 2/3)$.

Proof of the Claim. The set $\tilde{J}$ is a union of disjoint open intervals $\tilde{J}_1, \dots, \tilde{J}_k$ indexed respecting cyclic order so that $t \in \tilde{J}_1$. Similarly, J is the union of disjoint open intervals $J_1, \dots, J_l$ also indexed respecting cyclic order so that $dt \in J_1$.

Our aim is to show that $d \cdot \tilde{J}_i = J_i$ where the subscripts are taken modulo k and l, respectively (hence, l divides k). We proceed by induction. The set

$$\tilde{X} = \cup_{C \in \tilde{\mathcal{C}}} C$$

is the preimage under multiplication by d of

$$X = \cup_{C \in \mathcal{C}} C .$$

Thus each component of $\mathbb{R}/\mathbb{Z} \setminus \tilde{X}$ must map onto a component of $\mathbb{R}/\mathbb{Z} \setminus X$. In particular, $\tilde{J}_i$ maps onto a component of $\mathbb{R}/\mathbb{Z} \setminus X$. It follows that $d \cdot \tilde{J}_1 = J_1$. Now suppose that $d \cdot \tilde{J}_i = J_i$. By the definition of $\tilde{J}$, the right endpoint $\tilde{b}_i$ of $\tilde{J}_i = (\tilde{a}_i, \tilde{b}_i)$ and the left endpoint $\tilde{a}_{i+1}$ of $\tilde{J}_{i+1} = (\tilde{a}_{i+1}, \tilde{b}_{i+1})$ are such that $(\tilde{a}_{i+1}, \tilde{b}_i)$ is the component $I_{\tilde{C}}(t)$ of $\mathbb{R}/\mathbb{Z} \setminus \tilde{C}$, for some $\tilde{C} \in \tilde{\mathcal{C}}$. Since multiplication by d maps $\tilde{C}$ onto C in a consecutive preserving fashion, it follows that $(d\tilde{a}_{i+1}, d\tilde{b}_i)$ is a component of $\mathbb{R}/\mathbb{Z} \setminus d \cdot C$. Moreover, by the inductive hypothesis, $(d\tilde{a}_{i+1}, d\tilde{b}_i)$ contains J_i and thus dt, since $d\tilde{b}_i - \epsilon \in d \cdot \tilde{J}_i = J_i$. Therefore J contains $d\tilde{a}_{i+1} + \epsilon$, for $\epsilon > 0$ small enough. Since $(d\tilde{b}_i, d\tilde{a}_{i+1})$ is contained in the complement of J, the left endpoint of J_{i+1} is $d\tilde{a}_{i+1}$ and $d \cdot \tilde{J}_{i+1}$ must coincide with J_{i+1}, and the proof of the claim is now complete.

We now show that if L is a λ-unlinked class, then $d \cdot L$ is also a λ-unlinked class. Suppose that s lies in the λ-unlinked class of t and that ds does not belong to the λ-unlinked class of dt. Then there exists a λ-class A such that $ds \notin I_A(dt)$. Consider the collection $\mathcal{C} = \{A\}$. In the notation of the claim, $J = I_A(dt)$ and both λ-equivalent arguments s, t must lie in $\tilde{J}$. Since $d \cdot \tilde{J} = J$, we obtain a contradiction with $ds \notin I_A(dt) = J$. Thus if L is a λ-unlinked class, then $d \cdot L$ is contained in a λ-unlinked class. For the converse, suppose that ds and dt are λ-unlinked equivalent and that none of the preimages $s, s + 1/d, \dots, s + (d-1)/d$ of s are λ-unlinked equivalent to t. Then there exist λ-classes $A_1, \dots, A_{d-1}$ such that $s + i/d \notin I_{A_i}(t)$. Let $\mathcal{C} = \{d \cdot A_1, \dots, d \cdot A_{d-1}\}$. In the notation of the claim, $ds \in J$ but no preimage of ds lies in $\tilde{J}$, which is a contradiction. Thus $d \cdot L$ is a λ-unlinked class.

Suppose that L maps onto $d \cdot L$ in a one-to-one fashion. After choosing a point $t \in L$, there exists a collection $\mathcal{C}$ of λ-classes such that $\tilde{J}$ maps in a one-to-one fashion onto J, where $\tilde{J}$ and J are the sets considered in the claim of above. The claim shows that multiplication by d is cyclic order preserving on $\tilde{J}$. In particular, $L \mapsto d \cdot L$ is cyclic order preserving. □

In view of the previous result, there is a basic distinction among $\lambda_{\mathbb{Q}}$-unlinked classes; that is, there are **preperiodic, periodic and wandering $\lambda_{\mathbb{Q}}$-unlinked classes.**

Examples. Consider the cubic polynomial $f(z)$ of Example 1 in Section 3. As suggested by Figure 4, the rays with irrational arguments $\pm s, \pm t$ seem to land at the critical point c_+, where $t = 0.1016458427...$ and $s = 0.2316874906....$ Moreover, the figure suggests that, for all $\theta \notin \{\pm s, \pm t\}$, there exists a pair of rational rays landing at a common point that separates the external ray at angle θ from the rays at angles $\{\pm s, \pm t\}$. So $\{\pm s, \pm t\}$ forms a wandering $\lambda_{\mathbb{Q}}(f)$-unlinked class.

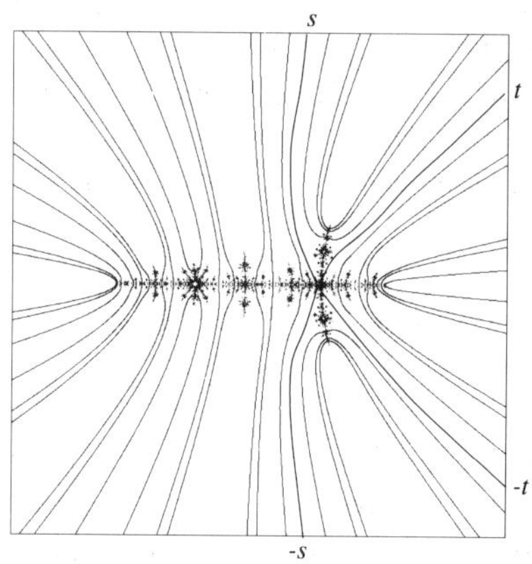

FIGURE 4. The rays at irrational angles $\{\pm s, \pm t\}$ seem to land at the same point and to be separated from any other ray by a pair of rational rays. Thus $\{\pm s, \pm t\}$ forms a $\lambda_{\mathbb{Q}}(f)$-unlinked class; as their arguments are irrational, this class wanders.

Our next two lemmas show that a non-trivial $\lambda_{\mathbb{R}}$-class is a $\lambda_{\mathbb{Q}}$-class or is contained in a wandering $\lambda_{\mathbb{Q}}$-unlinked class, where $\lambda_{\mathbb{R}}$ is the extension of a formal rational lamination $\lambda_{\mathbb{Q}}$.

LEMMA 4.7. *Let $\lambda_{\mathbb{Q}}$ be a formal rational lamination. If two arguments $s, t \in \mathbb{R}/\mathbb{Z}$ are $\overline{\lambda_{\mathbb{Q}}}$-related and not $\lambda_{\mathbb{Q}}$-equivalent, then s and t are irrational, and*

(i) $t = s$, or

(ii) both arguments t and s lie in the same $\lambda_{\mathbb{Q}}$-unlinked class L. Moreover, (s, t) or (t, s) is a connected component of $\mathbb{R}/\mathbb{Z} \setminus L$.

Remark. In the previous lemma, it may occur that (i) holds but not (ii). In fact, consider the polynomial g of Example 2 in Section 3. Let $t = s \in \mathbb{R}/\mathbb{Z} \setminus \mathbb{Q}/\mathbb{Z}$ be the argument of an external ray that lands in the boundary of a bounded periodic Fatou component U. The $\lambda_{\mathbb{Q}}(g)$-unlinked class of $t = s$ is the set L formed by the arguments of irrational external rays landing at ∂U. Obviously, (s, s) is not a connected component of $\mathbb{R}/\mathbb{Z} \setminus L$; hence (ii) does not hold.

PROOF OF LEMMA 4.7. We start by showing that $\overline{\lambda_{\mathbb{Q}}}$ contains purely irrational or purely rational pairs. Recall that we do not know that $\lambda_{\mathbb{Q}}$ is the rational lamination of a polynomial. Hence we are forced to prove, in an abstract setting, the analogue of Lemma 3.7.

Claim. If the ordered pair (s, t) is in $\overline{\lambda_{\mathbb{Q}}}$ and $s \in \mathbb{Q}/\mathbb{Z}$, then $t \in \mathbb{Q}/\mathbb{Z}$.

Proof of the Claim. We may assume that s is periodic under multiplication by d and, after passing to an iterate, we may also assume that s is, in fact, fixed.

By the unlinked property, there exists a sequence of ordered pairs $(s_n, t_n) \in \lambda_{\mathbb{Q}}$ converging to (s,t) such that $\{s_n, t_n\}$ is contained in one of the two intervals that form $\mathbb{R}/\mathbb{Z} \setminus \{s,t\}$. Without loss of generality, we suppose that $s_n \searrow s$ and that $t_n \nearrow t$. Hence $ds_n \searrow s$ and $dt_n \nearrow dt$. It is not difficult to check that if dt lies in the interval (s,t) or in the interval $(t,s]$, then for some n,m, we have that $\{s_n,t_n\}$ and $\{ds_m, dt_m\}$ are linked, which is a contradiction with the unlinked and invariant properties of $\lambda_{\mathbb{Q}}$. Thus $dt = t$ and $t \in \mathbb{Q}/\mathbb{Z}$.

Since $\lambda_{\mathbb{Q}}$ is closed in $\mathbb{Q}/\mathbb{Z} \times \mathbb{Q}/\mathbb{Z}$, we conclude that $\overline{\lambda_{\mathbb{Q}}} \setminus \lambda_{\mathbb{Q}}$ consists of pairs with both coordinates irrational. Consider a pair $(s,t) \in \overline{\lambda_{\mathbb{Q}}} \setminus \lambda_{\mathbb{Q}}$ such that $t \neq s$.

Now note that s,t are $\lambda_{\mathbb{Q}}$-unlinked equivalent. For otherwise, there exists a $\lambda_{\mathbb{Q}}$-class A which is linked with $\{s,t\}$. But the pair (s,t) is the limit of a sequence $(s_n,t_n) \in \lambda_{\mathbb{Q}}$. For n large, $\{s_n,t_n\}$ and A are linked, which is a contradiction with the unlinked property (Q5) of $\lambda_{\mathbb{Q}}$. Hence s,t lie in a $\lambda_{\mathbb{Q}}$-unlinked class L. Also, it follows that the interval (s,t) or the interval (t,s) is a connected component of $\mathbb{R}/\mathbb{Z} \setminus L$ □

LEMMA 4.8. *Let λ be a formal rational lamination or a $\mathbb{R}$eal lamination. If s,t are irrational angles contained in a λ-unlinked class L such that the interval (s,t) is a connected component of $\mathbb{R}/\mathbb{Z} \setminus L$, then L is a wandering λ-unlinked class.*

REMARK 4.9. If (s,t) is a connected component of $\mathbb{R}/\mathbb{Z} \setminus L$, then (ds,dt) is a connected component of $\mathbb{R}/\mathbb{Z}\setminus d{\cdot}L$. In fact, there exists $s_n \searrow s$ and $t_n \nearrow t$ such that s_n and t_n are λ-equivalent and consecutive in their common λ-class A_n (i.e., (s_n,t_n) is a connected component of $\mathbb{R}/\mathbb{Z} \setminus A_n$). By the consecutive preserving property (Q4) and (R4) of formal rational laminations and $\mathbb{R}$eal laminations, (ds_n, dt_n) is contained in $\mathbb{R}/\mathbb{Z} \setminus d \cdot L$. Thus, (ds, ts) is a connected component of $\mathbb{R}/\mathbb{Z} \setminus d \cdot L$.

PROOF OF LEMMA 4.8. There are two cases according to whether there exists an n such that $d^n s = d^n t$ or not.

First we deal with the case for which $d^n s = d^n t$, for some n. Since the interval (s,t) is a connected component of $\mathbb{R}/\mathbb{Z}\setminus L$, it follows that $\mathbb{R}/\mathbb{Z}\setminus d{\cdot}L$ contains (ds,dt). Furthermore, $\mathbb{R}/\mathbb{Z} \setminus d^n L$ contains $(d^n s, d^n t)$. Therefore $d^n L = \{d^n s\} \subset \mathbb{R}/\mathbb{Z} \setminus \mathbb{Q}/\mathbb{Z}$ is a wandering λ-unlinked class.

Next we deal with the case for which $d^n s \neq d^n t$, for all $n \geq 0$. We suppose that L is eventually periodic and, after some work, we will obtain a contradiction. Iterating L we may suppose that L is periodic and by passing to an iterate of the multiplication by d map we may assume that L is fixed. By the previous remark, for all n, the non-trivial interval $(d^n s, d^n t)$ is a connected component of $\mathbb{R}/\mathbb{Z} \setminus L$. Since $s,t \notin \mathbb{Q}/\mathbb{Z}$, the collection $\{(d^n s, d^n t)\}$ is a pairwise disjoint collection of non-trivial intervals. Disjointness implies that, for n large enough, the length of $(d^n s, d^n t)$ is strictly smaller than $1/d$. Thus, for n large enough, the length of the disjoint intervals $\{(d^n s, d^n t)\}$ is an increasing function of n, that is clearly impossible. We conclude that L is not fixed, which is a contradiction. □

Our next result will be used in several occasions to show that multiplication by d acts injectively on most of the members of a pairwise unlinked collection of finite subsets of $\mathbb{R}/\mathbb{Z}$.

LEMMA 4.10. *Let $\{\Theta_1, \cdots \Theta_l\}$ be a pairwise unlinked and disjoint collection of finite subsets of $\mathbb{R}/\mathbb{Z}$. Assume that the cardinality δ_k of Θ_k is at least two and that $d \cdot \Theta_k$ is a singleton, for $k = 1, \dots, l$. Then $\sum(\delta_k - 1) \leq d - 1$. In particular, $l \leq d - 1$.*

PROOF OF LEMMA 4.10. The collection $\{\Theta_1, \cdots \Theta_l\}$ cuts $\mathbb{R}/\mathbb{Z}$ into $[\sum(\delta_k - 1)] + 1$ pieces as follows. Two points are in the same piece J if they lie in the same component of $\mathbb{R}/\mathbb{Z} \setminus \Theta_k$, for all k. Since the length of each piece is a multiple of $1/d$, we conclude that the inequality $\sum(\delta_k - 1) \leq d - 1$ holds. □

The last ingredient in the proof of Proposition 4.3 is our next result which gives a uniform upper bound for the cardinality of wandering $\lambda_{\mathbb{Q}}$-unlinked classes. It is based on a result found in [**K2**] which is applied to obtain Theorem A.2 of the Appendix. For more estimates of this kind we refer the reader to [**BL**].

LEMMA 4.11. *Let λ be a formal rational lamination or a $\mathbb{R}$eal lamination. If L is a wandering λ-unlinked class, then L contains at most $d2^{d-1}$ elements. Moreover, for n large enough, $m_d : d^n \cdot L \to d^{n+1} \cdot L$ is one-to-one.*

PROOF OF LEMMA 4.11. Consider a finite collection of integers $k \geq 0$ such that a point in $d^{k+1} \cdot L$ has a preimage $\Theta_k \subset d^k \cdot L$ of cardinality $\delta_k > 1$. Since the collection $\Theta = \{\Theta_k\}$ is pairwise unlinked, we can apply the previous lemma and conclude that there are at most $d - 1$ values of k for which $d^k \cdot L \mapsto d^{k+1} \cdot L$ is not injective. Therefore, for n large enough, $m_d : d^n \cdot L \to d^{n+1} \cdot L$ is one-to-one. Moreover, it follows that $L \mapsto d^n \cdot L$ is at most 2^{d-1}-to-one. By Lemma 4.6, any finite subset A of $d^n \cdot L$ satisfies the hypothesis of Theorem A.2. Thus $d^n \cdot L$ has at most d elements, and therefore L has at most $2^{d-1}d$ elements. □

PROOF OF PROPOSITION 4.3. Concerning (R2), by lemmas 4.7 and 4.8, a $\lambda_{\mathbb{R}}$-class is either a $\lambda_{\mathbb{Q}}$-class or contained in a wandering $\lambda_{\mathbb{Q}}$-unlinked class. Lemma 4.11 states that wandering $\lambda_{\mathbb{Q}}$-unlinked classes are finite.

Concerning (R3), it suffices to show that each wandering $\lambda_{\mathbb{Q}}$-unlinked class L is a $\lambda_{\mathbb{R}}$-class. Since wandering classes are finite, connected components of $\mathbb{R}/\mathbb{Z} \setminus L$ are open intervals. Let (s, t) be one of these components. By the definition of L, there exists $s_n \searrow s$ and $t_n \nearrow t$ such that (s_n, t_n) is a component of $\mathbb{R}/\mathbb{Z} \setminus A_n$, where A_n is a $\lambda_{\mathbb{Q}}$-class. Hence s, t are $\overline{\lambda_{\mathbb{Q}}}$-related. In view of the transitivity of $\lambda_{\mathbb{R}}$ together with finiteness of L, the elements of L are $\lambda_{\mathbb{R}}$-equivalent. Since unlinked classes map onto unlinked classes, (R3) follows. By Remark 4.9, (ds, dt) is a component of $\mathbb{R}/\mathbb{Z} \setminus d \cdot L$ and (R4) follows as well.

It is easy to check that $\lambda_{\mathbb{Q}}$-classes are pairwise unlinked. Therefore, (R5) also holds.

It remains to show (R1), that is, that $\lambda_{\mathbb{R}}$ is closed. Two $\lambda_{\mathbb{R}}$-equivalent arguments s, t are connected through a chain of $\overline{\lambda_{\mathbb{Q}}}$-related arguments. In view of the bound of $d2^{d-1}$ for the cardinality of wandering unlinked classes, there is always a choice of such a chain of length not greater than $d2^{d-1}$. This uniform bound implies that $\lambda_{\mathbb{R}}$ is closed in $\mathbb{R}/\mathbb{Z} \times \mathbb{R}/\mathbb{Z}$. □

4.3. Extension of $\mathbb{R}$eal Laminations to $\mathbb{C}$. In several of the constructions available in the literature [**Th, D, BL**], unlinked equivalence relations in $\mathbb{R}/\mathbb{Z}$ are extended to $\mathbb{C}$ as in the following definition (see also Figure 5).

Recall that $\mathbb{D}$ denotes the open unit disk in $\mathbb{C}$.

DEFINITION 4.12. *Let $\lambda_{\mathbb{R}}$ be a $\mathbb{R}$eal lamination. After identification of $\mathbb{R}/\mathbb{Z}$ with $\partial\mathbb{D}$ we say that the* **extension** *$\lambda_{\mathbb{C}}$ of $\lambda_{\mathbb{R}}$ to $\mathbb{C}$ is the equivalence relation in $\mathbb{C}$ whose non-trivial equivalence classes are the Euclidean convex hulls of the non-trivial $\lambda_{\mathbb{R}}$-classes $A \subset \partial\mathbb{D}$.*

Sometimes it is more convenient to "draw" the convex hulls with respect to the hyperbolic metric in $\mathbb{D}$. This leads to a picture topologically equivalent to the one obtained with the Euclidean metric.

FIGURE 5. The convex hull with respect to the hyperbolic metric in $\mathbb{D}$ of some equivalence classes of $\lambda_{\mathbb{Q}}(g)$, where g is the cubic polynomial in Example 2 of Section 3.

According to a result by Moore [**Mo**], the quotient of a sphere S^2 by an equivalence relation $\sim$ is again a sphere if the following hold:

- each $\sim$-class is not the whole sphere S^2;
- $\sim$ is a closed equivalence relation;
- each $\sim$-class is connected and has connected complement.

Applying this result to the obvious extension of $\lambda_{\mathbb{C}}$ to a relation in the Riemann sphere, we obtain the following corollary.

COROLLARY 4.13. *If $\lambda_{\mathbb{C}}$ is the extension of a $\mathbb{R}$eal lamination, then $\mathbb{C}/\lambda_{\mathbb{C}}$ is homeomorphic to $\mathbb{C}$.*

If $\lambda_{\mathbb{R}} = \lambda(f)$ is the landing relation of a polynomial f with locally connected filled Julia set $K(f)$, then $\overline{\mathbb{D}}/\lambda_{\mathbb{C}}$ is homeomorphic to $K(f)$ (see Douady [**D**]). Usually, $\overline{\mathbb{D}}/\lambda_{\mathbb{C}}$ is called the *pinched disk model* of $K(f)$ and $\mathbb{C}/\lambda_{\mathbb{C}}$ is called the *pinched plane.*

In our context we have an abstract equivalence relation $\lambda_{\mathbb{R}}$ only related to the dynamics of the multiplication by d map m_d in $\mathbb{R}/\mathbb{Z}$. We wish to produce a dynamical system on $\mathbb{C}$. From a topological viewpoint, the desired dynamical system $P : \mathbb{C} \to \mathbb{C}$ will be just like a polynomial g with filled Julia set homeomorphic to $\overline{\mathbb{D}}/\lambda_{\mathbb{C}}$ and such that the Fatou critical points of g have finite forward orbit. More precisely:

PROPOSITION 4.14. *Let $\lambda_{\mathbb{C}}$ be the extension of a $\mathbb{R}$eal lamination. We let $q : \mathbb{C} \to \mathbb{C} \cong \mathbb{C}/\lambda_{\mathbb{C}}$ denote the quotient map. Then there exists a degree d branched covering $P : \mathbb{C} \to \mathbb{C}$ such that the following hold:*

(i) The following diagram commutes

$$\begin{array}{ccc} \mathbb{C}\setminus\mathbb{D} & \xrightarrow{(\cdot)^d} & \mathbb{C}\setminus\mathbb{D} \\ {\scriptstyle q}\downarrow & & \downarrow{\scriptstyle q} \\ q(\mathbb{C}\setminus\mathbb{D}) & \xrightarrow[P]{} & q(\mathbb{C}\setminus\mathbb{D}). \end{array}$$

(ii) If U is a bounded component of $\mathbb{C}\setminus q(\partial\mathbb{D})$, then U is a Jordan domain which is eventually periodic under iterations of P.

(iii) If U is a period p bounded component of $\mathbb{C}\setminus q(\partial\mathbb{D})$, then $P^{\circ p}:\overline{U}\to\overline{U}$ is topologically conjugate to one of the following maps:

$$\begin{array}{rccc} e^{2\pi i\theta}(\cdot): & \overline{\mathbb{D}} & \to & \overline{\mathbb{D}} \\ & z & \mapsto & e^{2\pi i\theta}z, \end{array}$$

for some $\theta\in\mathbb{R}\setminus\mathbb{Q}$, or

$$\begin{array}{rccc} (\cdot)^\delta: & \overline{\mathbb{D}} & \to & \overline{\mathbb{D}} \\ & z & \mapsto & z^\delta, \end{array}$$

for some integer $\delta>1$.

(iv) There exists a collection of homeomorphisms $h_U:\overline{\mathbb{D}}\to\overline{U}$ such that whenever U is NOT a period p component such that $P^{\circ p}:\overline{U}\to\overline{U}$ is topologically conjugate to an irrational rotation, then for some integer $\delta_U\geq 1$, the following diagram commutes

$$\begin{array}{ccc} \overline{\mathbb{D}} & \xrightarrow{(\cdot)^{\delta_U}} & \overline{\mathbb{D}} \\ {\scriptstyle h_U}\downarrow & & \downarrow{\scriptstyle h_{P(U)}} \\ \overline{U} & \xrightarrow[P]{} & P(\overline{U}). \end{array}$$

If U is a period p component such that $P^{\circ p}:\overline{U}\to\overline{U}$ is topologically conjugate to an irrational rotation $z\mapsto e^{2\pi i\theta}z$, then the following diagram commutes

$$\begin{array}{ccc} \overline{\mathbb{D}} & \xrightarrow{e^{2\pi i\theta/p}\cdot(\cdot)} & \overline{\mathbb{D}} \\ {\scriptstyle h_U}\downarrow & & \downarrow{\scriptstyle h_{P(U)}} \\ \overline{U} & \xrightarrow[P]{} & P(\overline{U}). \end{array}$$

We postpone the proof of this proposition until Subsection 4.4.

DEFINITION 4.15. *Let $\lambda_\mathbb{C}$ be the extension of a $\mathbb{R}$eal lamination $\lambda_\mathbb{R}$. A degree d branched covering $P:\mathbb{C}\to\mathbb{C}$ is called a* **topological realization of** $\lambda_\mathbb{R}$ *if (i) through (iv) of Proposition 4.14 hold. If $\lambda_\mathbb{R}$ is the extension of a formal rational lamination $\lambda_\mathbb{Q}$, then we simply say that P is a* **topological realization of** $\lambda_\mathbb{Q}$.

From the proof of the proposition we will conclude that, if ω is a critical point of P (i.e., P is not one-to-one about ω), then either $\omega=h_U(0)$ for some component U such that $P:U\to P(U)$ is not one-to-one, or $\omega=q(A)\in q(\partial\mathbb{D})$ for some $\lambda_\mathbb{R}$-class $A\subset\mathbb{R}/\mathbb{Z}\cong\partial\mathbb{D}$ such that $A\mapsto d\cdot A$ is not one-to-one.

Remark. From the definition it follows that a topological realization of $\lambda_{\mathbb{R}}$ is unique up to topological conjugacy.

In the context of topological realizations, it makes sense to introduce the analogues of the Julia sets, Fatou sets, filled Julia sets, etc.

DEFINITION 4.16. *Let $P : \mathbb{C} \to \mathbb{C}$ be a topological realization of a $\mathbb{R}$eal lamination $\lambda_{\mathbb{R}}$. Denote by $q : \mathbb{C} \to \mathbb{C} \cong \mathbb{C}/\lambda_{\mathbb{C}}$ the quotient map. We say that:*
(a) $q(\partial\mathbb{D})$ is the **topological Julia set** *$J^{top}(P)$ of P;*
(b) $\mathbb{C} \setminus J^{top}(P)$ is the **topological Fatou set** *$F^{top}(P)$ of P;*
(c) $q(\overline{\mathbb{D}})$ is the **topological filled Julia set** *$K^{top}(P)$ of P;*
(d) $\mathbb{C} \setminus K^{top}(P)$ is the **basin of infinity** *of P;*
(e) $R^t_{top} = \{q(re^{2\pi it}) : r \in (1, \infty)\}$ is the **topological ray** *with argument $t \in \mathbb{R}/\mathbb{Z}$.*

By the definition of topological realizations, all of the critical points of P which are contained in the topological Fatou set have finite forward orbit. We now introduce **extended topological rays**. Consider a ray R^t_{top} landing at a point z which lies in ∂U, where U is a bounded topological Fatou component. For some $s \in \mathbb{R}/\mathbb{Z}$, the point $z = h_U(e^{2\pi is})$, where $h_U : \overline{\mathbb{D}} \to \overline{U}$ is the homeomorphism given by the above proposition. An **extended topological ray** $\hat{R}^t_{top}$ is the arc obtained as the union of R^t_{top}, $\{z\}$ and $h_U([0, 1]e^{2\pi is})$. Therefore, extended rays connect ∞ to the center of a Fatou component.

Note that if $f : \mathbb{C} \to \mathbb{C}$ is a polynomial with locally connected Julia set, then f is a topological realization of its landing relation $\lambda(f)$, provided that Fatou critical points have finite forward orbit under f. When this occurs, the topological definitions of above coincide with the usual definitions of polynomial dynamics. We have pointed out that the extension to $\mathbb{R}/\mathbb{Z}$ of the rational lamination of f might not coincide with $\lambda(f)$. In a more abstract setting, the next lemma states that this situation occurs whenever there exists a critical point with infinite forward orbit in the boundary of a periodic Fatou component of f.

LEMMA 4.17. *Let $\lambda_{\mathbb{R}}$ be a $\mathbb{R}$eal lamination and $P : \mathbb{C} \to \mathbb{C}$ a topological realization of $\lambda_{\mathbb{R}}$.*

The $\mathbb{R}$eal lamination $\lambda_{\mathbb{R}}$ coincides with the extension of the formal rational lamination $\lambda_{\mathbb{Q}} = \lambda_{\mathbb{R}} \cap (\mathbb{Q}/\mathbb{Z} \times \mathbb{Q}/\mathbb{Z})$ if and only if no critical point of P with infinite forward orbit is in the boundary ∂U of a bounded topological Fatou component U of P.

We postpone the proof of this lemma until Subsection 4.4. In the next corollary, using a result of Douady (e.g., see Lemma 18.8 in [**M1**]) and the previous lemma, we show that the topological realization of a formal rational lamination cannot have rotation domains.

COROLLARY 4.18. *Consider a formal rational lamination $\lambda_{\mathbb{Q}}$ with topological realization $P : \mathbb{C} \to \mathbb{C}$. If U is a period p bounded topological Fatou component of $P : \mathbb{C} \to \mathbb{C}$, then the return map $P^{\circ p} : U \to U$ is not injective, so there are no rotation domains and only the first type of the classification in Proposition 4.3 (iii) is realized.*

PROOF OF COROLLARY 4.18. By Proposition 4.3, if $P^{\circ p} : U \to U$ is injective, then it is an irrational rotation. Hence every point in ∂U has infinite orbit. By the previous lemma, ∂U does not contain critical points. Thus $m_d^{\mathrm{op}} : q^{-1}(\partial U) \to q^{-1}(\partial U)$ is a homeomorphism. This is a contradiction with a lemma of Douady (e.g., see Lemma 18.8 in [**M1**]) which states that multiplication by $\delta > 1$ in $\mathbb{R}/\mathbb{Z}$ does not act injectively on a closed invariant infinite set. □

4.4. Gaps. In this subsection, after introducing "gaps" and discussing a series of related results, we prove Proposition 4.14 and Lemma 4.17.

Given a Real lamination $\lambda_{\mathbb{R}}$ with extension $\lambda_{\mathbb{C}}$ to the complex plane, we let $q : \mathbb{C} \to \mathbb{C} \cong \mathbb{C}/\lambda_{\mathbb{C}}$ be the quotient map. The dynamics of $z \mapsto z^d$ in $\mathbb{C} \setminus \mathbb{D}$ projects to $P : q(\mathbb{C} \setminus \mathbb{D}) \to q(\mathbb{C} \setminus \mathbb{D})$. Our task is to extend P over bounded components of $\mathbb{C} \setminus q(\partial\mathbb{D})$, if any.

A bounded component U of $\mathbb{C} \setminus q(\partial\mathbb{D})$ is the image, under q, of a connected component V of the set formed by the points in $\mathbb{D}$ with trivial $\lambda_{\mathbb{C}}$-class.

Definition 4.19. *Consider a $\lambda_{\mathbb{R}}$-class $A \subset \mathbb{R}/\mathbb{Z} \cong \partial\mathbb{D}$ and denote the Euclidean convex hull of A by* Convex(A) $\subset \overline{\mathbb{D}}$. *The* **laminated subset** *of $\overline{\mathbb{D}}$ is the union of* Convex(A) *as A ranges over all $\lambda_{\mathbb{R}}$-classes.*

Since $\lambda_{\mathbb{R}}$ is closed, the laminated subset $\mathcal{L}$ is compact. Therefore, the complement $\mathbb{D} \setminus \mathcal{L}$ (possibly empty) is a union of disjoint convex open sets. Note that $\mathbb{D} \setminus \mathcal{L}$ coincides with the set of points in $\mathbb{D}$ with trivial $\lambda_{\mathbb{C}}$-class.

Definition 4.20. *The closure of a connected component of $\mathbb{D} \setminus \mathcal{L}$ is called a* **gap**.

It follows that every point in the boundary ∂G of a gap G is $\lambda_{\mathbb{C}}$-equivalent to a point in $\partial\mathbb{D}$. Thus $q(\mathring{G})$ is a connected component U of $\mathbb{C} \setminus q(\partial\mathbb{D})$. Conversely, if U is a bounded component of $\mathbb{C} \setminus q(\partial\mathbb{D})$, then $q^{-1}(U)$ is an open set contained in $\mathbb{D}$ formed by points which are not $\lambda_{\mathbb{C}}$-equivalent to points in $\partial\mathbb{D}$, hence points with trivial $\lambda_{\mathbb{C}}$-class. Therefore, $q^{-1}(U)$ is a connected set and coincides with the interior of a gap. We conclude that there is a one-to-one correspondence between gaps and bounded components of $\mathbb{C} \setminus q(\partial\mathbb{D})$. That is, we have proven the following lemma.

Lemma 4.21. *U is a bounded component of $\mathbb{C} \setminus q(\partial\mathbb{D})$ if and only if $U = q(\mathring{G})$ for some gap G.*

Remark. Our definition of gaps is tailored so that the above correspondence between gaps and bounded components of $\mathbb{C} \setminus q(\partial\mathbb{D})$ holds. The definitions of gaps in [**Th, BL**] differ from our definition. Not to every gap, in the sense of these authors, does there necessarily correspond a bounded component of $\mathbb{C} \setminus q(\partial\mathbb{D})$.

If U is a bounded component of $\mathbb{C} \setminus q(\partial\mathbb{D})$, then $\partial U = q(\partial G)$ for some gap G. Every point in ∂G is $\lambda_{\mathbb{C}}$-equivalent to some point in $\partial\mathbb{D}$. Thus our map P is already determined on $\partial U = q(\partial G) \subset q(\partial\mathbb{D})$, by Proposition 4.14 (i). To extend P over U, we must understand the boundary ∂U and the dynamics of P on it.

Lemma 4.22. *If G is a gap of a non-trivial Real lamination $\lambda_{\mathbb{R}}$, then $G \cap \partial\mathbb{D}$ is a Cantor set. Moreover:*

(a) The gap G is the Euclidean convex hull of $G \cap \partial\mathbb{D}$.

(b) The convex hull of $d \cdot (G \cap \partial\mathbb{D})$ is, again, a gap which we denote by $d \cdot G$.

(c) If $(s,t) \subset \mathbb{R}/\mathbb{Z} \cong \partial\mathbb{D}$ is a component of $\partial\mathbb{D} \setminus G$, then (ds, dt) is a component of $\partial\mathbb{D} \setminus d \cdot G$ or $ds = dt$.

(d) Each component $\partial\mathbb{D} \setminus G$ is an interval with $\lambda_{\mathbb{R}}$-equivalent endpoints $s, t \in \mathbb{R}/\mathbb{Z} \cong \partial\mathbb{D}$. The line segment through $\mathbb{D}$ which joins s and t is contained in the boundary of G.

(e) Every boundary point of G which lies in $\mathbb{D}$ is contained in a line segment that joins $\lambda_{\mathbb{R}}$-equivalent points.

(f) The boundary ∂G is a Jordan curve.

It follows that gaps are "infinite sided polygons". Thus we say that a **side** of a gap G is a closed line segment $L \subset \overline{\mathbb{D}}$ connecting $s, t \in \partial\mathbb{D}$ such that $L \subset \partial G$.

PROOF OF LEMMA 4.22. We first show that $G \cap \partial\mathbb{D}$ is a Cantor set. Invariance under multiplication by d and the non-triviality of $\lambda_{\mathbb{R}}$ imply that arguments with non-trivial $\lambda_{\mathbb{R}}$-classes are dense in $\mathbb{R}/\mathbb{Z}$. Thus $G \cap \partial\mathbb{D}$ is totally disconnected. By the definition of gaps, if (s,t) is a component of $\partial\mathbb{D} \setminus G$, then there exist $s_n \searrow s$ and $t_n \nearrow t$ such that s_n is $\lambda_{\mathbb{R}}$-equivalent to t_n. Since $\lambda_{\mathbb{R}}$ is closed, s and t are $\lambda_{\mathbb{R}}$-equivalent. The transitivity of $\lambda_{\mathbb{R}}$ implies that every point in $G \cap \partial\mathbb{D}$ is not isolated. Hence $G \cap \partial\mathbb{D}$ is a Cantor set. Note that we have also proven the first statement of (d).

Concerning (a), consider $z \in G$ and a $\lambda_{\mathbb{R}}$-class A. Let C_A denote the closure of the connected component of $\mathbb{D} \setminus \text{Convex}(A)$ which contains z. Then

$$G = \bigcap_{A\ \lambda_{\mathbb{R}}\text{-class}} C_A.$$

Therefore G may be rewritten as a countable nested intersection of closed sets C which have the property that C is the convex hull of $\partial\mathbb{D} \cap C$. It follows that G is the convex hull of $\partial\mathbb{D} \cap G$.

For statement (b), consider the dense subset S of $\partial\mathbb{D} \cap G$ consisting of two sided non-isolated points. Choose a reference point $t_0 \in S$. Note that if $s \in S$, then $\{t_0, s\}$ and A are unlinked for all non-trivial $\lambda_{\mathbb{R}}$-classes A (their convex hulls are disjoint). Thus $S \setminus \mathbb{Q}/\mathbb{Z}$ is a $\lambda_{\mathbb{R}}$-unlinked class. By Lemma 4.6, we conclude that $d \cdot (S \setminus \mathbb{Q}/\mathbb{Z})$ is a $\lambda_{\mathbb{R}}$-unlinked class. Therefore the convex hull of $d \cdot S \subset \overline{\mathbb{D}}$ and $\text{Convex}(A)$ are disjoint, for all non-trivial $\lambda_{\mathbb{R}}$-classes A. It follows that the convex hull of $d \cdot S$ is formed by points with trivial $\lambda_{\mathbb{C}}$-class. Then its closure, the convex hull of $d \cdot (G \cap \partial\mathbb{D})$, is contained in a gap, say $d \cdot G$. To complete the proof of (b) we must show that $d \cdot G$ is contained in the convex hull of $d \cdot (\partial\mathbb{D} \cap G)$. Consider a component (s,t) of $\partial\mathbb{D} \cap G$. The consecutive preserving property of $\lambda_{\mathbb{R}}$ implies that (ds, dt) is a component of $\partial\mathbb{D} \setminus \text{Convex}(d \cdot A)$, where A is the $\lambda_{\mathbb{R}}$-class that contains s and t. Thus (ds, dt) is a component of $\mathbb{R}/\mathbb{Z} \setminus d \cdot (\partial\mathbb{D} \cap G)$, provided that $ds \neq dt$. It follows that $d \cdot G$ can not contain points in (ds, dt). Hence $d \cdot G$ is contained in the convex hull of $d \cdot (\partial\mathbb{D} \cap G)$ and statement (b) follows. Note that we have also proven (c).

The remaining assertions of the lemma follow easily. □

The consequences that stem from this lemma which are relevant to the dynamics of P are collected in our next result.

COROLLARY 4.23. *Let G be a gap of a $\mathbb{R}$eal lamination $\lambda_{\mathbb{R}}$ with extension $\lambda_{\mathbb{C}}$ and quotient map $q : \mathbb{C} \to \mathbb{C} \cong \mathbb{C}/\lambda_{\mathbb{C}}$. Denote by $P : q(\mathbb{C} \setminus \mathbb{D}) \to q(\mathbb{C} \setminus \mathbb{D})$ the projection of $z \mapsto z^d$. Then:*

(a) $q(\partial G)$ is a topological circle which maps under P onto $q(\partial(d \cdot G))$.

(b) $P : q(\partial G) \to q(\partial(d \cdot G))$ is an orientation preserving covering map.

(c) If $d^p \cdot G = G$, then $P^{\circ p} : q(\partial G) \to q(\partial G)$ is topologically conjugate to an irrational rotation of $\mathbb{R}/\mathbb{Z}$ or to multiplication by some integer $\delta > 1$ in $\mathbb{R}/\mathbb{Z}$.

PROOF OF COROLLARY 4.23. For statement (a), according to Lemma 4.22, the boundary ∂G is a Jordan curve and the portion of a $\lambda_{\mathbb{C}}$-class which lies in ∂G is either a closed arc or a point. Hence if we remove two distinct points from the quotient topological space $q(\partial G)$, then we are left with two connected sets. From Section 2-5 of Hocking and Young [**HY**], we conclude that $q(\partial G)$ is a topological circle.

Concerning (b), that P is a local homeomorphism on $q(\partial G)$ follows from part (c) of Lemma 4.22 and the fact that multiplication by d is a local homeomorphism.

Concerning (c), assume that $P^{\circ p}(q(\partial G)) = q(\partial G)$. Let E be the collection of endpoints s, t where (s, t) is a component of $\partial\mathbb{D} \setminus G$. We claim that there is a finite subset F of E such that, under multiplication by d^p, every element of E eventually maps to F. In fact, if (s, t) is a component of $\partial\mathbb{D} \setminus G$, then there exists a minimal k such that the length of $(d^{kp}s, d^{kp}t)$ is greater or equal than $1/d$. The interval $(d^{kp}s, d^{kp}t)$ is also a component of $\partial\mathbb{D} \setminus G$. There can only be finitely many components of $\partial\mathbb{D} \setminus G$ with length not less than $1/d$. Every point of E eventually maps to the set F formed by the endpoints of these finitely many components. Since $q(E)$ is dense in $q(\partial G)$, there exists a point $t \in F$ such that its backward orbit $\mathcal{O}^-(q(t)) = \{q(s) \in q(\partial G) : P^{\circ np}(q(s)) = q(t) \text{ for some } n \geq 0\}$ is dense in $q(\partial G)$. The existence of such a backward orbit implies that $P^{\circ p} : q(\partial G) \to q(\partial G)$ has no wandering interval. It follows from one-dimensional real dynamics that $P^{\circ p}$ is topologically conjugate to an irrational rotation of $\mathbb{R}/\mathbb{Z}$ or to multiplication by some $\delta > 1$ in $\mathbb{R}/\mathbb{Z}$. □

A key fact in the understanding of the dynamics of P is the absence of wandering gaps.

LEMMA 4.24. *If G is a gap of the $\mathbb{R}$eal lamination $\lambda_{\mathbb{R}}$, then G is eventually periodic (that is, there exists integers $p \geq 1, l \geq 0$ such that $d^{p+l} \cdot G = d^l \cdot G$).*

PROOF OF LEMMA 4.24. Suppose that $d^n \cdot G$ are distinct for all n. It is suffices to show that $G \cap \partial\mathbb{D}$ is finite to obtain a contradiction with Lemma 4.22 which states that $G \cap \partial\mathbb{D}$ is a Cantor set. Two gaps may only have non-trivial intersection along a common side. So $\{d^n \cdot S\}$ are disjoint and unlinked, where S is the set consisting of the two sided non-isolated points of $G \cap \partial\mathbb{D}$. It follows that S is a $\lambda_{\mathbb{R}}$-unlinked class. By Lemma 4.11, S is finite. □

PROOF OF PROPOSITION 4.14. Our task is to construct a degree d branched covering $P : \mathbb{C} \to \mathbb{C}$. Recall that part (i) of the proposition determines the definition of P in $q(\mathbb{C} \setminus \mathbb{D})$. Thus we have to extend P over bounded components of $\mathbb{C} \setminus q(\partial\mathbb{D})$. Each bounded component corresponds to a gap.

The first step for extending P is to show that there exists a collection of homeomorphisms $h_G : \partial\mathbb{D} \to q(\partial G)$, where G is gap, and such that the following holds:

If $q(\partial G)$ is periodic of period p with return map topologically conjugate to a rotation by angle θ, then for $z \in \partial\mathbb{D}$, we have that $P \circ h_G(z) = h_{dG}(e^{2\pi i\theta/p} z)$. Otherwise, for $z \in \partial\mathbb{D}$, we have that $P \circ h_G(z) = h_{dG}(z^\gamma)$, for some $\gamma \geq 1$.

We begin by constructing the homeomorphism corresponding to the gaps in a periodic cycle $G, d \cdot G, \dots, d^p \cdot G = G$. Let δ_j be the degree of $P : q(\partial d^j \cdot G) \to q(\partial d^{j+1} \cdot G)$.

If the product $\delta = \delta_0 \cdots \delta_{p-1} > 1$, then there exists h_G such that $P^{\circ p} \circ h_G(z) = h_G(z^\delta)$, for $z \in \partial\mathbb{D}$, by Corollary 4.23. For $j = 0, \ldots, p-1$, we define $h_{d^j \cdot G}$ as the unique map such that the following diagram commutes:

$$\begin{array}{ccc} \partial\mathbb{D} & \xrightarrow{z^{(\delta_j \cdots \delta_{p-1})}} & \partial\mathbb{D} \\ {\scriptstyle h_{d^j \cdot G}}\downarrow & & \downarrow{\scriptstyle h_G} \\ d^j \cdot G & \xrightarrow[P^{\circ p-j}]{} & \partial G \end{array}$$

and $h_{d^j \cdot G}(1) = P^{\circ j} \circ h_G(1)$. Note that with this normalization we obtain that $h_{d^p \cdot G} = h_G$. Moreover $P \circ h_{d^j \cdot G}(z) = h_{d^{j+1} \cdot G}(z^{\delta_j})$.

Now if $\delta = 1$, then there exists h_G such that $P^{\circ p} \circ h_G(z) = h_G(e^{2\pi i \theta} z)$, for some irrational θ. In this case, let $h_{d^j \cdot G}$ be the unique map such that the following diagram commutes

$$\begin{array}{ccc} \partial\mathbb{D} & \xrightarrow{e^{2\pi i \frac{(p-j)\theta}{p}} z} & \partial\mathbb{D} \\ {\scriptstyle h_{d^j \cdot G}}\downarrow & & \downarrow{\scriptstyle h_G} \\ d^j \cdot G & \xrightarrow[P^{\circ p-j}]{} & \partial G. \end{array}$$

Now that we have constructed the homeomorphisms h_G corresponding to periodic cycles of gaps, we may extend this collection to preperiodic gaps by successive liftings.

Having accomplished the construction of the collection h_G, we can now extend P as follows:
Extend $h_G : \partial\mathbb{D} \to q(\partial G)$ to a collection of homeomorphisms $H_U : \overline{\mathbb{D}} \to \overline{U}$, where U is the bounded component of $\mathbb{C} \setminus q(\partial\mathbb{D})$ such that $\partial U = q(\partial G)$. Next consider a bounded component U of $\mathbb{C} \setminus q(\partial\mathbb{D})$. Let $G = q^{-1}(\overline{U})$, and let $\overline{V} = q(d \cdot G)$. It follows that, for $z \in \partial\mathbb{D}$, the map $H_V^{-1} \circ P \circ H_U(z)$ is either z^γ, for some $\gamma \geq 1$, or αz, for some $\alpha \in \partial\mathbb{D}$. In the former case, let $Q_U(z) = z^\gamma$, for $z \in \overline{\mathbb{D}}$. In the latter, let $Q_U(z) = \alpha z$, for $z \in \overline{\mathbb{D}}$. For $z \in U$, define

$$P(z) = H_V \circ Q_U \circ H_U^{-1}(z).$$

Thus we have extended the action of P to an arbitrary bounded component U of $\mathbb{C} \setminus q(\partial\mathbb{D})$.

To see that $P : \mathbb{C} \to \mathbb{C}$ is a degree d branched covering it is enough to show that P is a local homeomorphism at all $z \in \mathbb{C}$ except for a finite set. We consider first the points $z \in \mathbb{C} \setminus q(\partial\mathbb{D})$ and then consider the points $z \in q(\partial\mathbb{D})$. For simplicity, we will use the notation of Definition 4.16. That is, we say that $q(\partial\mathbb{D})$ is the topological Julia set J^{top}, and its complement $\mathbb{C} \setminus q(\partial\mathbb{D})$ is the topological Fatou set F^{top}.

To start, we show that P is a local homeomorphism at all z in the topological Fatou set $F^{top} = \mathbb{C} \setminus q(\partial\mathbb{D})$ with the exception of a finite set. By Lemma 4.10, there are at most finitely many gaps G such that multiplication by d is not one-to-one on $\partial\mathbb{D} \cap G$. Therefore, there are at most finitely many bounded components, say $U_1, \ldots, U_l$, of the topological Fatou set, which are not mapped, under P, in a one-to-one fashion onto their corresponding image. By construction, P is a local homeomorphism at all $z \in F^{top}$ distinct from $H_{U_1}(0), \ldots, H_{U_l}(0)$.

Now we deal with points in the topological Julia set $J^{top} = q(\partial\mathbb{D})$. It follows from Lemma 4.10 that all but finitely many $\lambda_{\mathbb{R}}$-classes map in a one-to-one fashion onto their image, under multiplication by d. These finitely many exceptional classes correspond to the critical points of P in J^{top}. So it is enough to consider a $\lambda_{\mathbb{R}}$-class A such that $A \mapsto d\cdot A$ is one-to-one and show that P is a local homeomorphism around $z = q(\text{Convex}(A))$. For this we consider compact neighborhoods N of Convex(A) such that $N = q^{-1}(q(N))$. It suffices to establish that, for an appropriate choice of N, the continuous map $P : q(N) \to \mathbb{C}$ from a compact to a Hausdorff space is injective, thus a homeomorphism from $q(N)$ onto its image. In fact, we may choose N so that:

(a) $z \mapsto z^d$ is injective on $N \cap (\mathbb{C} \setminus \mathbb{D})$,

(b) if G is a gap such that $G \cap N \neq \emptyset$, then either G is contained in N or its boundary ∂G shares a side with Convex(A),

(c) if U is a bounded component of F^{top}, then $P : q(N) \cap U \to \mathbb{C}$ is injective.

In the quotient, (a) implies that P is injective in the portion of $q(N)$ which is in the closure of the basin of infinity and (b) says that every bounded Fatou component U which has non-trivial intersection with $q(N)$ is either contained in $q(N)$ or $z \in \partial U$.

We claim that for such N, the map $P : q(N) \to \mathbb{C}$ is injective. Otherwise, there exist $z_1, z_2 \in q(N)$ such that $P(z_1) = P(z_2)$. In view of (a) we have that z_1 and z_2 lie in bounded topological Fatou components, say U_1 and U_2. Note that $P(U_1) = P(U_2)$. By (c), these components are distinct. If one component, say U_1, is completely contained in $q(N)$, then every point in $P(\partial U_2 \cap q(N))$ is the image of at least two points in $q(N)$, which is a contradiction with (a). Hence, from (b) it follows that, for $i = 1, 2$, the gap corresponding to U_i, denote it G_i, has a common side with Convex(A). Let s_i and t_i denote the endpoints of this side, labeled so that $G_i \cap \partial\mathbb{D}$ is contained in the interval $[s_i, t_i]$. Since $d \cdot G_1 = d \cdot G_2$, from Lemma 4.22 (c) it follows that $ds_1 = ds_2$. But $A \mapsto d \cdot A$ is injective. This contradiction implies that $P : q(N) \to \mathbb{C}$ is injective.

Thus P is a branched covering and the proof of the proposition is complete. □.

Proof of Lemma 4.17. Consider a Real lamination which coincides with the extension of $\lambda_{\mathbb{Q}} = \lambda_{\mathbb{R}} \cap (\mathbb{Q}/\mathbb{Z} \times \mathbb{Q}/\mathbb{Z})$. Let $P : \mathbb{C} \to \mathbb{C}$ be a topological realization of $\lambda_{\mathbb{Q}}$. Let U be a bounded topological Fatou component with a critical point ω in its boundary. We must show that ω has finite forward orbit. We proceed by contradiction and suppose that ω has infinite forward orbit. Then $q^{-1}(\omega)$ is an irrational $\lambda_{\mathbb{R}}$-class with convex hull which shares a side with the closure of $q^{-1}(U)$. Hence there exists a gap G with a side that has irrational endpoints.

Therefore it suffices to rule out the existence of a gap G of $\lambda_{\mathbb{R}}$ with a side L which has irrational endpoints a, b. Suppose that such a gap G and side L exist.

We first show that $da \neq db$. In fact, there exist sequences of $\lambda_{\mathbb{Q}}$-equivalent points a_n, b_n converging to a, b, respectively. Without loss of generality, $a_n, b_n \in (a, b)$. If $da = db$, then $d\cdot G$ must be the singleton $\{da = db\}$, since the $\lambda_{\mathbb{Q}}$-equivalent points da_n and db_n separate $da = db$ from the rest of $\mathbb{R}/\mathbb{Z}$. But $d \cdot G$ is a Cantor set. Therefore $da \neq db$.

Now that we know that $d^n a \neq d^n b$, we may iterate and assume that G is periodic. Moreover by passing to an appropriate iterate, we may assume that G is fixed. The non-trivial intervals $(d^n a, d^n b)$ are also components of $\mathbb{R}/\mathbb{Z} \setminus \partial G$. Recall that a and b are irrational, so the collection formed by these intervals is infinite. The length of $(d^n a, d^n b)$ must tend to zero, which is a contradiction.

For the converse, let $\tilde{\lambda}_{\mathbb{R}}$ be the extension of $\lambda_{\mathbb{R}} \cap \mathbb{Q}/\mathbb{Z} \times \mathbb{Q}/\mathbb{Z}$ to $\mathbb{R}/\mathbb{Z}$. Suppose that $\tilde{\lambda}_{\mathbb{R}}$ is strictly contained in $\lambda_{\mathbb{R}}$. Then the union of the convex hulls of the non-trivial $\tilde{\lambda}_{\mathbb{R}}$-classes strictly contains the corresponding union for $\lambda_{\mathbb{R}}$. Thus there exists a $\tilde{\lambda}_{\mathbb{R}}$-gap G strictly contained in a $\lambda_{\mathbb{R}}$-gap. Since both Real laminations agree in $\mathbb{Q}/\mathbb{Z} \times \mathbb{Q}/\mathbb{Z}$, the gap G must have a side L with irrational endpoints a, b. We claim that $d^n a = d^n b$, for some n. Otherwise, we may suppose that G is fixed and, as above, obtain a collection $\{(d^n a, d^n b)\}$ of non-trivial disjoint intervals. Hence, let n be the minimal integer (≥ 0) such that $d^{n+1} a = d^{n+1} b$. It follows that $\omega = q(d^n a) = q(d^n b)$ is a critical point in the boundary of $q(d^n \cdot G)$ with infinite forward orbit. □

We have simultaneously proven the next result.

COROLLARY 4.25. *Let $\lambda_{\mathbb{R}}$ be the extension of a formal rational lamination $\lambda_{\mathbb{Q}}$. Then every side of a gap G of $\lambda_{\mathbb{R}}$ has rational endpoints.*

5. Separation Lemmas

Thus far we have produced a topological realization $P : \mathbb{C} \to \mathbb{C}$ of a given formal rational lamination $\lambda_{\mathbb{Q}}$. That is, we have constructed a topological branched covering P, which topologically realizes $\lambda_{\mathbb{Q}}$. In order to find a true polynomial with rational lamination $\lambda_{\mathbb{Q}}$, we will "perturb" P, in a certain sense, to obtain a sequence of post-critically finite polynomials f_n converging to some polynomial f. The rational laminations $\lambda_{\mathbb{Q}}(f_n)$ will become more and more similar to $\lambda_{\mathbb{Q}}$ as n increases. To show that f has rational lamination $\lambda_{\mathbb{Q}}(f) = \lambda_{\mathbb{Q}}$, we will need to rule out parabolic periodic orbits of f. Loosely speaking, we will rule out "crashing" of periodic points of f_n, as n tends to infinity. The obstruction for this to happen will follow from the three separation lemmas of below.

Separation lemmas already play a crucial role in the literature. For example, see [**GM, K1, K2, Sch4**].

The first two lemmas establish that a pair of preperiodic or periodic points z_1 and z_2 in the topological Julia set of a topological realization P may be separated by a graph formed by (extended) preperiodic rays. The lemmas depend on the location of z_1 and z_2 with respect to the closure of bounded topological Fatou components. Namely, the case in which one point is not in the closure of any bounded topological Fatou component and the case in which both points are in the closure of a bounded topological Fatou component. (see Figure 6).

LEMMA 5.1. *Let $P : \mathbb{C} \to \mathbb{C}$ be a topological realization of a formal rational lamination $\lambda_{\mathbb{Q}}$. Consider two distinct eventually periodic points z_1 and z_2 in the topological Julia set of P. Suppose that z_1 is not in the boundary of any bounded topological Fatou component. Then, for all $n \in \mathbb{N}$, there exists an eventually periodic point z such that the following hold:*

(i) $P^{\circ n}(z)$ is NOT periodic.

(ii) z_1 and z_2 lie in distinct components of $\mathbb{C} \setminus \Gamma$ where Γ is the graph formed by the union of z with the topological rays landing at this point z.

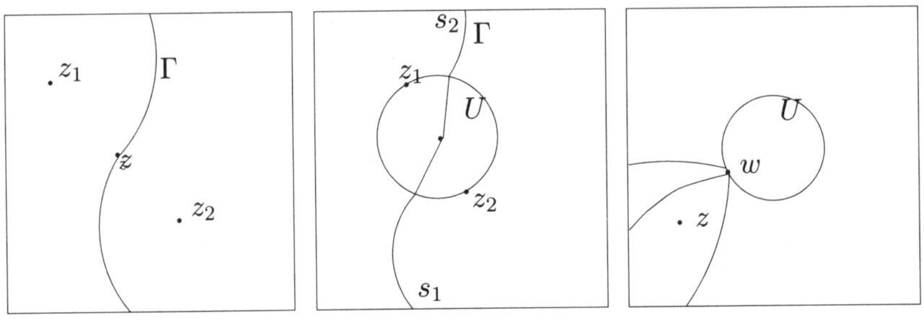

FIGURE 6. Lemma 5.1 (left), Lemma 5.2 (center) and Lemma 5.3 (right).

LEMMA 5.2. *Let $P : \mathbb{C} \to \mathbb{C}$ be a topological realization of a formal rational lamination $\lambda_{\mathbb{Q}}$. Consider two distinct eventually periodic points z_1 and z_2 in the boundary of a bounded topological Fatou component U of P. Then, for all $n \in \mathbb{N}$, there exists $s_1, s_2 \in \mathbb{Q}/\mathbb{Z}$ such that the following hold:*

(i) $d^n s_1, d^n s_2$ are NOT periodic.

(ii) The graph Γ formed by the union of the two extended topological rays with arguments s_1 and s_2 joining ∞ with the center of U separates z_1 from z_2 (i.e., z_1 and z_2 lie in different connected components of $\mathbb{C} \setminus \Gamma$).

PROOF OF LEMMA 5.1. Suppose that z_1 is periodic. By passing to an iterate, we may assume that z_1 is fixed and that the topological rays landing at z_1 are also fixed.

Denote by A_1 (resp. A_2) the arguments of the topological rays landing at z_1 (resp. z_2). It follows that $A_2 \subset (a, b)$, where (a, b) is a connected component of $\mathbb{R}/\mathbb{Z} \setminus A_1$. Since z_1 is not in the boundary of a topological Fatou component, the segment connecting $a \in \partial\mathbb{D}$ and $b \in \partial\mathbb{D}$ is not a side of a gap.

Claim. Given $\epsilon > 0$, there exist $\lambda_{\mathbb{Q}}$-equivalent rational angles $a_0 \in (a, a + \epsilon)$ and $b_0 \in (b - \epsilon, b)$.

Proof of the Claim. Otherwise, let V be the intersection of the convex hulls of $\{a, a + \epsilon, b\} \subset \partial\mathbb{D}$ and $\{b - \epsilon, b, a\} \subset \partial\mathbb{D}$, and note that the interior is disjoint from the convex hulls of $\lambda_{\mathbb{Q}}$-classes. Thus V is contained in a gap and V contains the segment connecting a and b. This contradicts the fact that this segment is not a side of a gap.

Thus there exists $\lambda_{\mathbb{Q}}$-equivalent points $a_0, b_0 \in (a, b)$ such that multiplication by d is a bijection from $(a, a_0) \cup (b_0, b)$ onto its image. It follows that $a_n \in (a, a_0)$ and $b_n \in (b_0, b)$ are $\lambda_{\mathbb{Q}}$-equivalent, where a_n and b_n are inductively defined as $da_n = a_{n-1}$ and $db_n = b_{n-1}$. For N large enough, the common landing point z of $R^{a_N}_{top}$ and $R^{b_N}_{top}$ is the desired point.

By taking successive and appropriate preimages of the sequences a_n and b_n constructed above, the general case (i.e., z_1 preperiodic) follows. □

PROOF OF LEMMA 5.2. For U periodic, in view of Proposition 4.14 and Corollary 4.18, the return map $P^{\circ p} : \mathbb{C} \to \mathbb{C}$ is topologically conjugate to $z \mapsto z^\delta$, for some $\delta > 1$, via $h_U : \overline{\mathbb{D}} \to \overline{U}$. Moreover, z_1 and z_2 are the endpoints of "internal rays" with arguments t_1 and t_2, respectively; that is, $z_j = h_U(e^{2\pi i t_j})$. By the dynamics of $z \to z^\delta$, there are arguments $\tau_1 \in \mathbb{Q}/\mathbb{Z}$ and $\tau_2 \in \mathbb{Q}/\mathbb{Z}$ separating t_1 from t_2 such that $\delta^n \tau_1$ and $\delta^n \tau_2$ are NOT periodic under multiplication by δ.

The "internal rays" $h_U([0,1]e^{2\pi i \tau_1})$ and $h_U([0,1]e^{2\pi i \tau_2})$ have endpoints w_1 and w_2, respectively, which lie in ∂U. At least one topological external ray lands at each point w_1 and w_2, say the ones with arguments s_1 and s_2, respectively. These are the desired arguments.

For U preperiodic, Proposition 4.14 and Corollary 4.18 also describe the dynamics of P in ∂U. It also follows that the points z_1 and z_2 in ∂U are separated by preperiodic points w_1 and w_2. Moreover, we may choose w_1 and w_2 such that $f^{\circ n}(w_1)$ and $f^{\circ n}(w_2)$ are not periodic. Hence, to complete the proof of the lemma, just let s_1 and s_2 be arguments of topological rays landing at w_1 and w_2, respectively. □

As a consequence of Corollary 4.25, which rules out sides of gaps with irrational endpoints, we may also separate eventually periodic points from bounded topological Fatou components:

LEMMA 5.3. *Let $P : \mathbb{C} \to \mathbb{C}$ be a topological realization of a formal rational lamination $\lambda_{\mathbb{Q}}$. Let U be a bounded topological Fatou component. Consider an eventually periodic point z which is not in ∂U. Then there exists an eventually periodic point $w \in \partial U$ such that:*

The graph Γ formed by the union of w and all the topological rays landing at w separates z from U.

PROOF OF LEMMA 5.3. By Corollary 4.25, every side of a gap of $\lambda_{\mathbb{R}}$ has rational endpoints. Let A be the arguments of the topological rays landing at z. Let G be the gap that projects onto $\overline{U}$. There exists a side L of G separating $A \subset \partial\mathbb{D}$ from the interior of G. It follows that $q(L) = w$ is the desired point. □

6. Critically Marked Topological Realizations

Among other results, Poirier [**P**] established a procedure to "mark" the critical points of a post-critically finite polynomial $f : \mathbb{C} \to \mathbb{C}$, thus generalizing the work of Bielefield, Fisher and Hubbard [**BFH**] which deals with critically prerepelling polynomials. In short, Poirier constructs a pair $\Theta = (\mathcal{F}, \mathcal{J})$ of finite collections of finite sets formed by rational angles. To each critical point c in the Fatou set, Poirier assigns a finite set $\mathcal{F}_c$ consisting of arguments of rational external rays which land in the boundary of the Fatou component that contains c. The collection consisting of the sets $\mathcal{F}_c$ is $\mathcal{F}$; that is,

$$\mathcal{F} = \{\mathcal{F}_c : f'(c) = 0, c \in F(f)\}.$$

Similarly, to each critical point c in the Julia set, there corresponds a finite set $\mathcal{J}_c$ formed by arguments of rational rays which land at c. The collection consisting of the sets $\mathcal{J}_c$ is $\mathcal{J}$; that is,

$$\mathcal{J} = \{\mathcal{J}_c : f'(c) = 0, c \in J(f)\}.$$

The pair $\Theta = (\mathcal{F}, \mathcal{J})$ is called a **critical marking** of f. The construction involves a finite amount of choices. Hence, the same post-critically finite polynomial may have several critical markings. One of the main results in [**P**] is a complete description of the pairs of collections Θ which arise as a critical marking of a post-critically finite polynomial. Moreover, it is shown that two polynomials which admit the same marking are equal. After introducing the needed objects, we will state Poirier's results in a precise manner. We strongly encourage the reader to refer to the original paper for a detailed exposition.

Recall that our objective is to find a polynomial f with rational lamination equal to a given formal rational lamination $\lambda_{\mathbb{Q}}$. In Subsection 6.1, we generalize critical markings to the context of topological realizations. To a topological realization $P : \mathbb{C} \to \mathbb{C}$ of the formal rational lamination $\lambda_{\mathbb{Q}}$ we assign a pair $\Theta = (\mathcal{F}, \mathcal{J})$ of finite collections of finite subsets of $\mathbb{R}/\mathbb{Z}$. One fundamental difference with critical marking of post-critically finite polynomials is that we allow the presence of irrational arguments in $\mathcal{J}$. Since the map P may have critical points with infinite forward orbits in the topological Julia set, the presence of irrational arguments is necessary. (We say that z is a critical point of P if P is not locally one-to-one about z.) In subsections 6.1 and 6.2, we show that Θ has some properties similar to those described by Poirier in his characterization of critical markings of post-critically finite polynomials. This allows us to perturb Θ and obtain a sequence $\{\Theta_n\}$ converging, in a certain sense, to Θ. After applying Poirier's result, we obtain post-critically finite polynomials f_n with critical markings Θ_n. By passing to a subsequence, $\{f_n\}$ converges to some polynomial f. In Section 7, we show that the polynomial f is the desired polynomial with rational lamination $\lambda_{\mathbb{Q}}(f) = \lambda_{\mathbb{Q}}$.

6.1. Critical Markings. Consider a topological realization $P : \mathbb{C} \to \mathbb{C}$ of a formal rational lamination $\lambda_{\mathbb{Q}}$. Our aim is to construct collections

$$\mathcal{F} = \{\mathcal{F}_{c_1}, \dots, \mathcal{F}_{c_k}\},$$

where $c_1, \dots, c_k$ are the critical points of P which lie in the topological Fatou set, and

$$\mathcal{J} = \{\mathcal{J}_{c_{k+1}}, \dots, \mathcal{J}_{c_m}\},$$

where $c_{k+1}, \dots, c_m$ are the critical points of P which lie in the topological Julia set. Recall that P is a degree d branched covering. It follows that P has finitely many critical points.

Although the description of the construction below is self-contained we encourage the reader to refer to [**P**] for examples and a detailed exposition. In fact, our exposition mimics Poirier's.

If the critical point c belongs to the bounded topological Fatou component U, then the cardinality of $\mathcal{F}_c \subset \mathbb{Q}/\mathbb{Z}$ will coincide with the local degree of P at c. The arguments in $\mathcal{F}_c$ are chosen from the arguments of the topological rays landing at ∂U. The appropriate choice requires the definition of supporting arguments.

DEFINITION 6.1. *Consider a bounded topological Fatou component U of a topological realization $P : \mathbb{C} \to \mathbb{C}$. Let $t_1, \dots, t_k$ be all the arguments of the topological rays landing at $z \in \partial U$ indexed respecting cyclic order. The union of the rays $R^{t_1}_{top}, \dots, R^{t_k}_{top}$ and the point $\{z\}$ cuts the complex plane into k regions. The component U is contained in one these regions, say the one bounded by $R^{t_1}_{top}$, $R^{t_2}_{top}$ and $\{z\}$ (if $k = 1$, then $t_1 = t_2$). We say that t_1 is the* **left supporting argument of** *U* **at** *z and that t_2 is the* **right supporting argument of** *U* **at** *z.*

If t is a left or right supporting argument of U at some point z in ∂U, then we sometimes simply say that t is a left or right supporting argument of U.

We now construct $\mathcal{F}_c$ in an orderly manner. We start with a periodic critical point c and construct the corresponding $\mathcal{F}_{c'}$ for all critical points c' in the cycle of c.

Periodic Critical Points. Let c be a periodic critical point of period p and let U denote the corresponding topological Fatou component. Choose a periodic point z in ∂U such that its period divides p. Let t denote the left supporting argument for U at z. Let $\mathcal{F}_c$ be the set consisting of all the arguments of the left supporting rays of U which map, under P, onto R^{dt}_{top}. Note that the cardinality of $\mathcal{F}_c$ coincides with the local degree of P at c. For future reference, we call t the preferred supporting argument associated to c. Now if there exists another critical point c' in the orbit of c, then we construct $\mathcal{F}_{c'}$ as follows. Since $c' = P^{\circ q}(c)$, it follows that the Fatou component $V = P^{\circ q}(U)$ is supported by $s = d^q t$. Again, for future reference, we call s the preferred supporting argument associated to c'. Let $\mathcal{F}_{c'}$ be the set consisting of all the arguments of the left supporting rays of V which map, under P, onto R^{ds}_{top}.

Next we construct $\mathcal{F}_c$ for strictly preperiodic critical points c in F^{top}.

Preperiodic Critical Points in the Topological Fatou Set. Let c be a strictly preperiodic critical point which lies in the Fatou component U. Consider the minimal number n such that $c' = P^{\circ n}(c)$ is a critical point. We may assume that $\mathcal{F}_{c'}$ is already chosen and that the preferred supporting argument of c' is t. Let $\mathcal{F}_c$ be the set consisting of all the arguments of the topological rays in $P^{-n}(R^t_{top})$ that are left supporting arguments of U. Choose any element of $\mathcal{F}_c$ and call it the preferred supporting argument associated to c.

Finally, we deal with the critical points in J^{top}.

Critical Points in the Topological Julia Set. Let c be a Julia critical point. There are two cases. If the forward orbit of c does not contain other critical points, then choose an external ray R^t landing at c. Let $\mathcal{J}_c$ be the arguments of the external rays in $P^{-1}(R^{dt}_{top})$ which land at c; call t the preferred argument associated to c. If n is the minimal number such that $P^{\circ n}(c)$ is a critical point c', then we may assume that c' has already associated $\mathcal{J}_{c'}$ with preferred argument t. Let $\mathcal{J}_c$ be the set consisting of all the arguments of the external rays in $P^{-n}(R^t_{top})$ that land at c. We arbitrarily choose an element of $\mathcal{J}_c$ and call it the preferred argument associated to c.

DEFINITION 6.2. *A pair $\Theta = (\mathcal{F}, \mathcal{J})$ obtained by the above procedure is called a* **critical marking of** P.

The next result is an immediate consequence of the construction, which we state without proof.

Notation. Throughout, we let $|A|$ denote the cardinality of a set A.

PROPOSITION 6.3. *Consider a topological realization $P : \mathbb{C} \to \mathbb{C}$ of a formal rational lamination $\lambda_{\mathbb{Q}}$. Let $\Theta = (\mathcal{F}, \mathcal{J})$ be a critical marking of P where:*

$$\mathcal{F} = \{\mathcal{F}_1, \ldots, \mathcal{F}_l\},$$

$$\mathcal{J} = \{\mathcal{J}_1, \ldots, \mathcal{J}_m\}.$$

Then:

(C0) $|\mathcal{F}_j| \geq 2$ and $|\mathcal{J}_k| \geq 2$. Also, $|d \cdot \mathcal{F}_j| = 1$ and $|d \cdot \mathcal{J}_k| = 1$.
(C1) $d - 1 = \sum(|\mathcal{F}_j| - 1) + \sum(|\mathcal{J}_k| - 1)$.
(C2) $\mathcal{F}$ and $\mathcal{J}^-$ are unlinked. That is, for $\epsilon > 0$ small enough, we have that

$$\mathcal{F}_1, \ldots, \mathcal{F}_l, \mathcal{J}_1 - \epsilon, \ldots, \mathcal{J}_m - \epsilon$$

are disjoint and pairwise unlinked.

(C3) $\mathcal{F}$ (resp. $\mathcal{J}$) is hierarchic. That is, for any two elements $\theta, \theta' \in \mathbb{R}/\mathbb{Z}$ that participate in $\mathcal{F}$ (resp. $\mathcal{J}$) such that $d^i\theta$ and $d^j\theta'$ lie in $\mathcal{F}_k$ (resp. $\mathcal{J}_k$), for some $i, j > 0$, we have that $d^i\theta = d^j\theta'$.

(C4) Given an element θ that participates in $\mathcal{F}$, there exists a periodic argument $d^i\theta$ which also participates in $\mathcal{F}$.

(C5) None of the arguments that participate in $\mathcal{J}$ are periodic.

6.2. Critical Marking and Itineraries. Recall that in the previous section we described how to obtain a critical marking Θ from a topological realization of a formal rational lamination $\lambda_{\mathbb{Q}}$. Here we are interested in reversing this procedure; that is, we will describe how to recover the formal rational lamination $\lambda_{\mathbb{Q}}$ from the critical marking Θ.

In this subsection we henceforth let $\lambda_{\mathbb{Q}}$ be a formal rational lamination, and let $P : \mathbb{C} \to \mathbb{C}$ be a topological realization of $\lambda_{\mathbb{Q}}$. Moreover, let $\Theta = (\mathcal{J}, \mathcal{F})$ be a critical marking of P, where

$$\mathcal{F} = \{\mathcal{F}_c : c \text{ is a critical point and } c \in F^{top}\},$$

$$\mathcal{J} = \{\mathcal{J}_c : c \text{ is a critical point and } c \in J^{top}\}.$$

To each critical point c there corresponds, a connected graph Γ_c as follows:

- If $c \in J^{top}$, then we let Γ_c be the graph formed by the union of the critical point c and all the topological rays R^t_{top} such that $t \in \mathcal{J}_c$.
- If $c \in F^{top}$, then we let Γ_c be the union of the critical point c and the extended rays with arguments in $\mathcal{F}_c$ connecting ∞ with c.

Following [**P**], the critical marking $\Theta = (\mathcal{F}, \mathcal{J})$ cuts the complex plane into Θ-regions.

DEFINITION 6.4. *We say that two points z, w belong to the same* **Θ-region** *if and only if z, w lie in the same connected component of $\mathbb{C} \setminus \Gamma_c$, for all critical points c.*

We next consider the corresponding situation in the circle at infinity.

DEFINITION 6.5. *We say that s and t are in the same* **Θ-unlinked class** *L if and only if s and t lie in the same component of $\mathbb{R}/\mathbb{Z} \setminus \mathcal{F}_c$, for all critical points $c \in F^{top}$, as well as in the same component of $\mathbb{R}/\mathbb{Z} \setminus \mathcal{J}_c$, for all critical points $c \in J^{top}$.*

Each Θ-unlinked class L is a finite union of intervals in $\mathbb{R}/\mathbb{Z}$ with total length $1/d$. Hence the critical marking $\Theta = (\mathcal{F}, \mathcal{J})$ chops the circle $\mathbb{R}/\mathbb{Z}$ into exactly d unlinked classes $L_1, \dots, L_d$ and the complex plane $\mathbb{C}$ into exactly d Θ-regions $U_1, \dots, U_d$. The correspondence is: $t \in L_i$ if and only if $R^t_{top} \subset U_i$.

Each L_i may be written as a union of open intervals as follows:

$$L_i = (a_1, b_1) \cup \dots \cup (a_l, b_l).$$

Next we introduce the following sets:

$$L_i^- = (a_1, b_1] \cup \dots \cup (a_l, b_l],$$

$$L_i^+ = [a_1, b_1) \cup \dots \cup [a_l, b_l).$$

Multiplication by d maps each L_i bijectively onto the complement of a finite subset of $\mathbb{R}/\mathbb{Z}$. Also, multiplication by d maps each $L_i^{\pm}$ bijectively onto $\mathbb{R}/\mathbb{Z}$.

Moreover, these maps preserve cyclic order (see Lemma II.1.1 of [**P**]). Furthermore, $P:\mathbb{C}\to\mathbb{C}$ maps each region U_i onto the complement of a finite union of rays and extended rays in one-to-one manner.

DEFINITION 6.6. *For $t\in\mathbb{R}/\mathbb{Z}$, we let $A^-_\Theta(t)=L^-_i$ if $t\in L^-_i$. We call $A^-_\Theta(t)$ the* **left address** *of t. Similarly, we let $A^+_\Theta(t)=L^+_i$ if $t\in L^+_i$. We call $A^+_\Theta(t)$ the* **right address** *of t. Now we let*

$$\begin{array}{rccl} S^\pm_\Theta: & \mathbb{R}/\mathbb{Z} & \to & \{L_1,\dots,L_d\}^{\mathbb{N}\cup\{0\}} \\ & t & \mapsto & (L_{j_0},L_{j_1},\dots) \end{array}$$

where $L^\pm_{j_n}=A^\pm(d^nt)$, for all $n\ge 0$. We say that $S^+_\Theta(t)$ is the **right symbol sequence** *of t and that $S^-_\Theta(t)$ is the* **left symbol sequence** *of t.*

We wish to reconstruct $\lambda_\mathbb{Q}$ from Θ. Consider a $\lambda_\mathbb{Q}$-class A. Let $z\in J^{top}$ be the landing point of the topological rays with arguments in A. For the sake of simplicity, suppose the forward orbit of the landing point z is disjoint from all the graphs Γ_c. Then all the topological rays landing at $f^{\circ n}(z)$ are contained in the same Θ-region, say U_{j_n}. Thus the arguments $d^n\cdot A$ of these rays are contained in L_{j_n}. We conclude that, under a mild assumption, the symbol sequences of $\lambda_\mathbb{Q}$-equivalent points coincide. The general case is more subtle. The exact relation between itineraries and $\lambda_\mathbb{Q}$-equivalence is given in the definition and proposition below.

DEFINITION 6.7. *We say that rational angles t and t' are Θ pre-equivalent, denoted $t\sim t'$, if and only if one of the following holds:*

(i) $S^-_\Theta(t)=S^-_\Theta(t')$.

(ii) The first $n\ge 0$ symbols of $S^-_\Theta(t)$ and of $S^-_\Theta(t')$ coincide, and the arguments d^nt, d^nt' lie in $\mathcal{J}_c$, for some critical point c.

The equivalence relation $\Lambda_\mathbb{Q}(\Theta)$ in $\mathbb{Q}/\mathbb{Z}$ **generated by** *Θ is the smallest transitive relation which contains $\sim$.*

The next proposition shows that $\Lambda_\mathbb{Q}(\Theta)$ coincides with $\lambda_\mathbb{Q}$.

PROPOSITION 6.8. *If Θ is the critical marking of a topological realization $P:\mathbb{C}\to\mathbb{C}$ of a formal rational lamination $\lambda_\mathbb{Q}$, then*

$$\Lambda_\mathbb{Q}(\Theta)=\lambda_\mathbb{Q}.$$

The proof of this proposition relies on a series of lemmas. The next two lemmas are translations of Lemma II.3.3 and Corollary II.3.5 of [**P**] to our context. We omit the proofs and refer the reader to [**P**] for them.

The left addresses of topological rays landing at a common point z either coincide or z is a Julia critical point.

LEMMA 6.9. *Consider two arguments t,s in $\mathbb{R}/\mathbb{Z}$. If R^t_{top} and R^s_{top} land at the same point z, then $A^-_\Theta(t)=A^-_\Theta(s)$ or z is a critical point.*

If two topological rays land at a common point, then their image must also land at a common point. The next lemma gives a sufficient condition for the converse.

LEMMA 6.10. *Consider two arguments t,s in $\mathbb{R}/\mathbb{Z}$. If R^{dt}_{top} and R^{ds}_{top} land at a common point and $A^-_\Theta(t)=A^-_\Theta(s)$, then R^t_{top} and R^s_{top} land at a common point.*

The previous two lemmas will allow us to show that $\lambda_{\mathbb{Q}} \subset \Lambda_{\mathbb{Q}}(\Theta)$. For the converse, we need to know that $\Lambda_{\mathbb{Q}}(\Theta)$ has some of the properties that formal rational laminations have.

LEMMA 6.11. *If A is a $\Lambda_{\mathbb{Q}}(\Theta)$-class, then A is a finite subset of $\mathbb{Q}/\mathbb{Z}$. Moreover, if A and A' are distinct $\Lambda_{\mathbb{Q}}(\Theta)$-classes, then A and A' are unlinked.*

PROOF OF LEMMA 6.11. For the first assertion, we begin by showing that if A is the $\Lambda_{\mathbb{Q}}(\Theta)$-class A of a period p argument t, then all the elements of A are periodic of period p. Consider the subset S of $\mathbb{R}/\mathbb{Z}$ that consists of all $s \in \mathbb{R}/\mathbb{Z}$ such that $A^-_\Theta(d^j s) = A^-_\Theta(d^j t)$, for $j = 0, \dots, p-1$. Note that if $s \in A$, then $d^{pk}s \in S$, for all integers $k \geq 0$. The map $S \to d^p S$ is cyclic order preserving, since multiplication by d is cyclic order preserving on Θ-unlinked classes. Thus the map $S \to d^p S$ extends to a monotone increasing circle map. By standard results on iteration of circle maps, under iterations of multiplication by d^p, every element of S converges to a periodic point. But every periodic point of multiplication by d^p is repelling. Hence all the elements of A are periodic of the same period. It follows that A is finite. Finiteness of an arbitrary $\Lambda_{\mathbb{Q}}(\Theta)$-class follows, since $\Lambda_{\mathbb{Q}}(\Theta)$-equivalent arguments map, under multiplication by d, to $\Lambda_{\mathbb{Q}}(\Theta)$-equivalent arguments.

For the second assertion, it suffices to show that if distinct $t, s, \hat{t}, \hat{s} \in \mathbb{Q}/\mathbb{Z}$ are such that $t \sim s$ and $\hat{t} \sim \hat{s}$, then either $\{t, s\}, \{\hat{t}, \hat{s}\}$ are unlinked or $t \sim \hat{t}$ (see Definition 6.7).

Recall that $m_d : L^-_i \mapsto \mathbb{R}/\mathbb{Z}$ is a cyclic order preserving bijection. Therefore, if the subsets $\{t, s\} \subset L^-_i$ and $\{\hat{t}, \hat{s}\} \subset L^-_j$ are linked, then $i = j$ and $\{dt, ds\}, \{d\hat{t}, d\hat{s}\}$ are also linked.

Consider distinct elements $t, s, \hat{t}, \hat{s} \in \mathbb{Q}/\mathbb{Z}$. Suppose that $t \sim s$, $\hat{t} \sim \hat{s}$ and that $\{t, s\}, \{\hat{t}, \hat{s}\}$ are linked. To establish the assertion, we claim that $t \sim \hat{t}$. In fact, according to Definition 6.7 there are three possibilities:

Case 1. If $S^-_\Theta(t) = S^-_\Theta(s)$ and $S^-_\Theta(\hat{t}) = S^-_\Theta(\hat{s})$, then either $S^-_\Theta(t) = S^-_\Theta(\hat{t})$ or there exists a smallest integer $n \geq 0$ such that $A^-_\Theta(d^n t) \neq A^-_\Theta(d^n \hat{t})$. The latter cannot occur because it implies that $\{d^n t, d^n s\}, \{d^n \hat{t}, d^n \hat{s}\}$ lie in distinct Θ-unlinked classes and, in view of our assumption, these sets are linked. Thus $S^-_\Theta(t) = S^-_\Theta(\hat{t})$ and $t \sim \hat{t}$.

Case 2. If $S^-_\Theta(t) = S^-_\Theta(s)$, and if $A^-_\Theta(d^j \hat{t}) = A^-_\Theta(d^j \hat{s})$, for $j = 0, \dots, n-1$, and $\{d^n \hat{t}, d^n \hat{s}\} \subset \mathcal{J}_c$, then $\{d^n \hat{t}, d^n \hat{s}\}$ and $\{d^n t, d^n s\}$ are linked. Since $\{d^n \hat{t}, d^n \hat{s}\} \subset \mathcal{J}_c$, and since $\{d^n t, d^n s\}$ is contained in some Θ-unlinked class L^-_i, we obtain a contradiction.

Case 3. If $A^-_\Theta(d^j \hat{t}) = A^-_\Theta(d^j \hat{s})$, for $j = 0, \dots, n-1$, and $\{d^n \hat{t}, d^n \hat{s}\} \subset \mathcal{J}_c$, and if $A^-_\Theta(d^j \hat{t}) = A^-_\Theta(d^j \hat{s})$ for $j = 0, \dots, m$ and $\{d^m \hat{t}, d^m \hat{s}\} \subset \mathcal{J}_{c'}$, then $m = n$, by the above argument. Moreover, $\{d^n \hat{t}, d^n \hat{s}\}$ and $\{d^m \hat{t}, d^m \hat{s}\}$ are linked. Recall that $\mathcal{J}_c, \mathcal{J}_{c'}$ are equal or unlinked. We conclude that $d^n t$ and $d^n \hat{t}$ are in $\mathcal{J}_c = \mathcal{J}_{c'}$. Therefore $t \sim \hat{t}$. □

The next lemma shows that topological rays R^t_{top} and $R^{t'}_{top}$ which land in the boundary ∂U of the same bounded topological Fatou component U such that their arguments t and t' have the same left or right symbol sequence must land at a common point. That is, t and t' are $\lambda_{\mathbb{Q}}$-equivalent.

LEMMA 6.12. *Let t and t' be elements of $\mathbb{Q}/\mathbb{Z}$. Denote by z and z' the landing points of R^t_{top} and $R^{t'}_{top}$, respectively. If $S^\pm_\Theta(t) = S^\pm_\Theta(t')$, and if both z and z' lie in ∂U, where U is a bounded topological Fatou component, then $z = z'$.*

PROOF OF LEMMA 6.12. Suppose that $S^-_\Theta(t) = S^-_\Theta(t')$ or that $S^+_\Theta(t) = S^+_\Theta(t')$. Also assume that z and z' land in ∂U, where U is a bounded topological Fatou component.

Suppose that U is a period p topological Fatou component with a critical point $c \in U$. The general case follows along the same lines. Let Γ_0 be the graph formed by the union of all the extended topological rays which have arguments in $\mathcal{F}_c$ and join c to ∞. We have that Γ_0 chops the complex plane into $\delta > 1$ components, where δ is the cardinality of $\mathcal{F}_c$. Then the topological rays R^t_{top} and $R^{t'}_{top}$ must lie in the closure of one of these components, say V_0.

Consider the graph Γ_n which consists of the union of all the extended rays which join c to ∞ and map, under $P^{\circ p}$, onto one contained in Γ_{n-1}. The graph Γ_n chops the complex plane into $\delta\gamma^n$ connected regions, for some $\gamma > 1$.

The rays R^t_{top} and $R^{t'}_{top}$ must lie in the closure of a component of $\mathbb{C} \setminus \Gamma_n$, say V_n.

Since the dynamics of $P^{\circ p}$ on $\overline{U}$ is topologically conjugate to $z \mapsto z^\gamma$ it is not difficult to check that $(\bigcap_{n \geq 0} \overline{V}_n) \cap \partial U$ is a singleton and the lemma now follows. □

PROOF OF PROPOSITION 6.8. We begin by showing that $\lambda_\mathbb{Q} \subset \Lambda_\mathbb{Q}(\Theta)$. Suppose that t and s are both periodic and $\lambda_\mathbb{Q}$-equivalent. By Lemma 6.9 and the invariance of $\lambda_\mathbb{Q}$ under multiplication by d, the left addresses of $d^n t$ and $d^n s$ coincide, for all n. It follows that t is $\Lambda_\mathbb{Q}(\Theta)$-equivalent to s.

Since every rational argument is eventually periodic under multiplication by d, we proceed by induction. Suppose that t, s are $\lambda_\mathbb{Q}$-equivalent and that dt, ds are $\Lambda_\mathbb{Q}(\Theta)$-equivalent. We must show that t, s are $\Lambda_\mathbb{Q}(\Theta)$-equivalent. By Lemma 6.9:

(a) $A^-_\Theta(t) = A^-_\Theta(s)$ or

(b) R^t_{top} and R^s_{top} land at a critical point c.

In case (a), the definition of $\Lambda_\mathbb{Q}(\Theta)$ implies that t and s are $\Lambda_\mathbb{Q}(\Theta)$-equivalent. In case (b), respecting cyclic order, index the elements of $\mathcal{J}_c$ by $\theta_1, \dots, \theta_l$. Let C be the $\lambda_\mathbb{Q}$-class of t and s (i.e., the arguments of the external rays landing at c). Note that $\mathcal{J}_c = \{\theta_1, \dots \theta_l\} \subset C$. The arguments in $(\theta_i, \theta_{i+1 (mod\, l)}] \cap C$ have the same left address. By case (a), $(\theta_i, \theta_{i+1 (mod\, l)}] \cap C$ is contained in a $\Lambda_\mathbb{Q}(\Theta)$-class. By Definition 6.7 (ii), all the elements $\theta_1, \dots \theta_l$ are $\Lambda_\mathbb{Q}(\Theta)$-equivalent. By transitivity of $\Lambda_\mathbb{Q}(\Theta)$, the $\lambda_\mathbb{Q}$-class C is contained in a $\Lambda_\mathbb{Q}(\Theta)$-class. In particular, t and s are $\Lambda_\mathbb{Q}(\Theta)$-equivalent.

We now show that $\Lambda_\mathbb{Q}(\Theta) \subset \lambda_\mathbb{Q}$. Assume that t and s are such that Definition 6.7 (ii) holds. The construction of Θ implies that $\mathcal{J}_c$ is formed by $\lambda_\mathbb{Q}$-equivalent arguments. By Lemma 6.10, t and s are $\lambda_\mathbb{Q}$-equivalent. Assume that t and s are distinct arguments such that Definition 6.7 (i) holds. Suppose t and s are not $\lambda_\mathbb{Q}$-equivalent. Denote by z and w the landing points of R^t_{top} and R^s_{top}, respectively. Since topological Fatou components are separated (see Lemma 5.3) and $\Lambda_\mathbb{Q}(\Theta)$-classes are unlinked, without loss of generality, we may assume that we are in one of the following cases:

(I) Both z and w are in the boundary of the same topological Fatou component.

(II) One point, say z, is not in the boundary of a bounded topological Fatou component.

In case (I), Lemma 6.12 implies that t and s are $\lambda_\mathbb{Q}$-equivalent.

In case (II), there exist infinitely many $\lambda_\mathbb{Q}$-classes separating t from s. But we have shown that $\lambda_\mathbb{Q} \subset \Lambda_\mathbb{Q}(\Theta)$ and $\Lambda_\mathbb{Q}(\Theta)$-classes are unlinked. Thus all of these separating classes are contained in the same $\Lambda_{\mathbb{Q}}(\Theta)$-class as that in which t and s are in. This contradicts the fact that $\Lambda_{\mathbb{Q}}(\Theta)$-classes are finite. □

The coincidence of $\lambda_{\mathbb{Q}}$ and $\Lambda_{\mathbb{Q}}(\Theta)$, and Lemma 6.12 imply that the critical marking Θ has the next two extra properties as stated in the following result.

LEMMA 6.13. *(C6) Consider a periodic argument t which participates in $\mathcal{F}$ and an argument $t' \in \mathbb{Q}/\mathbb{Z}$. If $S^+_\Theta(t) = S^+_\Theta(t')$, then $t = t'$.*

(C7) Consider preperiodic arguments $t \in \mathcal{J}_c$ and $t' \in \mathcal{J}_{c'}$. If $S^-_\Theta(d^i t) = S^-_\Theta(t')$, for some i, then $d^i t \in \mathcal{J}_{c'}$.

7. Combinatorial Continuity

Given a formal rational lamination $\lambda_{\mathbb{Q}}$, our goal is to find a polynomial f for which $\lambda_{\mathbb{Q}} = \lambda_{\mathbb{Q}}(f)$. The polynomial f will be the limit of a sequence $\{f_n\}$ of post-critically finite polynomials. We obtain this sequence after applying Poirier's theorem below to an appropriate sequence $\{\Theta_n\}$ (see Theorem 3.9 in [**P**]).

THEOREM 7.1 (Poirier). *Consider a pair $\Theta = (\mathcal{F}, \mathcal{J})$ of finite collections:*

$$\mathcal{F} = \{\mathcal{F}_1, \dots, \mathcal{F}_l\},$$
$$\mathcal{J} = \{\mathcal{J}_1, \dots, \mathcal{J}_m\}$$

which consist of finite subsets of $\mathbb{Q}/\mathbb{Z}$. If conditions (C0) through (C7) listed below hold for Θ, then there exists a unique post-critically finite polynomial f which admits Θ as a critical marking.

(C0) $|\mathcal{F}_j| \geq 2$ and $|\mathcal{J}_k| \geq 2$. Also, $|d \cdot \mathcal{F}_j| = 1$ and $|d \cdot \mathcal{J}_k| = 1$.

(C1) $d - 1 = \sum(|\mathcal{F}_j| - 1) + \sum(|\mathcal{J}_k| - 1)$.

(C2) $\mathcal{F}$ and $\mathcal{J}^-$ are unlinked; that is, for $\epsilon > 0$ small enough,

$$\mathcal{F}_1, \dots, \mathcal{F}_l, \mathcal{J}_1 - \epsilon, \dots, \mathcal{J}_m - \epsilon$$

are disjoint and pairwise unlinked.

(C3) $\mathcal{F}$ (resp. $\mathcal{J}$) is hierarchic. That is, for any two elements $\theta, \theta' \in \mathbb{Q}/\mathbb{Z}$ which participate in $\mathcal{F}$ (resp. $\mathcal{J}$) and such that $d^i\theta$ and $d^j\theta'$ lie in $\mathcal{F}_k$ (resp. $\mathcal{J}_k$), for some $i, j > 0$, we have that $d^i\theta = d^j\theta'$.

(C4) Given an element θ that participates in $\mathcal{F}$, there exists $i \geq 0$ such that $d^i\theta$ is a periodic argument which also participates in $\mathcal{F}$.

(C5) None of the arguments which participate in $\mathcal{J}$ are periodic.

(C6) Consider a periodic argument t which participates in $\mathcal{F}$ and an argument $t' \in \mathbb{Q}/\mathbb{Z}$. If $S^+_\Theta(t) = S^+_\Theta(t')$, then $t = t'$.

(C7) Consider preperiodic arguments $t \in \mathcal{J}_l$ and $t' \in \mathcal{J}_k$. If $S^-_\Theta(d^i \cdot t) = S^-_\Theta(t')$, for some i, then $d^i t \in \mathcal{J}_k$.

We introduced the right and left symbol sequences $S^\pm_\Theta$ assuming that Θ is a critical marking of a topological realization. Nevertheless, the definition easily generalizes to collections Θ for which (C0)–(C3) hold.

We now give a sketch of how to find the desired polynomial f. By Proposition 6.3 and Lemma 6.13, we know that (C0)–(C7) hold for a critical marking Θ of the topological realization $P : \mathbb{C} \to \mathbb{C}$ of the formal rational lamination $\lambda_{\mathbb{Q}}$. We may have some irrational arguments participating in Θ. We "perturb" Θ in order to find a sequence Θ_n converging to Θ, in a certain sense that is precised later. All of the angles participating in Θ_n will be rational. Moreover, each Θ_n will fulfill the requirements of Theorem 7.1 for being a critical marking of a post-critically finite polynomial f_n. By passing to a subsequence we may assume that $\{f_n\}$ converges to a polynomial f, as $n \to \infty$. The rational laminations $\lambda_{\mathbb{Q}}(f_n)$ of f_n become "similar" to $\lambda_{\mathbb{Q}}$, as n increases. The separation lemmas of Section 5 will enable us to show that f has no parabolic cycles. We may then apply a result of Goldberg and Milnor [**GM**] in order to pass to the limit and conclude that $\lambda_{\mathbb{Q}} = \lambda_{\mathbb{Q}}(f)$.

7.1. Convergence. We now introduce the appropriate topological space to "perturb" the critical marking of a topological realization.

Consider a formal rational lamination $\lambda_{\mathbb{Q}}$ with topological realization $P : \mathbb{C} \to \mathbb{C}$. Let $\Theta = (\mathcal{F}, \mathcal{J})$ be a critical marking of P, where

$$\mathcal{F} = \{\mathcal{F}_1, \dots, \mathcal{F}_l\},$$

$$\mathcal{J} = \{\mathcal{J}_1, \dots, \mathcal{J}_m\}.$$

Define $\mathcal{A}_\Theta$ as the set that consists of all $\Theta' = (\mathcal{F}', \mathcal{J}')$, where

$$\mathcal{F}' = \{\mathcal{F}'_1, \dots, \mathcal{F}'_{l'}\} \quad \text{and} \quad \mathcal{J}' = \{\mathcal{J}'_1, \dots, \mathcal{J}'_{m'}\},$$

such that:

(a) conditions (C0) through (C3) of Proposition 6.3 hold;

(b) $\mathcal{F}' = \mathcal{F}$;

(c) if $\mathcal{J}_k$ is a member of $\mathcal{J}$ such that $\mathcal{J}_k \subset \mathbb{Q}/\mathbb{Z}$, then $\mathcal{J}_k$ is also a member of $\mathcal{J}'$.

We endow $\mathcal{A}_\Theta$ with the "compact-unlinked" topology which we proceed to describe. For each closed set $X \subset \mathbb{R}/\mathbb{Z}$, we let $V_X \subset \mathcal{A}_\Theta$ be the set that consists of all $\Theta' = (\mathcal{F}', \mathcal{J}') \in \mathcal{A}_\Theta$ such that:

X is contained in a component of $\mathbb{R}/\mathbb{Z} \setminus \mathcal{J}'_i$ whenever $\mathcal{J}'_i$ is a set in the collection $\mathcal{J}'$ distinct from the sets which participate in $\mathcal{J}$ and that are formed by rational arguments.

The topology in $\mathcal{A}_\Theta$ which has sub-basis formed by all such V_X is called the "compact-unlinked" topology.

Roughly, critical markings are close in the compact-unlinked topology if the corresponding partitions of $\mathbb{R}/\mathbb{Z}$ into unlinked classes are close.

LEMMA 7.2. *If Θ is a critical marking of P, then there exists a sequence $\{\Theta_n\}$ in $\mathcal{A}_\Theta$ that converges to Θ such that (C0) through (C7) of Poirier's theorem 7.1 hold for all Θ_n, and all the arguments participating in Θ_n are rational.*

PROOF OF LEMMA 7.2. Let $\mathcal{J}_1, \dots, \mathcal{J}_k$ be the sets in $\mathcal{J}$ which consist of irrational angles. Given $\epsilon_1, \dots, \epsilon_k > 0$, we let $\mathcal{J}'$ be the collection obtained from $\mathcal{J}$ replacing $\mathcal{J}_1, \dots, \mathcal{J}_k$ with $\mathcal{J}_1 - \epsilon_1, \dots, \mathcal{J}_k - \epsilon_k$. Also, let $\mathcal{F}' = \mathcal{F}$, and let $\Theta' = (\mathcal{F}', \mathcal{J}')$.

It is possible to choose arbitrarily small $\epsilon_1, \dots, \epsilon_k > 0$ such that:

(i) Conditions (C0)–(C2) of Theorem 7.1 hold for Θ'.

(ii) $\mathcal{J}_i - \epsilon_i$ is a subset of $\mathbb{Q}/\mathbb{Z}$ formed by strictly preperiodic arguments with eventual period p_i.

(iii) The eventual period p_i of $\mathcal{J}_i - \epsilon_i$ is different from the periods of the periodic arguments participating in $\mathcal{F}' = \mathcal{F}$, as well as from the eventual periods of the remaining sets of the collection $\mathcal{J}'$.

(iv) Let p be the period of an argument participating in $\mathcal{F}$. Let s and t be periodic arguments with period p. If $A^+_{\Theta'}(t) = A^+_{\Theta'}(s)$, then $A^+_\Theta(t) = A^+_\Theta(s)$.

(v) Let $t, t' \in \mathbb{Q}/\mathbb{Z}$ be such that there exist θ, θ' participating in $\mathcal{J}$ with forward orbit containing t, t', respectively. If $A^-_{\Theta'}(t) = A^-_{\Theta'}(t')$, then $A^-_{\Theta}(t) = A^-_{\Theta}(t')$.

It suffices to check that (C3)–(C7) hold for Θ'. In effect:

(C3) follows from (iii) and the fact that (C3) holds for Θ.

(C4) follows from $\mathcal{F}' = \mathcal{F}$.

(C5) follows from (ii) and the fact that (C5) holds for Θ.

According to Poirier (see Lemma 3.10 in [**P**]), it suffices to check (C6) for t and t' periodic of the same period. Thus (C6) follows from (iv) and the fact that (C6) holds for Θ.

For (C7), let t, t' be arguments participating in $\mathcal{J}'$ such that $S^-_{\Theta'}(d^i t) = S^-_{\Theta'}(t')$. It follows that the eventual period of t and t' coincides. By (iii), t, t' participate in $\mathcal{J}$. Since (C7) holds for $\mathcal{J}$, we have that $d^i t = t'$, by (v). $\square$.

Now apply Poirier's theorem to obtain a sequence $\{f_n\}$ of post-critically finite polynomials with critical markings $\{\Theta_n\}$ of the previous lemma. By compactness of the connectedness locus (see Theorem 2.1), we may pass to a convergent subsequence and obtain the following result.

COROLLARY 7.3. *Consider $P : \mathbb{C} \to \mathbb{C}$ a topological realization of a rational lamination. Let Θ be a critical marking of P. Then there exists post-critically finite polynomials f_n with critical markings $\Theta_n \in \mathcal{A}_\Theta$ such that $f_n \to f$ and $\Theta_n \to \Theta$, as $n \to \infty$.*

A post-critically finite polynomial f_n with critical marking Θ_n has rational lamination $\lambda_\mathbb{Q}(f_n) = \Lambda_\mathbb{Q}(\Theta_n)$ (see Proposition 6.8). Intuitively, the rational laminations $\lambda_\mathbb{Q}(f_n)$ become more and more similar to $\lambda_\mathbb{Q}$ as n increases. We make this precise by restricting our attention to finite subsets of $\mathbb{Q}/\mathbb{Z}$:

LEMMA 7.4. *Let Θ be a critical marking of the topological realization $P : \mathbb{C} \to \mathbb{C}$ of $\lambda_\mathbb{Q}$. Consider a sequence of post-critically finite polynomials f_n with critical markings Θ_n in $\mathcal{A}_\Theta$ which converge to Θ. Let Q_m be the subset of $\mathbb{Q}/\mathbb{Z}$ formed by the arguments t such that $d^m t$ is periodic of period not greater than m.*

Then, given a positive integer m, there exists N such that, for all $t, s \in Q_m$, we have:

(i) For all $n \geq N$, the arguments t and s are $\lambda_\mathbb{Q}(f_n)$-equivalent if and only if t and s are $\lambda_\mathbb{Q}$-equivalent.

(ii) For all $n \geq N$, the external rays $R^t_{f_n}$ and $R^s_{f_n}$ land in the boundary of the same bounded Fatou component of f_n if and only if the topological rays R^t_{top} and R^s_{top} land in the boundary of the same bounded topological Fatou component of P.

PROOF OF LEMMA 7.4. For the first statement, given Q_m, by the definition of the topology in $\mathcal{A}_\Theta$, there exists N such that: for all $n > N$, two elements t and t' of Q_m have the same left Θ-address if and only if t and t' have the same left Θ_n-address. Moreover, the elements of the collection $\mathcal{J}$ which are subsets of $\mathbb{Q}/\mathbb{Z}$ also participate in the critical markings of f_n, for all n. Thus, for $n > N$, we have that $\Lambda_\mathbb{Q}(\Theta_n) = \lambda_\mathbb{Q}(f_n)$ must coincide with $\Lambda_\mathbb{Q}(\Theta) = \lambda_\mathbb{Q}$ over the set Q_m.

For the second, given a bounded topological Fatou component U, we consider the closed set $q^{-1}(\partial U) \subset \mathbb{R}/\mathbb{Z}$ formed by the argument of the topological rays landing at ∂U. Denote by $U_1, \dots, U_k$ the bounded topological Fatou components which contain in their boundaries a landing point of some topological ray with argument in Q_m (i.e., $q^{-1}(\partial U_i) \cap Q_m \neq \emptyset$). Note that, for $i = 1, \dots, k$, the sets

$q^{-1}(\partial U_i)$ and $\mathcal{J}_c$ are unlinked, for all critical points c with infinite forward orbit. The definition of the topology of $\mathcal{A}_\Theta$ guarantees that, for n greater than some N, and for all $t, t' \in C = q^{-1}(\partial U_1) \cup \cdots \cup q^{-1}(\partial U_k)$, we have that:

$S^-_\Theta(t) = S^-_\Theta(t')$ if and only if $S^-_{\Theta_n}(t) = S^-_{\Theta_n}(t')$.

Thus for $n > N$, two arguments $t, s \in C$ are Θ_n pre-equivalent, denoted $t \sim_n s$, if and only if $t, s \in C$ are Θ pre-equivalent, denoted $t \sim s$.

Suppose $t, s \in Q_m$ are such that the external rays $R^t_{f_n}$ and $R^s_{f_n}$ land at distinct points in the boundary of a bounded Fatou component U_n, for some $n > N$. We claim that the topological rays R^t_{top} and R^s_{top} also land in the boundary of the same bounded topological Fatou component. For otherwise, in view of separation Lemma 5.3, there exists topological rays with rational arguments τ, τ' **in** C separating R^t_{top} from R^s_{top}. Since $\tau, \tau' \in C$, the corresponding external rays $R^\tau_{f_n}$, $R^{\tau'}_{f_n}$ also land at a common point and separate $R^t_{f_n}$ from $R^s_{f_n}$. This contradicts the assumption that $R^t_{f_n}$ and $R^s_{f_n}$ land at distinct points in ∂U_n.

Conversely, suppose that R^t_{top} and R^s_{top} land at distinct points in the boundary of a bounded topological Fatou component U. We fix $n > N$ and proceed by contradiction to show that the external rays $R^t_{f_n}$ and $R^s_{f_n}$ also land in the boundary of the same bounded Fatou component. The argument that follows is illustrated in Figure 7. If $R^t_{f_n}$ and $R^s_{f_n}$ do not land in the boundary of the same Fatou component, then there exist $R^{t'}_{f_n}$ and $R^{s'}_{f_n}$ landing at a common point and separating $R^t_{f_n}$ from $R^s_{f_n}$, by Lemma 5.1 or Lemma 5.3. In view of separation Lemma 5.3, there exist topological rays, with arguments $\tau, \tau' \in C$, landing at a common point and separating the ray with argument t' from the rays with arguments t, s, s'. Similarly, there exist topological rays, with arguments $\sigma, \sigma' \in C$, landing at a common point and separating the ray with argument s' from the rays with arguments t, s, t'. Since $R^{t'}_{f_n}$ and $R^{s'}_{f_n}$ land at a common point, the f_n-external rays with arguments $\sigma, \sigma', \tau, \tau'$ also land at a common point. Recall that these arguments belong to C, thus the topological rays with arguments $\sigma, \sigma', \tau, \tau'$ must also land at a common point and separate R^t_{top} from R^s_{top}. This contradicts the fact that R^t_{top} and R^s_{top} land at distinct points in the boundary of U. □

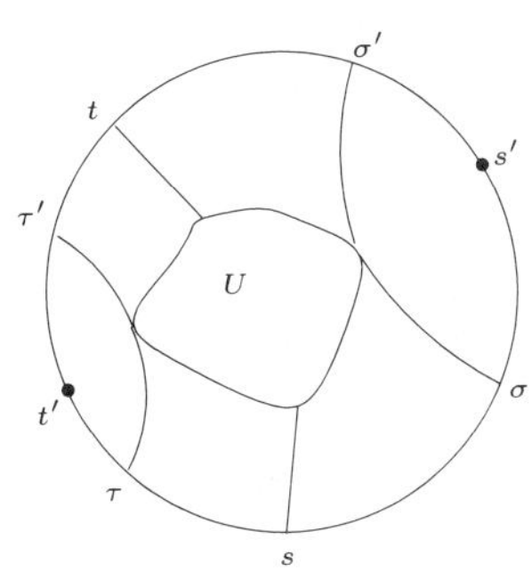

FIGURE 7. Illustrates part of the proof of Lemma 7.4.

COROLLARY 7.5. *Under the hypothesis of Lemma 7.4, given a positive integer p, there exists N such that the periodic classes of $\lambda_\mathbb{Q}$ with period dividing p coincide with the periodic classes of $\lambda_\mathbb{Q}(f_n)$ with period dividing p, for all $n > N$.*

PROOF OF COROLLARY 7.5. Note that every periodic point of P is either topologically attracting or topologically repelling. Moreover, $P : \overline{\Delta} \to \mathbb{C}$ is "weakly polynomial-like" of degree d, where $\Delta = q(\{|z| < 2\})$, in the sense of Definition 3.6 of [**GM**]. By Lemma 3.7 of [**GM**], the number of periodic points of P of period dividing p is d^p. Thus the number of periodic points of f_n, as well as of P, with period dividing p is d^p, for all n. Moreover, the number of Fatou periodic points of f_n with period dividing p also coincides with the number of Fatou periodic points of P with period dividing p, for all n. Thus, the number of $\lambda_{\mathbb{Q}}$-classes with period dividing p coincides with the number of $\lambda_{\mathbb{Q}}(f_n)$-classes with period dividing p, for all n.

Now note that there exists m such that Q_m contains all the classes of $\lambda_{\mathbb{Q}}$ with period diving p, where Q_m is the subset of $\mathbb{Q}/\mathbb{Z}$ introduced in the statement of Lemma 7.4. Hence we may consider an integer N such that assertion (i) of the previous lemma holds. Consequently, every $\lambda_{\mathbb{Q}}$-class with period dividing p is a $\lambda_{\mathbb{Q}}(f_n)$-class with period dividing p. The corollary now follows, by the cardinality considerations of above. □

7.2. Combinatorial Continuity. In Section 4, given a formal rational lamination $\lambda_{\mathbb{Q}}$ we constructed a topological realization $P : \mathbb{C} \to \mathbb{C}$. In Section 6, we obtained a critical marking Θ for the topological realization P. We then found a sequence $\{\Theta_n\}$ of critical markings of post-critically finite polynomials $\{f_n\}$ such that Θ_n converge to Θ and f_n converge to a polynomial f (see Corollary 7.3). To prove that f is our desired polynomial, we need to understand some properties of a sequence of external rays $\{R^t_{f_n}\}$, where t is a rational argument.

Consider a sequence of (extended) external rays $\{S_n\} \subset \mathbb{C}$. We say that

$$\limsup S_n$$

is the set of points $z \in \mathbb{C}$ such that every neighborhood of z intersects infinitely many S_n.

LEMMA 7.6. *Consider a sequence $\{f_n\}$ of post-critically finite polynomials converging to f.*

If $t \in \mathbb{Q}/\mathbb{Z}$ is periodic and $z \in (\limsup R^t_{f_n}) \cap J(f)$, then z is periodic under f.

If $t \in \mathbb{Q}/\mathbb{Z}$ is not periodic and $z \in (\limsup R^t_{f_n}) \cap J(f)$, then z is not periodic under f.

Moreover, assume that there exists an extended ray $\hat{R}^t_{f_n}$, for all n.

If $t \in \mathbb{Q}/\mathbb{Z}$ is periodic and $z \in (\limsup \hat{R}^t_{f_n}) \cap J(f)$, then z is periodic under f.

If $t \in \mathbb{Q}/\mathbb{Z}$ is not periodic and $z \in (\limsup \hat{R}^t_{f_n}) \cap J(f)$, then z is not periodic under f.

PROOF OF LEMMA 7.6. Suppose that $t \in \mathbb{Q}/\mathbb{Z}$ is periodic of period p. If $z \in \limsup R^t_{f_n}$ then, there exists $z_n \in R^t_{f_n}$ converging to z. Recall that $\limsup J(f_n) \supset J(f)$. Thus if z also lies in the Julia set $J(f)$, then the Euclidean distance

$$dist(z_n, J(f_n)) \to 0.$$

Let ρ_n be hyperbolic distance in the basin of infinity $\Omega(f_n)$. The dynamics of f_n lifts to multiplication by d acting on the upper half plane. It follows that $\rho_n(z_n, f_n^{\circ p}(z_n)) = \log d^p$. By the standard comparison between the hyperbolic and Euclidean metrics, we conclude that $dist(z_n, f_n^{\circ p}(z_n)) \to 0$ (see, for example, Theorem 4.3 of [**CG**]). Hence $z = f^{\circ p}(z)$.

Now, when $t \in \mathbb{Q}/\mathbb{Z}$ is strictly preperiodic, we first consider the case in which dt is a periodic argument. If $w \in (\limsup R^t_{f_n}) \cap J(f)$, then we proved that $f(w)$ must be periodic. There are two possibilities: either w is the unique periodic preimage of $f(w)$ or w is strictly preperiodic. We claim that only the latter occurs. We proceed by contradiction. Suppose that w is periodic. Let $w_n \in R^t_{f_n}$ be such that $w_n \to w$, and let $t' \in \mathbb{Q}/\mathbb{Z}$ be the unique periodic preimage of t. Consider $w'_n \in R^{t'}_f$ such that $f_n(w'_n) = f_n(w_n)$. Since t' is periodic, w'_n must converge to the unique periodic preimage w of the periodic point $f(w)$. Thus f is not locally injective at w. But w is a periodic point in the Julia set $J(f)$, which is obviously impossible. The assertion for strictly preperiodic t follows, by induction.

For extended rays the proof is fairly similar. In fact, let $\hat{R}^t_{f_n}$ be periodic extended rays. Denote by U_n the bounded Fatou component of f_n such that $\hat{R}^t_{f_n} \cap K(f_n)$ is contained in $\overline{U}_n$. Consider a point $z \in (\limsup \hat{R}^t_{f_n}) \cap J(f)$. In view of the above, it suffices to consider the case for which z is the limit of $z_n \in U_n$. The period of U_n is bounded by the period q of t. By passing to a subsequence, we may assume that the period of U_n is p, for all n. By passing to a further subsequence, we may assume that the return map $f_n^{\circ p} : U_n \to U_n$ has degree $\delta > 1$, for all n. Therefore, the return map is conformally conjugate to $z \mapsto z^\delta$ acting on $\mathbb{D}$. Let c_n denote the critical point of $f_n^{\circ p}$ in U_n. It follows that, the distance between z_n and $f^{\circ q}(z_n)$ is $\log \delta^{q/p}$, in the hyperbolic metric of $U_n \setminus \{c_n\}$. Moreover, the Euclidean distance between z_n and $J(f_n)$ tends to 0, a n tends to ∞. Again, by the standard comparison between the hyperbolic and Euclidean metrics, $z = f^{\circ q}(z)$. Now, replacing $R^t_{f_n}$ by $\hat{R}^t_{f_n}$ in the previous paragraph, the assertion for t preperiodic follows. □

The last ingredient in the proof of the main theorem is the next lemma by Goldberg and Milnor. We refer the reader to their original paper for the proof.

LEMMA 7.7 (Goldberg and Milnor [**GM**]). *Consider a polynomial f_0 and an external ray $R^t_{f_0}$ which lands at a repelling periodic point z_0. Then R^t_f lands at the analytic continuation of z_0, for all f in a sufficiently small neighborhood V of f_0. Moreover if $f_n \to f$, then $\limsup R^t_{f_n} = \overline{R}^t_f$.*

PROOF OF THEOREM 1.1. We will show that the polynomial f of Corollary 7.3 is such that $\lambda_{\mathbb{Q}}(f) = \lambda_{\mathbb{Q}}$. We claim that f has no parabolic cycles. In fact, if z_0 is a parabolic point of f, then there exist periodic points z_n and w_n of f_n such that both sequences $\{z_n\}$ and $\{w_n\}$ converge to z_0. By Corollary 7.5, we may pass to subsequences and assume that z_n is the landing point of a ray with argument t and w_n is the landing point of a ray with argument s, for all n. By Lemma 7.4, the topological external rays with arguments s and t land at distinct periodic points of P. By separations lemmas 5.1 and 5.2, there exists a graph Γ formed by (extended) strictly preperiodic topological rays separating R^t_{top} from R^s_{top}, say τ, σ are the arguments of the rays which participate in Γ. By Lemma 7.4, for n large enough, the graph Γ_n separates z_n from w_n, where Γ_n is the union of the closure of the (extended) f_n-rays with arguments σ and τ. It follows that the parabolic point z_0 lies in $\limsup \Gamma_n$, which contradicts Lemma 7.6.

We now claim that f has no irrationally neutral cycles. In fact, if z_0 is a period p irrationally neutral point, then z_0 is the limit of a sequence $\{z_n\}$ of period p repelling periodic points z_n of f_n. By Corollary 7.5, we may assume that there is an argument t, which is independent of n, such that $R^t_{f_n}$ lands at z_n. But Goldberg and Milnor's lemma implies that $R^t_{f_n}$ must land at the analytic continuation of the repelling periodic point w_0, where w_0 is the landing point of R^t_f, which is impossible.

We now show that $\lambda_{\mathbb{Q}} \subset \lambda_{\mathbb{Q}}(f)$. Suppose that two arguments t, s are $\lambda_{\mathbb{Q}}$ equivalent. Then, for n large enough, $R^t_{f_n}$ and $R^s_{f_n}$ land at a common point z_n. If R^t_f and R^s_f land at two distinct points z and w, respectively, then $\limsup(R^t_{f_n} \cup R^s_{f_n})$ contains a non-trivial connected set $X \subset K(f)$ joining z and w. Note that $f^{\circ l}(X) \subset \limsup(R^{d^l t}_{f_n} \cup R^{d^l s}_{f_n})$ is a non-trivial subset of $K(f)$, where $l \geq 0$ may chosen so that the arguments $d^l t$ and $d^l s$ are periodic. This is a contradiction with Goldberg and Milnor's lemma which states that the "limsup" of a periodic ray is the closure of a periodic ray.

Finally, we show that $\lambda_{\mathbb{Q}}(f) \subset \lambda_{\mathbb{Q}}$. Assume that, for some $t, s \in \mathbb{Q}/\mathbb{Z}$, the external rays R^t_f and R^s_f land at z. Again, we proceed by contradiction. Suppose that $R^t_{f_n}$ and $R^s_{f_n}$ land at distinct points, for n large enough. Then R^t_{top} and R^s_{top} land at distinct points. By the separation lemmas, there exists a graph Γ formed by (extended) preperiodic topological rays separating R^t_{top} from R^s_{top} such that, for some l,

(i) $P^{\circ l}(\Gamma)$ is formed by strictly (extended) preperiodic topological rays, and

(ii) $P^{\circ l}(R^t_{top})$ and $P^{\circ l}(R^s_{top})$ are periodic topological rays (i.e., $f^{\circ l}(z)$ is periodic).

By Lemma 7.4, the corresponding graph Γ_n separates $R^t_{f_n}$ from $R^s_{f_n}$. Since $\limsup R^t_{f_n} \supset R^t_f$, and since $\limsup R^s_{f_n} \supset R^s_f$, it follows that $z \in \limsup \Gamma_n$. But $f^{\circ l}(z)$ is periodic, which contradicts Lemma 7.6. □

Appendix A. Wandering Portraits

Here we upgrade the following result of [**K2**].

THEOREM A.1 ([**K2**]). *Let f be a monic polynomial of degree d. Then the number of smooth external rays landing at a point with infinite forward orbit is bounded by d, provided that the forward orbit of z is disjoint from the critical points of f.*

For quadratic polynomials, Theorem A.1 is a result due to Thurston [**Th**] known as the No Wandering Gap Theorem. The proof of the author's thesis [**K2**] generalizes the proof of Thurston using the language of Goldberg and Milnor's paper [**GM**]. It is not known if the bound of d is sharp or not. Levin and Blokh have developed a novel and a different set of ideas to tackle this problem (see [**L, BL**]). When several points are considered, these ideas lead to better bounds.

Our goal is to prove the next more abstract version of Theorem A.1.

THEOREM A.2. *Consider a finite set $A \subset \mathbb{R}/\mathbb{Z}$ such that the following hold:*
(i) $d^n \cdot A$ and $d^m \cdot A$ are disjoint and pairwise unlinked, for all $n \neq m$.
(ii) $A \mapsto d^n \cdot A$ is a cyclic order preserving bijection, for all n.
Then the cardinality of A is at most d.

In view of the Theorem A.1, to prove Theorem A.2 we only need to show that, given a finite set $A \subset \mathbb{R}/\mathbb{Z}$ as in the statement of the latter, we may find a polynomial f such that the rays with arguments in A land at a point z with infinite orbit. To obtain such a polynomial f, we construct a "critical portrait" Θ "compatible" with A and apply results that may be found in [**BDK**] or [**G**].

A.1. Critical Portraits. Critical markings are inspired by critical portraits. Critical portraits were introduced in Fisher's thesis (see [**BFH**]). They abstractly capture the location of the critical points of a polynomial.

DEFINITION A.3. *A collection $\Theta = \{\Theta_1, \ldots, \Theta_m\}$ of finite subsets of $\mathbb{R}/\mathbb{Z}$ is called a* **critical portrait** *of degree d if (CP1), (CP2) and (CP3) hold, where:*
(CP1) $|\Theta_j| \geq 2$ and $|d \cdot \Theta_j| = 1$, for $j = 1, \ldots, m$.
(CP2) $\Theta_1, \ldots, \Theta_m$ are pairwise unlinked.
(CP3) $\sum(|\Theta_j| - 1) = d - 1$.

PROPOSITION A.4. *Consider a finite set $A \subset \mathbb{R}/\mathbb{Z}$ such that the following hold:*
(i) $d^n \cdot A$ and $d^m \cdot A$ are disjoint and pairwise unlinked, for all $n \neq m$.
(ii) $A \mapsto d^n \cdot A$ is a cyclic order preserving bijection, for all $n \geq 0$.
Then there exists a critical portrait $\Theta = \{\Theta_1, \ldots, \Theta_l\}$ such that, for all i and all n sufficiently large, we have that Θ_i and $d^n A$ are disjoint and unlinked.

The proof of the proposition may be found in Subsection A.2 below. We now show that the proposition implies Theorem A.2.

Roughly, given a critical portrait Θ, according to Goldberg [**G**] (see Proposition 3.8 [**G**], also compare with [**BDK**]) there always exists a polynomial f with all critical points escaping such that the "location" of its critical points is given by Θ.

More precisely, let $\Theta = \{\Theta_1, \ldots, \Theta_m\}$ be a critical portrait. Following Goldberg (also compare with Subsection 6.2), we say that s, t are Θ-unlinked if the collection of subsets $\{\Theta_1, \ldots, \Theta_m, \{s, t\}\}$ is disjoint and pairwise unlinked. According to Lemma 3.5 of [**G**], if $\{\Theta_1, \ldots, \Theta_m, \{d^n s, d^n t\}\}$ is a disjoint and pairwise unlinked collection, for all n, then the external rays R^s_f and R^t_f are smooth and land at a common point, provided that Θ is a critical portrait for f (in the sense of Goldberg, see page 99 of [**G**]). It follows that, if Θ is the critical portrait given by the proposition above, then there exists a polynomial f such that the smooth external rays of f with arguments in $d^n \cdot A$ land at a common point, for all n large enough. Since the iterates of A are pairwise unlinked, it is not difficult to conclude that the landing point, of the external rays with arguments in $d^n \cdot A$, cannot be periodic (unless the cardinality of A is one). Thus, assuming Proposition A.4, we have proven the following.

COROLLARY A.5. *Consider a finite set $A \subset \mathbb{R}/\mathbb{Z}$ such that the following hold:*
(i) $d^n \cdot A$ and $d^m \cdot A$ are disjoint and pairwise unlinked for all $n \neq m$.
(ii) $A \mapsto d^n \cdot A$ is a cyclic order preserving bijection, for all n.
Then there exists a polynomial f and an integer N such that the external rays with arguments in $d^N \cdot A$ are smooth and land at a point z with infinite forward orbit.

By this corollary and Theorem A.1, the proof of Theorem A.2 follows immediately.

A.2. Compatible Critical Portraits. The proof of Proposition A.4 relies on the construction of critical portraits which are unlinked with finitely many finite subsets of $\mathbb{R}/\mathbb{Z}$.

LEMMA A.6. *Let $M(\alpha) \equiv j \cdot \alpha + \alpha_0 \pmod 1$ be a linear self-covering of $\mathbb{R}/\mathbb{Z}$ with $j \geq 2$. Let $B_1, \dots B_m$ be a collection of finite subsets of $\mathbb{R}/\mathbb{Z}$ such that the following hold:*
(i) $B_1, \dots, B_m$ are disjoint and pairwise unlinked.
(ii) $M(B_1), \dots, M(B_m)$ are disjoint and pairwise unlinked.
(iii) M maps B_i onto $M(B_i)$ in a cyclic order preserving fashion.

Then there exists $\{\theta, \theta'\}$ such that $M(\theta) = M(\theta')$, and the sets $\{\theta, \theta'\}$ and B_i are disjoint and unlinked, for all i.

PROOF. Consider the collection $\mathcal{I}$ of open intervals in $\mathbb{R}/\mathbb{Z}$ consisting of all the components of $\mathbb{R}/\mathbb{Z} \setminus B_i$, for all $i = 1, \dots, m$. There exists at least one element $I = (s_0, s_1)$ of this collection which has length greater than $1/j$ and does not contain elements of $\mathcal{I}$ of length greater than $1/j$.

Consider the set

$$H = \bigcup_{J \subsetneq I, J \in \mathcal{I}} \overline{J} = [h_0, h_1] \cup \dots \cup [h_{2n}, h_{2n+1}]$$

and let $H_k = (h_{2k}, h_{2k+1})$. Note that the extreme points of H_k must lie in B_i, for some i. Also, remark that if $\{\theta, \theta'\} \subset I \setminus H$, then $\{\theta, \theta'\}$ and B_i are unlinked, for all i.

We now need the following.

Claim. If $s_0 + 1/j \in H_r$, for some r, then $h_{2r+1} - 1/j \in I \setminus H$ (see Figure 8).

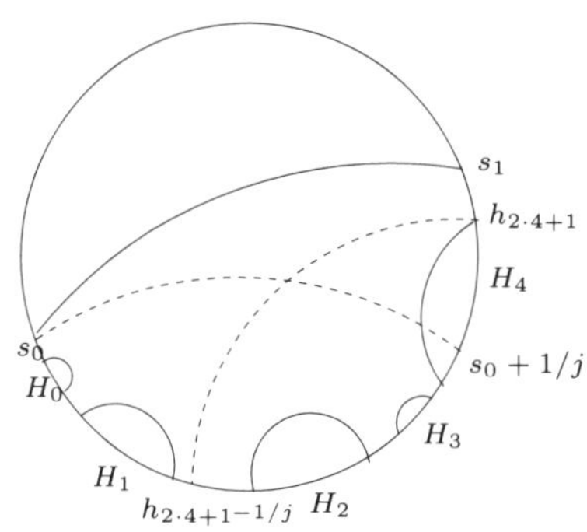

FIGURE 8. Claim in the proof of Lemma A.6.

Proof of Claim. Suppose that

$$s_0 + 1/j \in H_r \text{ and that } h_{2r+1} - 1/j \in H_l.$$

Then, under M, the interval $[h_{2l}, h_{2r}]$ is mapped in a one-to-one fashion onto its image, since this interval has length less than $1/j$. Therefore,

$$M(h_{2r}) \in \mathbb{R}/\mathbb{Z} \setminus M(H_l).$$

On the other hand, $h_{2r+1} - 1/j \in H_l$ implies that

$$M(h_{2r+1}) = M(h_{2r+1} - 1/j) \in M(H_l).$$

Hence, $\{M(h_{2l}), M(h_{2l+1})\}$ and $\{M(h_{2r}), M(h_{2r+1})\}$ are linked, which is a contradiction. The claim now follows.

From the claim we conclude that, for $\epsilon > 0$ sufficiently small, either $\{s_0+\epsilon, s_0+\epsilon+1/j\}$ or $\{h_{2r+1}+\epsilon, h_{2r+1}+\epsilon-1/j\}$ is unlinked with B_i, for all i. Thus the proof of the lemma is now complete. □

PROOF OF PROPOSITION A.4. For a fixed integer $n \in \mathbb{N}$, we construct a critical portrait Θ_n formed by $d-1$ pairs $\{\theta_i(n), \theta'_i(n)\}$ which are disjoint and unlinked with $A, \dots, d^n A$. Suppose that we have already constructed pairs $\{\theta_i(n), \theta'_i(n)\}$, for $i = 1, \cdots, k < d-1$, such that each pair is pairwise unlinked with $A, \dots, d^n A$. These k pairs chop $\mathbb{R}/\mathbb{Z}$ into $k+1$ "pieces". We say that two arguments t, s lie in the same "piece" if and only if t, s lie in the same component of $\mathbb{R}/\mathbb{Z} \setminus \{\theta_i(n), \theta'_i(n)\}$, for all $i = 1, \dots, k$. The total length of each "piece" is a multiple of $1/d$. Thus there exists a "piece" L with total length j/d strictly greater than $1/d$. Consider an order preserving map $\pi : \overline{L} \to S = \mathbb{R}/\mathbb{Z}$ with constant derivative d/j. Multiplication by d "projects" to the circle S as a j linear covering M_j of $S = \mathbb{R}/\mathbb{Z}$. Let $B_1, \dots, B_m$ be the collection obtained as the image under π of the sets of the collection $A, \dots, d^n \cdot A$ that are contained in L. Apply the previous lemma to obtain a pair $\{\gamma, \gamma + 1/j\} \subset \pi(L) \subset S$, which is pairwise unlinked with $B_1, \dots, B_m$. The preimage $\pi^{-1}(\{\gamma, \gamma+1/j\})$ may be chosen as $\{\theta_{k+1}(n), \theta'_{k+1}(n)\}$.

By passing to a subsequence, we may assume that $\theta_i(n)$ and $\theta'_i(n)$ converge to θ_i and θ'_i, respectively. Let $\tilde{\Theta}_i = \{\theta_i, \theta'_i\}$ and, after taking appropriate unions of non-disjoint elements of the collection $\{\tilde{\Theta}_1, \cdots, \tilde{\Theta}_{d-1}\}$, we obtain the desired critical portrait $\Theta = \{\Theta_1, \dots, \Theta_l\}$. □

References

[BDK] P. Blanchard, R. L. Devaney, and L. Keen, *The dynamics of complex polynomials and automorphism of the shift*, Invent. Math. **104** (1991), 545–580.

[BFH] B. Bielefield, Y. Fisher, and J. H. Hubbard, *The classification of critically preperiodic polynomials as dynamical systems*, Journal AMS **5** (1992), 721–762.

[BH] B. Branner and J. H. Hubbard, *The iteration of cubic polynomials, Part I: the global topology of parameter space*, Acta Math. **160** (1988), 143–206.

[Be] A. F. Beardon, *Iteration of Rational Functions*, Grad. Texts Math. *132*, Springer-Verlag, 1991.

[Bl] P. Blanchard, *Complex analytic dynamics on the Riemann sphere*, Bull. A.M.S. **11** (1984), 85–141.

[BL] A. Blokh and G. Levin, *On invariant laminations in the disc*, preprint, April 99.

[CG] L. Carlesson and T. W. Gamelin, *Complex Dynamics*, Springer-Verlag, 1993.

[DH0] A. Douady and J. H. Hubbard, *Itération des polynomes quadratiques complexes*, C. R. Acad. Sci. Paris **294** (1982), 123–126.

[DH1] A. Douady and J. H. Hubbard, *Étude dynamique des polynomes complexes, I & II*, Publ. Math. Orsay, 1984–85.

[D] A. Douady, *Description of compact sets in C*, pp. 429–465 of " Topological Methods in Modern Mathematics", ed. Goldberg and Phillips, Publish or Perish, 1993.

[Fa] D. Faught, *Local connectivity in a family of cubic polynomials*, Thesis, Cornell University, 1992.

[GM] L. R. Goldberg and J. Milnor, *Fixed point portraits of polynomial maps, Part II: Fixed Point Portraits*, Ann. Sci. École Norm. Sup. Paris 4e série, **26** (1993), 51–98.

[G] L. R. Goldberg, *On the Multiplier of a Repelling Fixed Point*, Invent. Math., (1994), 85–108.

[HY] J. Hocking and G. Young, *Topology*, Dover, 1988.

[K1] J. Kiwi, *Non Accessible Critical Points of Cremer Polynomials*, Ergodic Theory and Dynamical Systems, to appear.

[K2] J. Kiwi, *Rational Rays and Critical Portraits of Complex Polynomials*, Thesis, SUNY at Stony Brook, 1997 (available as SUNY-IMS Preprint #1997/15).

[La1] P. Lavaurs, *Une description combinatoire de l'involution definie par M sur les rationneles à dénominateur impair*, C.R. Acad. Sc. Paris (4) **303** (1986), 143–146.

[La2] P. Lavaurs, *Systemes dynamiques holomorphes: explosion de points périodiques paraboliques*, Thesis, Université Paris-Sud, Orsay, 1989.

[L] G. Levin, *On backward stability of holomorhic dynamical systems*, Fundamenta Mathematicae, **158** (1998), 97–107.

[McM] C. T. McMullen, *Complex Dynamics and Renormalization*, Ann. Math Studies **135**, Princeton University Press, 1994.

[M1] J. Milnor, *Dynamics in One Complex Variable: Introductory Lectures*, Vieweg, 1999.

[M2] J. Milnor, *Periodic Orbits, External Rays and the Mandelbrot Set: An Expository Account*, pp. 277–331 of "Astérisque", ed. Flexor, Sentenac and Yoccoz, Société Mathématique de France, 2000.

[Mo] R. L. Moore, *Concerning Upper Semi-continuous Collection of Continua*, Trans. A.M.S. **27** (1925), 412–428.

[P] A. Poirier, *On Postcritically Finite Polynomials, Parts I, II*, Stony Brook IMS Preprint #1993/5 and 7.

[Pe] C. L. Petersen, *Local connectivity of some Julia sets containing a circle with an irrational rotation*, Acta. Math. **177**, 163–224.

[Sch1] D. Schleicher, *Internal Addresses in the Mandelbrot Set and Irreducibility of Polynomials*, Thesis, Cornell University, 1994.

[Sch2] D. Schleicher, *On Fibers and Local Connectivity of Compact Sets*, Stony Brook IMS Preprint #1998/12.

[Sch3] D. Schleicher, *I. On Fibers and Local Connectivity of Mandelbrot and Multibrot Sets; II. On Fibers and Renormalization of Julia sets and Multibrot Sets*, Stony Brook IMS Preprint #1998/13.

[Sch4] D. Schleicher, *Rational Parameter Rays of the Mandelbrot Set*, pp. 405–443 of "Astérisque", ed. Flexor, Sentenac and Yoccoz, Société Mathématique de France, **261**, 2000.

[St] N. Steinmetz, *Rational Iteration: Complex Analytic Dynamical Systems*, de Gruyter, 1993.

[Th] W. P. Thurston, *On the Combinatorics of Iterated Rational Maps*, manuscript, 1985.

Facultad de Matemáticas, Pontificia Universidad Católica, Casilla 306, Correo 22, Santiago, Chile.

E-mail address: jkiwi@mat.puc.cl

Contemporary Mathematics
Volume **269**, 2001

ACTIONS OF DISCRETE GROUPS ON COMPLEX PROJECTIVE SPACES

José Seade
AND
Alberto Verjovsky

24 May, 1999

0. Introduction

Holomorphic dynamics can be divided, roughly speaking, into three parts: iteration theory, actions of discrete groups and foliations, which includes actions of continuous Lie groups. Iteration theory and holomorphic foliations are being studied in all dimensions, while discrete groups actions have been essentially focused on Kleinian groups acting on the Riemann sphere, which is a fascinating area of mathematics. There are also very interesting results in higher dimensions, for example in [DM], about discrete subgroups of $U(1,n)$ acting on the complex hyperbolic space. In [SV] we introduced the concept of a *Complex Kleinian Group*; by this we mean a discrete subgroup of $PSL(n+1,\mathbb{C})$ acting on the projective space $P^n_{\mathbb{C}}$ so that its region of discontinuity is non-empty, i.e., its limit set is not all of $P^n_{\mathbb{C}}$. That seems to be a whole branch of mathematics waiting to be explored. This article is essentially an exposition of [SV]; we have tried to present the main ideas of that article, removing most of the technical difficulties. Appropriate references are given in the text.

In section 1 we present some well known facts about conformal Kleinian groups, i.e. discrete groups of conformal transformations of the round spheres, acting with non-empty region of discontinuity. We include in this section a slight generalization of [SV], of an important theorem of Flaminio and Spatzier, concerning the ergodicity of the action of a Kleinian group on frames on the sphere over the limit set. This generalizes classical results of Sullivan and others, about the non-existance

---

1991 *Mathematics Subject Classification.* Primary: 32H, 58D, 58E20, 58F23, 53C. Secondary: 20H10, 22E40, 30D05, 30F40, 30F35, 32L25, 32J18, 32G05, 32M05, 57S20, 57S25,.

Key words and phrases. Fuchsian and Kleinian groups, conformal groups, conformally flat manifolds, Möbius groups, complex manifolds, twistor spaces, hyperbolic geometry, complex solenoids or laminations, Patterson-Sullivan measures, isometries, minimal sets, ergodicity, Hausdorff dimension.

Research partially supported by CONACYT (Mexico) 1026P-E, PAPIIT, UNAM, IN 104 596 IM, Academia Mexicana de Ciencias and the London Royal Society.

of invariant fields of lines over the limit set. In section 2 we introduce the complex Kleinian groups and give several types of examples and constructions of these groups. This includes the subgroups of $U(1,n)$, which act on $P^n_{\mathbb{C}}$ preserving a ball; they are to complex Kleinian groups what Fuchsian groups are to the classical Kleinian groups. In sections 3 and 4 we discuss two interesting constructions of complex Kleinian groups. The construction in section 3 shows that the conformal dynamics of Kleinian groups acting on spheres, embedds in the holomorphic dynamics of complex Kleinian groups acting on complex projective spaces. This is done by using twistor theory. If N is a $2n$-dimensional, oriented, Riemannian manifold, its twistor space is the total space of the fibre bundle over N with fiber, at each $x \in N$, the set of complex structures on T_xN compatible with the metric and orientation. If N is the round sphere S^{2n}, its twistor space is a complex manifold with a canonical projective embedding. In particular, the twistor space of S^4 is $P^3_{\mathbb{C}}$. The idea in Penrose's twistor theory programme is that there is a fantastic interplay between the conformal geometry on the sphere (and more generally on a locally conformally flat Riemannian manifold) and the complex geometry on its twistor space. Here we push this idea forward to dynamics. In section 4 we generalize the construction of the classical Schottky groups acting on the Riemann sphere $S^2 \cong P^1_{\mathbb{C}}$ to higher dimensional projective spa ces of odd-dimension (the even dimensional case is presently being studied). We consider an arbitrary configuration $\{(L_1, M_1), ..., (L_r, M_r)\}, r > 2$, of pairs of projective n-spaces in $P^{2n+1}_{\mathbb{C}}$, which are all of them pairwise disjoint. Given arbitrary neighbourhoods $U_1, ..., U_r$ of the $L_i\,'s$, pairwise disjoint, we show that there exists, for each $i = 1, ..., r$, projective transformations T_i of $P^{2n+1}_{\mathbb{C}}$, which interchange the interior with the exterior of a compact tubular neighbourhood N_i of L_i contained in U_i, leaving invariant the boundary $E_i = \partial(N_i)$. The $E_i\,'s$ are *mirrors*, they play the same role in $P^{2n+1}_{\mathbb{C}}$ as circles play in S^2 to define the classical Schottky groups. Each mirror E_i is a $(2n+1)$-sphere bundle over $P^n_{\mathbb{C}}$. The group of automorphisms of $P^{2n+1}_{\mathbb{C}}$ generated by the $T_i\,'s$ is a complex Kleinian group Γ. The region of discontinuity $\Omega(\Gamma)$ is a fibre bundle over $P^n_{\mathbb{C}}$ with fibre S^{2n+2} minus a Cantor set $\mathcal{C}$. The limit set Λ is the complement of $\Omega(\Gamma)$ in $P^{2n+1}_{\mathbb{C}}$; it is the set of accumulation points of the Γ-orbit of an L_i, any one of them, and it is a product $\mathcal{C} \times P^n_{\mathbb{C}}$. The action of Γ on this set of projective lines is minimal in the sense that the Γ-orbit of every point x_o in $P^{2n+1}_{\mathbb{C}}$ accumulates to (at least a point in) each one of the projective lines in Λ. This set is transversally *projectively self-similar*, i.e., Λ corresponds to a Cantor set in the Grassmannian $G_{2n+1,n}$, which is dynamically-defined. Hence Λ is a *solenoid* (or *lamination*) by projective spaces, which is transversally Cantor and projectively self-similar. Each of these groups Γ contains a subgroup $\check{\Gamma}$ of index two, which is a free group of rank $r-1$ and acts *freely* on $\Omega(\Gamma)$. The quotient $\Omega(\Gamma)/\check{\Gamma}$ is a compact complex manifold, which is a fibre bundle over $P^n_{\mathbb{C}}$ with fibre the connected sum of $(r-1)$ copies of $S^{2n+1} \times S^1$. These manifolds have a canonical projective structure, *i.e.*, they have an atlas $\{(\mathcal{U}_i, \phi_i)\}$ whose changes of coordinates are restrictions of complex projective transformations, c. f. [Gu]. However, these manifolds are never Kähler, due to cohomological reasons. When $n = 1$ and the configuration $\{(L_1, M_1), ..., (L_r, M_r)\}$ consists of twistor lines of the fibration $p\colon P^3_{\mathbb{C}} \to S^4$, then

Γ and $\check{\Gamma}$ descend to conformal Schottky groups on S^4. In this case $\Omega(\Gamma)/\check{\Gamma}$ is the twistor space of the manifold $S^4/p(\check{\Gamma})$, which is a Schottky manifold [Ku2]; $\Omega(\Gamma)/\check{\Gamma}$ is a *flat twistor space* [Pe2]. Finally, in section 5 we compare the deformations of our Schottky groups with the deformations of the complex manifolds that one gets as quotients of the action of the group on its region of discontinuity. For this we estimate an upper bound for the Hausdorff dimension of the limit set of the complex Schottky groups. We use this to show that, with the appropriate conditions for the Schottky group $\check{\Gamma}$, the Kuranishi space $\mathfrak{K}$ of versal deformations of the complex manifold $M_{\check{\Gamma}} := \Omega(\check{\Gamma})/\check{\Gamma}$, is smooth near the reference point determined by $M_{\check{\Gamma}}$. Furthermore, we estimate the dimension of $\mathfrak{K}$ and we prove that every small deformation of $M_{\check{\Gamma}}$ actually corresponds to a small deformation of the group $\check{\Gamma}$ in the projective group $PSL(2n+2,\mathbb{C})$, in analogy with the classical Teichmüller and moduli theory for Riemann surfaces.

1. CONFORMAL KLEINIAN GROUPS

Let $SL(2,\mathbb{C})$ be the group of 2×2 complex matrices with determinant 1. This group acts linearly on $\mathbb{C}^2$, so it acts on the complex projective line $P^1_{\mathbb{C}}$. Moreover, if we identify $P^1_{\mathbb{C}}$ with the Riemann sphere $S^2 \cong \mathbb{C}\cup\infty := \widehat{\mathbb{C}}$, then the action of $SL(2,\mathbb{C})$ on $P^1_{\mathbb{C}}$ is via the Möbius transformations: $z \mapsto (az+b)(cz+d)^{-1}$. Two such matrices define the same Möbius transformation iff they differ by multiplication by ± 1, so the group $PSL(2,\mathbb{C}) \cong SL(2,\mathbb{C})/\{\pm I\}$ can be identified with the group of all Möbius transformations acting on $P^1_{\mathbb{C}}$. This coincides with the group of holomorphic automorphisms of the Riemann sphere; it also coincides with $Conf_+(S^2)$, the group of orientation preserving conformal automorphisms on S^2.

More generally, let $Conf(S^n)$ be the group of all conformal automorphisms of the n-sphere, and let $Conf_+(S^n)$ be the index-two subgroup of orientation preserving maps. $Conf(S^n)$ is generated by inversions on all possible $(n-1)$-spheres in the unit sphere $S^n \subset \mathbb{R}^{n+1}$; $Conf_+(S^n)$ consists of the elements obtained by an even number of inversions. S^n bounds the unit ball D^{n+1} and given a $(n-1)$-sphere $S^{n-1}_\delta \subset S^n$, there exists a unique n-sphere $S^n_\delta \subset \mathbb{R}^{n+1}$ intersecting S^n orthogonally at S^{n-1}_δ. The inversion in $\mathbb{R}^{n+1}$ on this S^n_δ, takes the unit ball D^{n+1} into itself, preserving the hyperbolic metric on D^{n+1}. This identifies $Conf(S^n) \cong Iso(\mathbb{H}^{n+1})$, the group of isometries of the hyperbolic $(n+1)$-space. Hence $Conf_+(S^n)$ acts transitively on $\mathbb{H}^{n+1}$ and the isotropy group is $SO(n+1)$, the group of orientation preserving maps in $\mathbb{R}^{n+1}$ generated by reflections on hyperplanes through the origin. Therefore $Conf_+(S^n)$ is diffeomorphic to $SO(n+1)\times\mathbb{H}^{n+1}$.

We recall that one has the isomorphisms (see [A1, A2]):

i) $Iso_+(\mathbb{H}^2) \cong PSL(2,\mathbb{R}) \cong \text{Möb}(2,\mathbb{R})$;

ii) $Conf_+(S^2) \cong Iso_+(\mathbb{H}^3) \cong PSL(2,\mathbb{C}) \cong \text{Möb}(2,\mathbb{C})$;

iii) $Conf_+(S^4) \cong Iso_+(\mathbb{H}^5) \cong PSL(2,\mathcal{H}) \cong \text{Möb}(2,\mathcal{H})$;

where $\mathcal{H}$ is the space of quaternions and $\text{Möb}(2,K)$, $K=\mathbb{R},\mathbb{C},\mathcal{H}$, denotes the group of Möbius transformations $z\mapsto(az+b)(cz+d)^{-1}$ with either real, complex or quaternionic coefficients.

We recall that if Γ is a discrete subgroup of $Conf_+(S^n)$, then its **limit set**

$\Lambda(\Gamma) \subset S^n$ is the set of accumulation points of some (any) Γ-orbit; by definition, $\Lambda(\Gamma)$ is closed, hence compact. The complement $\Omega(\Gamma) = P^1_{\mathbb{C}} - \Lambda(\Gamma)$ is the **region of discontinuity** of Γ. A **(conformal, hyperbolic or Möbius) Kleinian group** is a discrete subgroup Γ of $Conf_+(S^n)$, whose limit set is not all of S^n, *i.e.,* its region of discontinuity is not empty. The limit set of an infinite discrete group Γ of $Iso_+(\mathbb{H}^n)$ is necessarily non-empty, and it may consist of one, two or an infinite number of points. If the limit set Λ consists of one or two points, Γ is said to be *elementary.* We only consider here non-elementary groups. The set Λ is obviously an invariant set for Γ, hence its complement Ω is an open invariant set, which can be empty (if Γ is not Kleinian). The dynamics of the action of Γ on Λ is entirely different to the dynamics on Ω, where the action is properly discontinuous, while on Λ every orbit is dense, i.e., the action of Γ on Λ is **minimal**. Moreover, if we assume that Γ is *geometrically-finite*, i.e., that it has a finite-sided polyhedron as a fundamental domain in H^n (see [Ra, Ni, FSp]), then [Pa, Su1, FSp] there is a unique (up to scaling) finite invariant measure μ on $\Lambda := \Lambda(\Gamma)$, called *the geometric*, or *the Patterson-Sullivan measure*, such that for all $\gamma \in \Gamma$, $\gamma^*(\mu) = |\gamma\prime|^\delta \mu$, where δ is the Hausdorff dimension of Λ and $\gamma^*(\mu)(A) = \mu(\gamma(A))$. The support of μ is Λ and the action there is **ergodic**. In particular, if Λ is the whole sphere at infinity, then μ is the Lebesgue measure on the sphere, up to scaling.

1.1 Examples.

i) Let M^n be a (complete) hyperbolic manifold, so it is of the form $\mathbb{H}^n/\Gamma$, where Γ is its fundamental group. Γ is a discrete subgroup of $Iso_+(\mathbb{H}^n)$ which may or may not be Kleinian, i.e., its limit set can be the whole sphere at infinity or a proper subset of it. For instance, if M is compact, then the limit set of Γ is the whole sphere S^{n-1} at infinity so it is not Kleinian there. However, in all cases Γ can be regarded as a subgroup of $Iso_+(\mathbb{H}^{n+1})$ via the canonical inclusion $Iso_+(\mathbb{H}^n) \hookrightarrow Iso_+(\mathbb{H}^{n+1})$; in this case its action on the sphere S^n at infinity leaves invariant the equator, a $(n-1)$-sphere; this implies that the limit set is contained in the equator, so it has a non-empty region of discontinuity. Hence Γ is Kleinian as a subgroup of $Iso_+(\mathbb{H}^{n+1}) \cong Conf_+(S^n)$.

ii) There are complete (connected), open hyperbolic n-manifolds of infinite volume, whose fundamental group is a Kleinian group in $Iso_+(\mathbb{H}^n)$.

iii) The Schottky groups are remarkable examples of Kleinian groups. These groups are obtained by considering pairwise disjoint $(n-1)$-spheres $\mathbb{S}_1, ..., \mathbb{S}_r$ in S^n. Each sphere $\mathbb{S}_i$ plays the role of a mirror: it divides S^n in two diffeomorphic components, and one has an involution T_i of S^n interchanging these components, the inversion on $\mathbb{S}_i$. The Schottky group is defined to be the group of conformal transformations generated by these involutions. For $r > 2$ the limit set is a Cantor set, whose complement is non-empty, so they are Kleinian. It is worth noting that we can deform these groups slightly by moving a little the generating spheres and letting them be tangent, or even overlap slightly. We can get in this way groups which are still Kleinian, though their limit set is no longer Cantor, see [Bo].

1.2 Definition. *A discrete subgroup Γ of $Iso_+(\mathbb{H}^n)$ is said to be Zariski-dense in*

$Iso_+(\mathbb{H}^n)$ if its Zariski closure is $Iso_+(\mathbb{H}^n)$ (see [FSp]*).*

One has the following theorem by L. Flaminio and R. Spatzier:

1.3 Theorem [FSp]. *Let Γ be a geometrically-finite discrete subgroup of $Iso_+(\mathbb{H}^n)$ which is Zariski-dense. Let μ be the Patterson-Sullivan measure on the limit set Λ of Γ. Then, every Γ-invariant measurable distribution on $\Lambda \subset S^{n-1}$ by subspaces of dimension d is μ-almost everywhere trivial, i.e., either $d = n-1$ or $d = 0$. Furthermore, the restriction to Λ of the bundle of orthonormal n-frames on S^n is locally $U_x \times SO(n)$, with U_x an open set in Λ, so we can endow it with the measure $\widetilde{\mu}$ which is (locally) the product of μ with the invariant measure measure on $SO(n)$. Then Γ acts on this bundle, via the differential, and this action is ergodic with respect to $\widetilde{\mu}$.*

In the sequel we will actually need the following slight generalization of this theorem, proved in [SV]:

1.4 Theorem. *Let Γ be a geometrically-finite and Zariski-dense discrete subgroup of $Iso(\mathbb{H}^{n+1})$. Let μ be the Patterson-Sullivan measure on the limit set Λ. Let $0 < p < n$. Let $G_{\Lambda,p}$ be the the restriction to Λ of the Grassmannian fibre bundle, $G_{n,p}(S^n)$, of p-dimensional subspaces of TS^n . Then Γ acts, via the differential, minimally on $G_{\Lambda,p}$. Furthermore, Γ acts ergodically on $G_{\Lambda,p}$ with respect to the measure $\widetilde{\mu}$ which is (locally) the product of μ with the homogeneous measure on $G_{n,p} = SO(n)/(SO(p) \times SO(n-p))$.*

The proof of the ergodicity in (1.4) follows easily from (1.3) above. Then one shows that the action of Γ on $G_{\Lambda,p}$ carries fibers isometrically into fibers. Then the minimality of the action follows from the ergodicity together with the following general theorem (2.3.1) in [SV]; only the first statement below is needed for (1.4), but the second statement will be needed later in the text:

Theorem 1.5. *Let X and Y be compact metric spaces, and let G be a compact group which acts minimally on X. Let $\pi : E \to X$ be a locally trivial fibre bundle with fibre Y, such that G acts on E as a skew-product, $g(x,y) = (g(x), F_{(g,x)}(y))$, where $F_{(g,x)} : Y \to Y$ is an isometry for each $(g,x) \in G \times X$. Let H be a minimal subset of E for the action of G. Then:*

i) The restriction of π to H, $\pi|_H : H \to X$, is a locally trivial fibre bundle, whose fibres are homogeneous spaces on which G acts transitively.

ii) If the space Y is the 2-sphere, then either the minimal set H is all of E, or else the bundle $\pi : E \to X$ admits a section.

2. COMPLEX KLEINIAN GROUPS

The classical Kleinian groups correspond to the case $n = 2$, i.e., when we consider groups of Möbius transformations acting holomorphically on the Riemann sphere. This motivates our definition below of *complex Kleinian groups.* For this we need to define first the appropriate concept of the limit set for an action of a discrete group, that we take from Kulkarni's paper [Kul]. This definition is non-standard and it is motivated by the following example: consider the dynamical system in $\mathbb{R}^2$ generated by the diffeomorphism $f(x,y) \mapsto (2x, \frac{1}{2}y)$. If we defined the limit set to be the set of accumulation points of the orbits, this would consist of 0 alone, while the action is clearly non-discontinuous on the axes. With the definition (2.2) below, the limit set consists of both axes and its complement is the region of discontinuity, as it ought to be.

2.1 Definition. *Let X be a locally compact Hausdorff space with a countable base for its topology. Let G be a group acting on X and let $\Omega \subset X$ be a G-invariant subset. The action on Ω is properly discontinuous if for every pair of compact subsets C and D of Ω, the cardinality of the set $\{\gamma \in G \,|\, \gamma(C) \cap D \neq \emptyset\}$, is finite.*

Let $\{A_\beta\}$ be a family of subsets of X where β runs over some infinite indexing set B. A point $x \in X$ is a *cluster point* of $\{A_\beta\}$ if every neighbourhood of x intersects A_β for infinitely many $\beta \in B$. Let $L_0(G)$ be the closure of the set of points in X with infinite isotropy group. Let $L_1(G)$ be the closure of the set of cluster points of the family $\{\gamma(x)\}_{\gamma \in G}$, where x runs over $X - L_0(G)$. Finally, let $L_2(G)$ be the closure of the set of cluster points of $\{\gamma(K)\}_{\gamma \in G}$, where K runs over all the compact subsets of $X - \{L_0(G) \cup L_1(G)\}$. We have:

2.2 Definitions-propositions. *(See* [Ku1]*.)*

i) Let X be as above and let G be a group of homeomorphisms of X. The **limit set** *of G in X is the set $\Lambda(G) := L_0(G) \cup L_1(G) \cup L_2(G)$. This set is closed in X and it is G-invariant (it can be empty).*

ii) The **region of discontinuity** *of G is $\Omega(G) \subset X := X - \Lambda(G)$. This set (which can be empty) is open, G-invariant, and G acts properly discontinuously on $\Omega(G)$.*

2.3 Definition. *A* **complex Kleinian group** *is a discrete subgroup Γ of $PSL(n+1,\mathbb{C})$, the group of automorphisms of the projective n-space $P^n_{\mathbb{C}}$, whose limit set is not all of $P^n_{\mathbb{C}}$.*

The classical definitions of the *limit set* and the *discontinuity set* coincide with the above definitions when G is a Möbius (or conformal) group. Furthermore, for conformal Kleinian groups, the limit set can be defined as the set of accumulation points of one (any) orbit, because the action is minimal, i.e., every orbit is dense in the limit set. This does not happen for complex Kleinian groups in general, as we show below (see [SV]). This motivatates the following questions, for which we only have partial answers:

2.4 Question. *When is the limit set of a complex Kleinian group (as defined above) the set of accumulation points of the orbits?; when is the action of a complex Kleinian group minimal on its limit set?.*

2.5 Examples of Complex Kleinian groups:

i) The Suspension or Cone construction. This allows us to construct a complex Kleinian group $\tilde{\Gamma} \subset PSL(n+1,\mathbb{C})$ out of a complex Kleinian group

$\Gamma \subset PSL(n,\mathbb{C})$ which acts on $P^{n-1}_{\mathbb{C}}$. We note that $P^{n-1}_{\mathbb{C}}$ is the space of lines in $\mathbb{C}^n$ and $P^n_{\mathbb{C}}$ can be thought of as being $\mathbb{C}^n$ union the hyperplane at infinity. If we extend the action of Γ on $P^{n-1}_{\mathbb{C}}$ to a linear (unimodular) action on $\mathbb{C}^n$, *i.e.*, to an action of a subgroup $\widehat{\Gamma} \subset SL(n,\mathbb{C})$ on $\mathbb{C}^n$, then we have an action of $\widehat{\Gamma}$ on $P^n_{\mathbb{C}}$. The limit set of $\widehat{\Gamma}$ is the complex cone, with vertex at 0, over the limit set of Γ in the $P^{n-1}_{\mathbb{C}}$ at infinity. One has an exact sequence:

$$0 \to \mathbb{Z}_n \to SL(n,\mathbb{C}) \xrightarrow{\mathcal{P}} PSL(n,\mathbb{C}) \to 1,$$

where $\mathcal{P}$ denotes projectivization of linear maps. One can always lift Γ to $SL(n,\mathbb{C})$ by taking its pull back under $\mathcal{P}$, so that Γ can always be extended to a Kleinian group $\widehat{\Gamma}$ on $P^n_{\mathbb{C}}$, whose restriction to the $P^{n-1}_{\mathbb{C}}$ at infinity is Γ. If we can actually lift Γ to a subgroup $\widetilde{\Gamma} \subset SL(n,\mathbb{C})$ that intersects the kernel of $\mathcal{P}$ only at the identity, then Γ itself can be considered as a Kleinian group in $P^n_{\mathbb{C}}$. The obstruction to lift $\Gamma \subset PSL(n,\mathbb{C})$ to an isomorphic group in $SL(n,\mathbb{C})$ is an element in $H^2(\Gamma,\mathbb{Z}_n)$, determined via the Schur multipliers [Ki]. If this obstruction vanishes, then Γ can be regarded as a Kleinian group on $P^n_{\mathbb{C}}$; This happens, for instance, if $H^2(\Gamma,\mathbb{Z}_n) \cong 0$.

If Γ is the fundamental group of a complete (non-necessarily compact), hyperbolic 3-manifold, so that $\Gamma \subset PSL(2,\mathbb{C})$, then the obstruction in question can be identified with the second Stiefel-Whitney class ω_2 of the 3-manifold, as pointed out by Thurston, see [Kr]. This class is always 0, because every oriented 3-manifold is parallelizable. Hence Γ can always be lifted isomorphically to $SL(2,\mathbb{C})$. Thus the fundamental group Γ of a hyperbolic 3-manifold acting on $\mathbb{H}^3$, whose action on the sphere at infinity is Kleinian, can be considered as a complex Kleinian group $\widetilde{\Gamma}$ acting on $P^2_{\mathbb{C}}$ and leaving the line at infinity invariant. The limit set of $\widetilde{\Gamma}$ is the cone, with vertex at 0, over the limit set of Γ on $P^1_{\mathbb{C}}$. By Ahlfors' Finiteness Theorem [A3], the quotient $\Omega(\Gamma)/\Gamma$ is a Riemann surface of finite type. Hence the quotient $\Omega(\widetilde{\Gamma})/\widetilde{\Gamma}$, is a complex line bundle over a Riemann surface of finite type; $\Omega(\widetilde{\Gamma})/\widetilde{\Gamma}$ is homotopically equivalent to $\mathbb{H}^3/\Gamma$.

It is worth noting that one has a canonical embedding $SL(n,\mathbb{C}) \to SL(n+k,\mathbb{C})$, for all $n > 1, k > 0$, given by $A \mapsto \begin{pmatrix} A & 0 \\ 0 & I_{k\times k}\end{pmatrix}$. Therefore, if $\Gamma \subset PSL(n,\mathbb{C})$ is a Kleinian group on $P^{n-1}_{\mathbb{C}}$ that can be suspended to a linear group in $SL(n,\mathbb{C})$, then Γ can be automatically suspended to a linear group in $SL(n+k,\mathbb{C})$; thus, via the cone construction, Γ can be regarded as a Kleinian group on $P^{n+k}_{\mathbb{C}}$. If Ω is the region of discontinuity of Γ in $P^{n-1}_{\mathbb{C}}$, then the region of discontinuity Ω^k of Γ in $P^{n+k}_{\mathbb{C}}$ is a $(k+1)$-dimensional complex bundle over $P^{n-1}_{\mathbb{C}}$. Hence, in particular, *the fundamental group of every complete, connected, open, 3-dimensional hyperbolic manifold can be regarded as a Kleinian group on* $P^n_{\mathbb{C}}$, *for all* $n > 1$. We note, however, that in all these cases the action on $P^n_{\mathbb{C}}$ is defined from an action on the previous $P^{n-1}_{\mathbb{C}}$, and it leaves $P^{n-1}_{\mathbb{C}}$ invariant.

ii) The Join. A related construction is the join of two Kleinian actions: let $\Gamma_1 \subset GL(n+1,\mathbb{C})$ be a discrete group acting on $\mathbb{C}^{n+1}$ which induces a Kleinian action on $P^n_{\mathbb{C}}$; let $\Gamma_2 \subset GL(m+1,\mathbb{C})$ be a discrete group acting on $\mathbb{C}^{m+1}$, which induces a Kleinian action on $P^m_{\mathbb{C}}$. Then $\Gamma_1 \times \Gamma_2$ acts on $\mathbb{C}^{n+m+2}$ and induces a Kleinian action on $P^{n+m+1}_{\mathbb{C}}$. The limit set and discontinuity set of this action are the corresponding projective joins of the limit sets and discontinuity sets in $P^n_{\mathbb{C}}$ and $P^m_{\mathbb{C}}$.

It is clear that in this case $P^n_{\mathbb{C}}$ and $P^m_{\mathbb{C}}$ are both invariant subsets for the action on $P^{n+m+1}_{\mathbb{C}}$. This motivates the following definitions:

2.6 Definition. *Let Γ be a complex Kleinian group on $P^n_{\mathbb{C}}$, $n > 1$.*

i) The action of Γ on $P^n_{\mathbb{C}}$ is **reducible** *if it is obtained either by suspending a complex Kleinian action on $P^{n-1}_{\mathbb{C}}$, or as the join of two Kleinian actions on $P^{n_1}_{\mathbb{C}}$ and $P^{n_2}_{\mathbb{C}}$, $n = n_1+n_2+1$. Otherwise we say that the action of Γ is* **irreducible**.

ii) The action of Γ on $P^n_{\mathbb{C}}$ is **(complex) algebraically-mixing** *if there are no proper complex compact submanifolds of $P^n_{\mathbb{C}}$ which are Γ-invariant.*

It is clear that algebraically-mixing implies irreducible.

iii) Subgroups of the affine group. It is clear that every affine transformation in $\mathbb{C}^n$ can be extended to infinity, so it defines a transformation of $P^n_{\mathbb{C}}$. Hence the affine group $Aff(\mathbb{C}^n)$ is a subgroup of $PSL(n+1,\mathbb{C})$, and any discrete subgroup of $Aff(\mathbb{C}^n)$ with non-empty region of discontinuity is a complex Kleinian group in the sense of this article. Moreover, by [KO], it is likely that for $n = 2$, the only complex Kleinian groups be either suspensions of groups in dimension 1, subgroups of $Aff(\mathbb{C}^2)$ or subgroups of the group $PU(1,2)$ that we explain below.

iv) Subgroups of $PU(1,n)$. This group is the projectivization of the group $U(1,n) \subset GL(n+1,\mathbb{C})$ consisting of the linear transformations that preserve the quadratic form $-||z_0||^2 + \cdots + ||z_n||^2$. This group preserves the open ball in $P^n_{\mathbb{C}}$ whose homogeneous coordinates satisfy $-||z_0||^2+\cdots+||z_n||^2 < 0$; this is in fact the largest group of holomorphic transformations of the ball. The discrete subgroups of this group have been studied by many authors as a mean to generalize the classical theory of automorphic forms and functions, and also as a mean to construct complex manifolds. So these groups can be regarded as complex Kleinian groups which are analogous to the Fuchsian groups [Ma1]. Deligne and Mostow [DM] studied discrete subgroups of $U(1,n)$ which are uniform, non-arithmetic and they have a non-empty region of discontinuity on the ball in $P^n_{\mathbb{C}}$.

v) The Schottky groups (see section 4 below; c. f. [No]**).** Consider an arbitrary configuration:

$$\mathcal{L} := \{(L_1, M_1), ..., (L_r, M_r)\},$$

of pairs of projective n-spaces in $P^{2n+1}_{\mathbb{C}}$, which are all of them pairwise disjoint. Given arbitrary neighbourhoods $U_1, ..., U_r$ of the L_i's, pairwise disjoint, we show that there exist, for each $i = 1, ..., r$, projective transformations T_i of $P^{2n+1}_{\mathbb{C}}$, which interchange the interior with the exterior of a compact tubular neighbourhood N_i of L_i contained in U_i, leaving invariant the boundary $E_i = \partial(N_i)$. The E_i's are *mirrors*, they play the same role in $P_{\mathbb{C}}^{2n+1}$ than $(n-1)$-spheres play in S^n to define the classical Schottky groups. Each mirror E_i is a $(2n+1)$-sphere bundle over $P_{\mathbb{C}}^n$. Choosing these mirrors appropriately, one has that the group of automorphisms of $P_{\mathbb{C}}^{2n+1}$ generated by the T_i's is a complex Kleinian group Γ. The region of discontinuity $\Omega(\Gamma)$ is a fibre bundle over $P_{\mathbb{C}}^n$ with fibre S^{2n+2} minus a Cantor set $\mathcal{C}$. The limit set Λ is the complement of $\Omega(\Gamma)$ in $P_{\mathbb{C}}^{2n+1}$; it is the set of accumulation points of the Γ-orbit of the L_i's, and it is a product $\mathcal{C} \times P_{\mathbb{C}}^n$. The action of Γ on this set of projective lines is minimal in the sense that the Γ-orbit of every point x_o in $P_{\mathbb{C}}^{2n+1}$ accumulates to (at least a point in) each one of the projective lines in Λ. This set is transversally *projectively self-similar*, i.e., Λ corresponds to a Cantor set in the Grassmannian $G_{2n+1,n}$, which is dynamically-defined. Hence Λ is a *solenoid* (or *lamination*) by projective spaces, which is transversally Cantor and projectively self-similar. Furthermore, the index two subgroup $\check{\Gamma} \subset \Gamma$ consisting of words of even lenght, is a free group of rank $r-1$ and acts *freely* on $\Omega(\Gamma)$. The quotient $\Omega(\Gamma)/\check{\Gamma}$ is a compact complex manifold, which is a fibre bundle over $P_{\mathbb{C}}^n$ with fibre the connected sum of $(r-1)$ copies of $S^{2n+1} \times S^1$. The diffeomorphism type of these manifolds is actually independent of the choice of the configuration $\mathcal{L}$, depending only on the (complex) dimension $2n+2$ and the number of mirrors used to define them. However, as complex manifolds they do have deformations, and we describe these in the last section of this article.

vi) Conformal vs. Complex Kleinian groups (see section 3 below).

If N^{2n} is an oriented Riemannian manifold, its twistor space is the total space of the bundle $p : \mathfrak{Z}(N) \to N$, whose fibre at each point $x \in N$ is the symmetric space $SO(2n)/U(n)$, of all complex structures on T_xN which are compatible with the metric and the orientation on N. $\mathfrak{Z}(N)$ is canonically an almost Hermitian manifold, whose complex structure is integrable whenever N is (locally) conformally flat, by [AHS, DV, OR]. Furthermore, $\mathfrak{Z}(N)$ is actually a projective manifold if N is the sphere S^{2n} with its usual metric. Moreover, $\mathfrak{Z}(S^4)$ is in fact $P_{\mathbb{C}}^3$.

The "Penrose twistor programme", in mathematical physics, shows that there is a remarkable interplay between the conformal geometry of (even dimensional) Riemannian manifolds and the complex geometry of their twistor spaces. In [SV] we showed that this idea can also be "pushed forward" to dynamics, hence proving that every conformal Kleinian group can be regarded, canonically, as a complex Kleinian group. In other words, the conformal dynamics of Kleinian groups on spheres embedds in the holomorphic dynamics of complex Kleinian groups on projective spaces. More precisely, we show that if Γ is a discrete group of orientation preserving conformal diffeomorphisms of N, then Γ has a canonical lifting to a group $\widetilde{\Gamma}$ of diffeomorphisms of $\mathfrak{Z}(N)$ that preserve the almost complex structure and carry fibres isometrically into fibres. Moreover, if the almost complex structure on $\mathfrak{Z}(N)$ is integrable, then the elements of $\widetilde{\Gamma}$ are indeed holomorphic transformations. The limit set of $\widetilde{\Gamma}$ is $p^{-1}(\Lambda(\Gamma))$, the inverse image of the limit set of Γ. We show that an interesting feature occurs, which does not happen for conformal Kleinian groups: the action of $\widetilde{\Gamma}$ on its limit set $\Lambda(\widetilde{\Gamma})$ may or may not be minimal and ergodic, with respect to the densities $\widetilde{\mu}_y$ defined from the Patterson-

Sullivan densities for Γ. More precisely, if Γ is any geometrically-finite, discrete subgroup of $Iso_+(\mathbb{H}^{2n+1}) \cong Conf_+(S^{2n}) \cong SO_0(2n+1,1)$ which is Zariski-dense, then the action of $\widetilde{\Gamma}$ is minimal and ergodic on its limit set $\widetilde{\Lambda}$; the same statements hold for subgroups of $Iso_+(\mathbb{H}^{2n})$, regarded as subgroups of $Iso_+(\mathbb{H}^{2n+1})$. However, if Γ is the fundamental group of a hyperbolic orbifold of dimension $m < 2n$, regarded as a subgroup of $Conf(S^{2n})$ via the inclusion $Iso(\mathbb{H}^m) \hookrightarrow Iso(\mathbb{H}^{2n+1})$, then the action of $\widetilde{\Gamma}$ on its limit set is neither minimal nor ergodic. The manifold $\mathfrak{Z}(S^{2n}) = SO(2n+1)/U(n)$ has a canonical projective embedding in $P_{\mathbb{C}}^{2^n-1}$, given via the spin representation [In]. This is the projective embedding of $\mathfrak{Z}(S^{2n})$ of smallest codimension, and the group of holomorphic automorphisms of $\mathfrak{Z}(S^{2n})$ extends uniquely to a subgroup of $PSL(2^n, \mathbb{C})$, by [Bl]. Therefore, every conformal Kleinian group on S^{2n} is canonically a complex Kleinian group on $P_{\mathbb{C}}^{2^n-1}$, for all $n \geq 1$. The case $n = 2$ is particularly interesting and we describe it in section 3 below.

vii) Addition of Kleinian groups. In a series of articles [Ka1-4], Kato has studied a class of 3-dimensional complex manifolds that he calls "L-manifolds". This means that the manifold in question contains a copy of $P_{\mathbb{C}}^1$ embedded in such a way that it has a neighbourhood which is biholomorphic to a neighbourhood of the canonical $P_{\mathbb{C}}^1$ in $P_{\mathbb{C}}^3$. One can use these neighbourhoods to glue together L-manifolds in the obviuos way; the result is a new complex L-manifold. This is the operation that Kato calls "the connected sum of L-manifolds". Then he shows that if M_1 and M_2 are two L-manifolds which are of the form U_i/Γ_i, $i = 1, 2$, where each U_i is an open subset of $P_{\mathbb{C}}^3$ and Γ_i is a discrete subgroup of $PSL(4, \mathbb{C})$ that acts freely on U_i, then the connected sum of M_1 and M_2, which is again an L-manifold, is the quotient of an open set $U \subset P_{\mathbb{C}}^3$ by a subgroup of $PSL(4, \mathbb{C})$. In other words, these complex Kleinian gro! ups have an addition property similar to that of Maskit's combination theorem.

The connected sum defined by Kato can actually be carried on to higher dimensional "L-manifolds", i.e. to complex manifolds of dimension $2n+1$ which contain a copy of $P_{\mathbb{C}}^n$, embedded so that it has a neighbourhood biholomorphic to the canonical $P_{\mathbb{C}}^n$ in $P_{\mathbb{C}}^{2n+1}$. Kato's addition of complex Kleinian groups can also be carried over to higher dimensions, thus providing many new complex Kleinian groups. In particular, the Schottky groups mentioned above can be "added" in this way.

It is worth noting that if a discrete group Γ acts freely on an open subset $\Omega \subset P_{\mathbb{C}}^N$, then the quotient Ω/Γ is a complex manifold with a projective structure in the sense of Gunning [Gu] and others.

3. COMPLEX KLEINIAN GROUPS ON $P_{\mathbb{C}}^3$: TWISTOR THEORY

If M is a closed, oriented 4-manifold endowed with a conformal metric, its twistor space $\mathfrak{Z}(M)$ (see, for instance, [At, AHS, BR, C1, C2, Pe1,2]), is the total space of the fibre bundle over M whose fibre, at each $x \in M$, is the set of all complex structures on T_xM which are compatible with the metric and orientation on M. So the fibre is $SO(4)/U(2)$, and it is not hard to see that this manifold is in fact the complex projective line $P^1_{\mathbb{C}}$. Thus one has the 2-sphere bundle,

$$p: \mathfrak{Z}(M) \to M\,,$$

called the Calabi-Penrose (or twistor) fibration, whose fibres are called the *twistor lines.* When $M = S^4$ with its canonical metric, the twistor space $\mathfrak{Z} := \mathfrak{Z}(S^4)$ turns out to be the complex projective space $P^3_{\mathbb{C}}$. Let us give an alternative (well known) construction of the twistor fibration $\mathfrak{Z} \to S^4$. We know that S^4 is the space of right quaternionic lines in $\mathcal{H}^2$. Multiplication on the right by i determines a complex structure on $\mathcal{H}^2 \cong \mathbb{R}^8$. In this way, each *right* quaternionic line L_q in $\mathcal{H}^2$ becomes a 2-dimensional complex space in $\mathbb{C}^4 \cong \mathcal{H}^2$. Moreover, given any $\alpha \in \mathcal{H}$, multiplication by α by the right preserves $L_q := \{q\lambda : \lambda \in \mathcal{H}\,,\, q \in \mathcal{H}^2 - \{(0,0)\}\,\}$, so each line L_q is covered by the complex lines $l_{q\alpha} := \{q\alpha\lambda : \lambda \in \mathbb{C}\,,\} \subset L_q$. If we identify each complex line $l_{q\alpha}$ to a point we obtain $P^3_{\mathbb{C}} = \mathfrak{Z}$, and if we identify each quaternionic line L_q to a point we obtain $P^1_{\mathcal{H}} = S^4$. This gives the 2-sphere bundle, $p: P^3_{\mathbb{C}} \to S^4$, where each fibre is a projective line $P^1_{\mathbb{C}}$.

It is well known that every $h \in \mathit{Conf}_+(S^4)$ has a canonical lifting to a holomorphic map $\widetilde{h}: \mathfrak{Z} = P^3_{\mathbb{C}} \to P^3_{\mathbb{C}}$. This is given through the identification in section 1:

$$\mathit{Conf}_+(S^4) \cong \mathit{M\ddot{o}b}(2,\mathcal{H}) \cong PSL(2,\mathcal{H}) \subset PSL(4,\mathbb{C})\,.$$

This also says that $\mathit{Conf}_+(S^4)$ actually lifts to $PSL(4,\mathbb{C})$ as a group, and it carries twistor lines into twistor lines.

We recall that $P^1_{\mathbb{C}}$ has the Fubini-Study metric [We], which coincides with the standard metric on S^2. This metric is essentially the angle between the complex lines in $\mathbb{C}^2$. For each line L_q, with the above complex structure, we consider the standard Hermitian metric. Then a transformation $A \in PSL(2,\mathcal{H})$ sends the right line L_q isometrically into the right line $L_{q'}$, because the vectors $\{q, qi, qj, qk\}$ form a real orthonormal basis in L_q, and their image in $L_{q'}$ is the basis $\{q', q'i, q'j, q'k\}$; therefore A preserves the angle between complex lines contained in the *same* right quaternionic line, so it preserves the Fubini-Study metric on the fibres of the twistor fibration. Since every biholomorphism of $P^3_{\mathbb{C}}$ is a projective linear transformation, we have that every $h \in \mathit{M\ddot{o}b}(2,\mathcal{H})$ lifts canonically to an automorphism $\widetilde{h}$ of $P^3_{\mathbb{C}}$. This lifting preserves the Calabi-Penrose fibration and it is an isometry on each fibre $P^1_{\mathbb{C}}$, with respect to the Fubini-Study metric. Furthermore, the map $\phi: h \mapsto \widetilde{h}$ injects $\mathit{M\ddot{o}b}(2,\mathcal{H})$ into the complex projective group $PSL(4,\mathbb{C})$. One has the following theorem.

3.1 Theorem. *Every conformal (discrete, Kleinian) group* Γ *in* S^4 *with limit set* $\Lambda(\Gamma)$, *is canonically a complex (discrete, Kleinian) group* $\widetilde{\Gamma} = \phi(\Gamma)$ *in* $P^3_{\mathbb{C}}$. *The limit set* $\widetilde{\Lambda} := \Lambda(\widetilde{\Gamma})$ *is* $p^{-1}(\Lambda)$, *where* p *is the Calabi-Penrose fibration* $p\colon P^3_{\mathbb{C}} \to S^4$. *If* Γ *is Kleinian, then* $\widetilde{\Lambda}$ *is in fact* $\Lambda \times P^1_{\mathbb{C}}$.

The proof of this theorem follows from the fact that the canonical lifting of $\mathit{M\ddot{o}b}(2,\mathcal{H})$ to $PSL(4,\mathbb{C})$ acts by isometries on the fibers of the twistor fibration (see [SV] for details).

Now consider a geometrically finite, discrete subgroup Γ of $Conf_+(S^4)$; Γ is said to be *Fuchsian* if it leaves invariant a 3-dimensional round sphere S^3_δ in S^4 where by *round sphere* we mean a sphere at infinity which is the boundary of a complete totally geodesic subspace of hyperbolic space $\mathbb{H}^n$. Every such group is automatically a Kleinian group in S^4, because its limit set is contained in S^3_δ. The fundamental group of every complete hyperbolic n-manifold with $n < 5$ is a Fuchsian group in S^4, hence also Kleinian, because the canonical inclusion $Iso(\mathbb{H}^n) \hookrightarrow Iso(\mathbb{H}^5)$ leaves invariant a hyperplane in $\mathbb{H}^5$.

As mentioned before, for conformal Kleinian groups one has that the action on the limit set is always minimal and ergodic with respect to the geometric measure. For Complex Kleinian groups this is not always the case. There is a dichotomy. For this we note that one has the geometric measure μ on its limit set $\Lambda \subset S^4$. Since, by (3.1) above, the limit set $\widetilde{\Lambda}$ of the canonical lifting $\widetilde{\Gamma}$ is a fibre bundle over Λ with fiber $P^1_{\mathbb{C}}$, one has a natural measure $\widetilde{\mu}$ on $\widetilde{\Lambda}$, which is locally the product of μ by the Lebesgue measure on $P^1_{\mathbb{C}}$. On has:

3.2 Theorem. *Let Γ be a conformal (discrete, Kleinian) group Γ in S^4 with limit set $\Lambda(\Gamma)$. Let $\widetilde{\Gamma}$ be its canonical lifting to $P^3_{\mathbb{C}}$. Then:*

i) If Γ is the fundamental group of a hyperbolic n-orbifold with $n < 4$, via the inclusion $Iso(\mathbb{H}^n) \hookrightarrow Iso(\mathbb{H}^5)$, then the action of $\widetilde{\Gamma}$ on $P^3_{\mathbb{C}}$ leaves invariant a closed subset of $P^3_{\mathbb{C}}$ contained properly in $\widetilde{\Lambda}(\Gamma)$. Thus its action on $\widetilde{\Lambda}(\Gamma)$ is not minimal nor ergodic with respect to the measure $\widetilde{\mu}$.

ii) If Γ is Zariski dense in either $Iso(\mathbb{H}^4)$ or $Iso(\mathbb{H}^5)$, then then the action of $\widetilde{\Gamma}$ is minimal on its limit set $\widetilde{\Lambda} \cong S^{n-1} \times P^1_{\mathbb{C}}$, and it is ergodic with respect to the measure $\widetilde{\mu}$. Hence the action of $\widetilde{\Gamma}$ on $P^3_{\mathbb{C}}$ is algebraically-mixing, i.e., there is no proper complex submanifold (nor sub-variety) of $P^3_{\mathbb{C}}$ which is $\widetilde{\Gamma}$-invariant.

Statement (i) is an easy application of twistor theory, see also [C1, C2]: if Γ is a conformal Kleinian group in S^4 that leaves invariant a maximal round sphere $S^2 \subset S^4$, then S^2 lifts to a holomorphic Legendrian rational curve in $P^3_{\mathbb{C}}$, which is $\widetilde{\Gamma}$-invariant and it is transversal to all the twistor lines that this curve meets. Since the action on the fibers is by isometries, this implies that it may not possibly be minimal nor ergodic. To prove the statement ii) in Theorem 3.2 we use theorems (1.4) and (1.5) above. In fact, to prove the minimality of the action, we first observe that Zorn's lemma implies that there exists a subset $H \subset \Lambda(\widetilde{\Gamma}) \subset \widetilde{\Lambda}$ where the action of $\widetilde{\Gamma}$ is minimal. By (1.5), either $H = \Lambda(\widetilde{\Gamma})$, or else this set is a fibre bundle over Λ and it admits a section. This section is therefore a Γ-invariant continuous family of almost complex structures $\{J_x\}_{x\in\Lambda}$. If $m = 5$, let E_x be the subspace of dimension 2 of T_xS^4 which is the eigenspace corresponding to the eigenvalue i of J_x. Then, the family $\{E_x\}_{x\in S^4}$ is a 2-dimensional Γ-invariant distribution. This contradicts theorem (1.4). If $m = 4$, for each $x \in \Lambda \subset S^3 \subset S^4$, let E_x be the 2-plane $E_x = T_x(S^3) \cap J_x(T_x(S^3))$. Then, the family $\{E_x\}_{x\in S^3}$ is a 2-dimensional Γ-invariant distribution. This contradicts theorem (1.4). □

3.2.1 Remark. The construction in this section is a special case of a more general construction: let $P^n_{\mathcal{H}}$ denote the quaternionic projective n-space consisting of *right* quaternionic lines in $\mathcal{H}^{n+1}$. Then one has the classical fibration $p : P_{\mathbb{C}}^{2n+1} \to P_{\mathcal{H}}^{n}$ with fibre $P_{\mathbb{C}}^{1}$. If $SL(n+1, \mathcal{H})$ denotes the group of $n \times n$ matrices with coefficients in $\mathcal{H}$ acting on the *left* on $\mathcal{H}^{n+1}$, then this action can be projectivized to obtain an action on $P_{\mathcal{H}}^{n}$. Then, $PSL(n+1, \mathcal{H}) = SL(n+1, \mathcal{H})/\{\pm I\}$ is the group of *quaternionic* projective transformations. It sends fibres of p into fibres of p.

Theorem 3.2 generalizes to Kleinian groups acting on even dimensional spheres of higher dimensions, see [SV]. One gets a canonical lifting $\widetilde{\Gamma}$ of $\Gamma \subset Conf_{+}(S^{2n})$ to the twistor space $\mathfrak{Z}$ of S^{2n}, and theorems 3.1 and 3.2 generalize in a straightforward way to this situation. The space $\mathfrak{Z}$ has a canonical embedding in the projective space $P_{\mathbb{C}}^{2^n-1}$, via the spin representation, and $\widetilde{\Gamma}$ has a unique extension to a complex Kleinian subgroup of $PSL(2^n, \mathbb{C})$. Therefore **the dynamics of conformal Kleinian groups embedds in the dynamics of complex Kleinian groups.**

4. COMPLEX SCHOTTKY GROUPS

Now we discuss briefly the generalization, mentioned in section 2 above, of the Schottky groups to higher dimensional complex projective spaces. We recall that in the conformal case, the Schottky groups are obtained by considering pairwise disjoint $(n-1)$-spheres $\mathbb{S}_1, ..., \mathbb{S}_r$ in S^n. Each sphere $\mathbb{S}_i$ plays the role of a mirror: it divides S^n in two diffeomorphic components, and one has an involution T_i of S^n interchanging these components, the inversion on $\mathbb{S}_i$. The Schottky group is defined to be the group of conformal transformations generated by these involutions. We are going to make a similar construction on $P_{\mathbb{C}}^{2n+1}$, $n > 0$.

Consider the subspaces of $\mathbb{C}^{2n+2} = \mathbb{C}^{n+1} \times \mathbb{C}^{n+1}$ defined by $\widehat{L}_0 := \{(a, 0) \in \mathbb{C}^{2n+2}\}$ and $\widehat{M}_0 := \{(0, b) \in \mathbb{C}^{2n+2}\}$. Let $\widehat{S}$ be the involution of $\mathbb{C}^{2n+2}$ defined by $\widehat{S}(a, b) = (b, a)$. This interchanges $\widehat{L}_0$ and $\widehat{M}_0$.

4.1 Lemma. *Let* $\Phi: \mathbb{C}^{2n+2} \to \mathbb{R}$ *be given by* $\Phi(a, b) = |a|^2 - |b|^2$. *Then:*

i) $\widehat{E}_{\widehat{S}} := \Phi^{-1}(0)$ *is a real algebraic hypersurface in* $\mathbb{C}^{2n+2}$ *with an isolated singularity at the origin* 0. *It is embedded in* $\mathbb{C}^{2n+2}$ *as a (real) cone over* $S^{2n+1} \times S^{2n+1}$, *with vertex at* $0 \in \mathbb{C}^{2n+2}$.

ii) $\widehat{E}_{\widehat{S}}$ *is invariant under multiplication by* $\lambda \in \mathbb{C}$, *so it is in fact a complex cone.* $\widehat{E}_{\widehat{S}}$ *separates* $\mathbb{C}^{2n+2} - \{(0, 0)\}$ *in two diffeomorphic connected components* U *and* V, *which contain respectively* $\widehat{L_0} - \{(0, 0)\}$ *and* $\widehat{M_0} - \{(0, 0)\}$. *These two components are interchanged by the involution* $\widehat{S}$, *for which* $\widehat{E}_{\widehat{S}}$ *is an invariant set.*

iii) Every linear subspace $\widehat{K}$ *of* $\mathbb{C}^{2n+2}$ *of dimension* $n+2$ *containing* $\widehat{L_0}$ *meets transversally* $\widehat{E}_{\widehat{S}}$ *and* $\widehat{M_0}$. *Therefore a tubular neighbourhood* V *of* $\widehat{M_0} - \{(0, 0)\}$ *in* $P_{\mathbb{C}}^{2n+1}$ *is obtained, whose normal disc fibres are of the form* $\widehat{K} \cap V$, *with* $\widehat{K}$ *as above.*

Proof. The first statement is clear because Φ is a quadratic form with $0 \in \mathbb{C}^{2n+2}$ as unique critical point. Clearly $\widehat{E}_{\widehat{S}}$ is invariant under multiplication by complex numbers, so it is a complex cone. That $\widehat{E}_{\widehat{S}} \cap S^{4n+3} = S^{2n+1} \times S^{2n+1} \subset \mathbb{C}^{2n+2}$, is because this intersection consists of all pairs (x,y) so that $|x| = |y| = \frac{1}{\sqrt{2}}$. That $\widehat{S}$ leaves $\widehat{E}_{\widehat{S}}$ invariant is obvious, and so is that $\widehat{S}$ interchanges the two components of $\mathbb{C}^{2n+2} - \{(0,0)\}$ determined by $\widehat{E}_{\widehat{S}}$, which must be diffeomorphic because $\widehat{S}$ is an automorphism. Finally, if $\widehat{K}$ is a subspace as in the statement (iii), then $\widehat{K}$ meets transversally $\widehat{E}_{\widehat{S}}$, because through every point in $\widehat{E}_{\widehat{S}}$ there exists an affine line in $\widehat{K}$ which is transverse to $\widehat{E}_{\widehat{S}}$. □

Let S be the linear projective involution of $P^{2n+1}_{\mathbb{C}}$ defined by $\widehat{S}$. Since $\widehat{E}_{\widehat{S}}$ is a complex cone, it projects to a codimension 1 real submanifold of $P^{2n+1}_{\mathbb{C}}$, that we denote by E_S .

4.2 Definition. *We call* E_S **the canonical mirror** *and* S **the canonical involution**. *More generally,* **mirrors** *in* $P^{2n+1}_{\mathbb{C}}$ *are defined to be the images of* E_S *under the action of* $PSL(2n+2,\mathbb{C})$.

A mirror is the boundary of a tubular neighbourhood of a $P^n_{\mathbb{C}}$ in $P^{2n+1}_{\mathbb{C}}$, so it is an S^{2n+1}-bundle over $P^n_{\mathbb{C}}$.

The above lemma 4.1 can be easily extended to the following proposition (see [SV] for details):

4.3 Proposition. *Let* $L \cong M \cong P^n_{\mathbb{C}}$ *be disjoint projective subspaces of* $P^{2n+1}_{\mathbb{C}}$. *Then:*

i) There exist involutions of $P^{2n+1}_{\mathbb{C}}$ *that interchange* L *and* M.

ii) Each of these involutions has a mirror, i.e., an invariant set $E = E_T \subset P^{2n+1}_{\mathbb{C}}$ *which separates* $P^{2n+1}_{\mathbb{C}}$ *in two connected components which are interchanged by* T. *Each component is diffeomorphic to a tubular neighbourhood of the canonical* $P^n_{\mathbb{C}} \subset P^{2n+1}_{\mathbb{C}}$.

iii) Given an arbitrary tubular neighbourhood U *of* L, *we can choose* T *so that the corresponding mirror* E_T *is contained in the interior of* U.

In fact one can obviously make stronger the last statement of (4.3):

4.4 Lemma. *Let* L *and* M *be as above. Given an arbitrary constant* λ, $0 < \lambda < 1$, *we can find an involution* T *interchanging* L *and* M, *with a mirror* E *such that if* U^* *is the open component of* $P^{2n+1}_{\mathbb{C}} - E$ *which contains* M *and* $x \in U^*$, *then* $d(T(x), L) < \lambda d(x, M)$, *where the distance* d *is induced by the Fubini-Study metric.*

We notice that the parameter λ in (4.4) gives control upon the degree of expansion and contraction of the generators of the groups, so one can estimate bounds on the Hausdorff dimension of the limit set (see section 5 below).

The previous discussion can be summarized in the following theorem (Compare [No]):

4.5 Theorem. *Let* $\mathcal{L} := \{(L_1, M_1), \dots, (L_r, M_r)\}$, $r > 1$, *be a set of* r *pairs of projective subspaces of dimension* n *of* $P^{2n+1}_{\mathbb{C}}$, *all of them pairwise disjoint. Then:*

i) There exist involutions $T_1, ..., T_r$ of $P^{2n+1}_{\mathbb{C}}$, such that each T_i, $i = 1, ..., r$, interchanges L_i and M_i, and the corresponding mirrors E_{T_i} are all pairwise disjoint.

ii) If we choose the $T_i's$ in this way, then the subgroup of $PSL(2n+2, \mathbb{C})$ that they generate is complex Kleinian.

iii) Moreover, given a constant $C > 0$, we can choose the $T_i's$ so that if $T := T_{j_1} \cdots T_{j_k}$ is a reduced word of length $k > 0$ (i.e., $j_1 \neq j_2 \neq \cdots \neq j_{k-1} \neq j_k$), then $T(N_i)$ is a tubular neighbourhood of the projective subspace $T(L_i)$ which becomes very thin as k increases: $d(x, T(L_i)) < C\lambda^k$ for all $x \in T(N_i)$, where N_i is the connected component of $P^{2n+1}_{\mathbb{C}} - E_{T_i}$ that contains L_i, for all $i = 1, ...r$.

A Kleinian group constructed as above will be called a **complex Schottky group**. The following result is proved in [SV]:

4.6 Theorem. *Let Γ be a complex Schottky group in $P^{2n+1}_{\mathbb{C}}$, generated by involutions $\{T_1, .., T_r\}$, $n \geq 1$, $r > 1$, as in (4.7) above. Let $\Omega(\Gamma)$ be the region of discontinuity of Γ and let $\Lambda(\Gamma) = P^{2n+1}_{\mathbb{C}} - \Omega(\Gamma)$ be the limit set. Then, one has:*

i) Let $W = P^{2n+1}_{\mathbb{C}} - \cup_{i=1}^{r} \overset{\circ}{N}_i$, where $\overset{\circ}{N}_i$ is the interior of the tubular neighbourhood N_i as in (4.7). Then W is a compact fundamental domain for the action of Γ on $\Omega(\Gamma)$. One has: $\Omega(\Gamma) = \bigcup_{\gamma \in \Gamma} \gamma(W)$.

ii) $\Lambda(\Gamma)$ is an intersection of nested sets: $\Lambda(\Gamma) = \cap_{i=1}^{\infty} \gamma_i(N_{j(i)})$, where $\{\gamma_i\}_{i=1}^{\infty}$ is a sequence of distinct elements of Γ and $j \colon \mathbb{N} \to \{1, ..., r\}$ is a function such that $\gamma_{i+1}(N_{j(i+1)})$ is a subset of $\gamma_i(N_{j(i)})$. Hence $\Lambda(\Gamma)$ is the set of accumulation points of the Γ-orbit of the union $L_1 \cup ... \cup L_r$.

iii) If $r = 2$, then $\Gamma \cong \mathbb{Z}/2\mathbb{Z} * \mathbb{Z}/2\mathbb{Z}$, the infinite dihedral group, and $\Lambda(\Gamma)$ is the union of two disjoint projective subspaces L and M of dimension n. In this case we say that Γ is **elementary**, *in analogy with Kleinian groups acting on $P^1_{\mathbb{C}}$.*

iv) If $r > 2$, then $\Lambda(\Gamma)$ is a complex solenoid (lamination), homeomorphic to $P^n_{\mathbb{C}} \times C$, where C is a Cantor set. Γ acts minimally on the set of projective subspaces in $\Lambda(\Gamma)$ considered as a closed subset of the Grassmannian $G_{2n+1,n}$.

v) If $r > 2$, let $\check{\Gamma} \subset \Gamma$ be the index 2 subgroup consisting of the elements which are reduced words of even length. Then $\check{\Gamma}$ is free of rank $r - 1$ and acts freely on $\Omega(\Gamma)$. The compact manifold with boundary $\check{W} = W \cup T_1(W)$ is a fundamental domain for the action of $\check{\Gamma}$ on $\Omega(\Gamma)$. We also call $\check{\Gamma}$ a **complex Schottky group**.

vi) Each element $\gamma \in \check{\Gamma}$ leaves invariant two copies, P_1 and P_2, of $P^n_{\mathbb{C}}$ in $\Lambda(\Gamma)$. For every $L \subset \Lambda(\Gamma)$, $\gamma^i(L)$ converges to P_1 (or to P_2) as $i \to \infty$ (or $i \to -\infty$).

In fact we prove that if $r > 2$, then Γ acts on a graph whose vertices have all valence either 2 or r. This graph is actually a tree, which can be compactified by adding its "ends". These form a Cantor set and the action of Γ can be extended to this compactification. The limit set $\Lambda(\Gamma)$ corresponds to the uncountable set of ends of this tree. We use this to prove statement $v)$ above.

4.7 Remarks. i) It is worth noting that one may let the mirrors overlap slightly and still obtain complex Kleinian groups, in the same way as in the case of subgroups of conformal transformations of $P^1_{\mathbb{C}}$ obtained by inversions on overlapping circles, c.f., [Bo]. Also, in theorem 4.6 we could take $r = \infty$, *i.e.*, an infinite countable set of disjoint mirrors.

ii) The action of $\check{\Gamma}$ in the Cantor set of projective subspaces is analogous to the action of a classical Fuchsian group of the second kind on its Cantor limit set. We also observe that, since each involution T_i is conjugate to the canonical involution defined in lemma 4.1, the laminations obtained in theorem 4.6 are transversally *projectively self-similar*. Hence one could try to apply results analogous to the results for (conformally) self-similar sets (for instance Bowen's formula [Bo]) to estimate the transverse Hausdorff dimension of the laminations obtained. Here by *transverse Hausdorff dimension* we mean the Hausdorff dimension of the Cantor set $\mathcal{C}$ of projective subspaces of $G_{2n+1,n}$ which conform the limit set. If $\widetilde{T}_i$, $i = 1, \dots, r$, denote the maps induced in the Grassmannian $G_{2n+1,n}$ by the linear projective transformations T_i, then $\mathcal{C}$ is dynamically-defined by the group generated by the set $\{\widetilde{T}_i\}$.

iii) The construction of Kleinian groups given in 4.6 actually provides families of Kleinian groups, obtained by varying the size of the mirrors that bound tubular neighbourhoods around the $L_i's$, c.f. section 5 below.

iv) We notice that the constructions and theorems above can be realized in an infinite dimensional complex projective space H and the associated projective space $P^\infty_{\mathbb{C}}$, see [SV].

5. QUOTIENT SPACES OF THE REGION OF DISCONTINUITY

We now discuss the nature of the quotients $\Omega(\Gamma)/\Gamma$ and $\Omega(\Gamma)/\check{\Gamma}$, for the complex Schottky groups of section 4. We have:

5.1 Theorem. *Let Γ be a complex Schottky group as in theorem 4.6, with $r > 2$. Then:*

i) The fundamental domain W of Γ is (the total space of) a locally trivial differentiable fibre bundle over $P^n_{\mathbb{C}}$ with fibre $S^{2n+2} - \overset{\circ}{D}_1 \cup \cdots \cup \overset{\circ}{D}_r$, where each $\overset{\circ}{D}_i$ is the interior of a smooth closed $(2n+2)$-disc D_i in S^{2n+2} and the D_i's are pairwise disjoint.

ii) $\Omega(\Gamma)$ fibres differentiably over $P^n_{\mathbb{C}}$ with fibre S^{2n+2} minus a Cantor set.

iii) If $\check{\Gamma}$ is the subgroup of index two as in theorem 4.6, which acts freely on $\Omega(\Gamma)$, then $\Omega(\Gamma)/\check{\Gamma}$ is a compact complex manifold that fibres differentiably over $P^n_{\mathbb{C}}$ with fibre $(S^{2n+1} \times S^1)\# \cdots \#(S^{2n+1} \times S^1)$, the connected sum of $r-1$ copies of $S^{2n+1} \times S^1$.

The key for proving (5.1) is an elementary result of projective geometry. We notice that if L is a copy of $P^n_{\mathbb{C}}$ in $P^{2n+1}_{\mathbb{C}}$ and if x is a point in $P^{2n+1}_{\mathbb{C}} - L$, then there is a unique copy K_x of $P^{n+1}_{\mathbb{C}}$ in $P^{2n+1}_{\mathbb{C}}$ that contains L and x. Then K_x intersects transversally every other copy of $P^n_{\mathbb{C}}$ embedded in $P^{2n+1}_{\mathbb{C}} - L$, and this intersection consists of a single point. Thus, given two disjoint copies L and M of

$P^n_{\mathbb{C}}$ in $P^{2n+1}_{\mathbb{C}}$, there is a canonical projection map

$$\pi := \pi_L : P^{2n+1}_{\mathbb{C}} - L \to M ,$$

which is a (holomorphic) submersion. Each fibre $\pi^{-1}(x)$ is diffeomorphic to $\mathbb{R}^{2n+2}$.

We also prove in [SV]:

5.2 Theorem. *Let M_Γ be the compact complex orbifold $M_\Gamma := \Omega(\Gamma)/\Gamma$, which has complex dimension $(2n+1)$. Then:*

i) The singular set of M_Γ, $Sing(M_\Gamma)$, is the disjoint union of r submanifolds analytically equivalent to $P^n_{\mathbb{C}}$, one contained in (the image in M_Γ of) each mirror E_i of Γ.

ii) Each component of $Sing(M_\Gamma)$ has a neighbourhood homeomorphic to the normal bundle of $P^n_{\mathbb{C}}$ in $P^{2n+1}_{\mathbb{C}}$ modulo the involution $v \mapsto -v$, for v a normal vector.

iii) M_Γ fibres over $P^n_{\mathbb{C}}$ with fibre a real analytic orbifold with r singular points, each having a neighbourhood (in the fibre) homeomorphic to the cone over the real projective space $P^{2n+1}_{\mathbb{R}}$.

We now let $\mathcal{L} := \{(L_1, M_1), \dots, (L_r, M_r)\}$ be a configuration of $P^n_{\mathbb{C}}$'s in $P^{2n+1}_{\mathbb{C}}$ as before, $r > 2$. Let Γ and Γ' be complex Schottky groups obtained from this same configuration, i.e., they are generated by sets $\{T_1, ..., T_r\}$ and $\{T'_1, ..., T'_r\}$ of holomorphic involutions of $P^{2n+1}_{\mathbb{C}}$ that interchange the L_i's with the M_i's. For each $i = 1, ..., r$, the composition $T'_i \circ T_i^{-1}$ preserves the subspaces L_i, M_i. It is an exercise to see that the subgroup of $PSL(n+2, \mathbb{C})$ of transformations that preserve these subspaces is the projectivization of a copy of $GL(n+1, \mathbb{C}) \times GL(n+1, \mathbb{C}) \subset GL(2n+2, \mathbb{C})$. Therefore, we can always find an analytic family $\{\Gamma_t\}$, $0 \le t \le 1$, of complex Schottky groups, with configuration $\mathcal{L}$, such that $\{\Gamma_0\} = \Gamma$ and $\{\Gamma_1\} = \Gamma'$. Furthermore, let $\mathcal{L} := \{(L_1, M_1), \dots, (L_r, M_r)\}$ and $\mathcal{L}' := \{(L'_1, M'_1), \dots, (L'_r, M'_r)\}$ be two configurations of $P^n_{\mathbb{C}}$'s in $P^{2n+1}_{\mathbb{C}}$ as before. Due to dimensional reasons, we can always move these configurations to obtain a differentiable family of pairs of disjoint n-dimensional subspaces $\{(L_{1,t}, M_{1,t}), \dots, (L_{r,t}, M_{r,t})\}$, with $0 \le t \le 1$, providing an isotopy between $\mathcal{L}$ and $\mathcal{L}'$. Thus one has a differentiable family Γ_t of complex Kleinian groups, where $\Gamma_0 = \Gamma$ and $\Gamma_1 = \Gamma'$. The same statements hold if we replace Γ and Γ' by their subgroups $\check{\Gamma}$ and $\check{\Gamma}'$, consisting of words of even length. So one has a differentiable family $\check{\Gamma}_t$ of Kleinian groups, where $\check{\Gamma}_0 = \check{\Gamma}$ and $\check{\Gamma}_1 = \check{\Gamma}'$. Hence the manifolds $\Omega(\Gamma_t)/\check{\Gamma}_t$ are all diffeomorphic. By section 6, these manifolds are (in general non-trivial) fibre bundles over $P^n_{\mathbb{C}}$ with fibre $\#^{(r-1)}(S^{2n+1} \times S^1)$, a connected sum of $(r-1)$-copies of $S^{2n+1} \times S^1$. If $n = 1$, given any configuration of r pairwise disjoint lines in $P^3_{\mathbb{C}}$, there exists an isotopy of $P^3_{\mathbb{C}}$ which carries the configuration into a family of r twistor lines. Hence $P^3_{\mathbb{C}}$ minus this configuration is diffeomorphic to the Cartesian product of S^4 minus r points with $P^1_{\mathbb{C}}$. Moreover, the attaching functions that we use to glue the boundary components of W, the fundamental domain of Γ, are all isotopic to the identity, because they live in $PSL(4, \mathbb{C})$, which is connected. Thus, if $n = 1$, then $\Omega(\Gamma_t)/\check{\Gamma}_t$ is diffeomorphic to a product $P^1_{\mathbb{C}} \times \#^{(r-1)}(S^3 \times S^1)$. Hence we have:

5.3 Proposition. *The differentiable type of the compact (complex) manifold $\Omega(\Gamma_t)/\check{\Gamma}_t$ is independent of the choice of configuration. It is a manifold of real dimension $(4n+2)$, which is a fibre bundle over $P^n_{\mathbb{C}}$ with fibre $\#^{(r-1)}(S^{2n+1}\times S^1)$; moreover, this bundle is trivial if $n=1$. We denote the corresponding manifold by M^n_r.*

The fact that the bundle is trivial when $n=1$ is interesting because, as pointed out in the introduction, when the configuration $\mathcal{L}$ consists of twistor lines in $P^3_{\mathbb{C}}$, the quotient $\Omega(\Gamma)/\check{\Gamma}$ is the twistor space of the conformally flat manifold $p(\Omega(\Gamma))/p(\check{\Gamma})$, which is a connected sum of the form $\#^{(r-1)}(S^3\times S^1)$. Hence, in this case the natural fibration goes the other way round, i.e., it is a fibre bundle over $\#^{(r-1)}(S^3\times S^1)$ with fibre $P^1_{\mathbb{C}}$.

Given a configuration $\mathcal{L}$ as above, let us denote by $[\mathcal{L}]_G$ its orbit under the action of the group $G=PSL(2n+2,\mathbb{C})$. These orbits are equivalence classes of such configurations. Let us denote by $\mathcal{C}^n_r$ the set of equivalence classes of configurations consisting of r pairs of $P^n_{\mathbb{C}}$'s as above. Then $\mathcal{C}^n_r$ is a Zariski open set of the moduli space $\mathfrak{M}^n_r$, of configurations of r unordered couples of projective subspaces of dimension n in $P^{2n+1}_{\mathbb{C}}$, which is obtained as the Mumford quotient [MFK] of the action of G on such configurations. By [Mu], $\mathcal{C}^n_r$ is a complex algebraic variety: the *moduli space of configurations* of r pairs of n-planes $P^n_{\mathbb{C}}$ in $P^{2n+1}_{\mathbb{C}}$. Similarly, we denote by $\mathfrak{G}^n_r$ the equivalence classes, or moduli space, of the corresponding Schottky groups, where two such groups are equivalent if they are conjugate by an element in $PSL(n+2,\mathbb{C})$. Given $\mathcal{L}:=\{(L_1,M_1),\dots,(L_r,M_r)\}$, and r-tuples of involutions $(T_1,\dots,T_r)$ and $(S_1,\dots,S_r)$ as above, i.e., interchanging L_i with M_i for all $i=1,..,r$ and having pairwise disjoint mirrors, we say that these r-tuples are *equivalent* if there exists $h\in G$ such that $hT_ih^{-1}=S_i$ for all i. Let $\mathfrak{T}_{\mathcal{L}}$ denote the set of equivalence classes of such r-tuples of involutions. It is clear that a conjugation h as above must leave $\mathcal{L}$ invariant. Hence, if r is big enough with respect to n, then h must be actually the identity, so the equivalence classes in fact consist of a single element.

5.4 Theorem. *There exists a holomorphic surjective map $\pi\colon \mathfrak{G}^n_r\to\mathcal{C}^n_r$ which is a C^∞ locally trivial fibration with fibre $\mathfrak{T}_{\mathcal{L}}$. Furthermore, let Γ,Γ' be complex Schottky groups as above and let $\Omega(\Gamma),\Omega(\Gamma')$ be their regions of discontinuity. Then the complex orbifolds $M_\Gamma:=\Omega(\Gamma)/\Gamma$ and $M_{\Gamma'}:=\Omega(\Gamma')/\Gamma'$ are biholomorphically equivalent if and only if Γ and Γ' are projectively conjugate, i.e., they represent the same element in $\mathfrak{G}^n_r$. Similarly, if $\check{\Gamma},\check{\Gamma}'$ are the corresponding index 2 subgroups, consisting of the elements which are words of even length, then the manifolds $M_{\check{\Gamma}}:=\Omega(\Gamma)/\check{\Gamma}$ and $M_{\check{\Gamma}'}:=\Omega(\Gamma')/\check{\Gamma}'$, are biholomorphically equivalent if and only if $\check{\Gamma}$ and $\check{\Gamma}'$ are projectively conjugate.*

Proof. The first statement in (5.4) is obvious, i.e., that we have a holomorphic surjection $\pi\colon\mathfrak{G}^n_r\to\mathcal{C}^n_r$ with kernel $\mathfrak{T}_{\mathcal{L}}$. The other statements are immediate consequences of the following lemma (5.5) of [SV].

5.5 Lemma. *Let U be a connected open set in $P^{2n+1}_{\mathbb{C}}$ that contains a subspace $L\subset P^{2n+1}_{\mathbb{C}}$ of dimension n, and let $h\colon U\to V$ be a biholomorphism onto an open* *set $V \subset P^{2n+1}_{\mathbb{C}}$. Suppose that V also contains a subspace M of dimension n. Then h extends uniquely to an element in $PSL(2n+2,\mathbb{C})$.*

Notice that if $n=1$, then (5.5) becomes Lemma 3.2 in [Ka1].

5.6 Corollary. *For $r>2$ sufficiently large, the manifold $\Omega(\Gamma)/\check{\Gamma}$ has non-trivial moduli.*

In fact, if the manifolds $\Omega(\Gamma)/\check{\Gamma}$ and $\Omega(\Gamma')/\check{\Gamma}'$ are complex analytically equivalent, then $\check{\Gamma}$ is conjugate to $\check{\Gamma}'$ in $PSL(2n+2,\mathbb{C})$, by (5.4), and the corresponding configurations $\mathcal{L}$ and $\mathcal{L}'$ are projectively equivalent. Now it is sufficient to choose r big enough to have two such configurations which are not projectively equivalent. This is possible because the action induced by the projective linear group G on the Grassmannian $G_{2n+1,n}$ is obtained from the projectivization of the action of $SL(2n+2,\mathbb{C})$ acting on the Grassmann algebra Λ^{n+1}, of $(n+1)$-vectors of $\mathbb{C}^{2n+2}$, restricted to the set of decomposable $(n+1)$-vectors $\mathcal{D}^{n+1}$. The set $\mathcal{D}^{n+1}$ generates the Grassmann algebra and $G_{2n+1,n}=(\mathcal{D}^{n+1}-\{0\})/\sim$, where $\sim$ is the equivalence relation of projectivization.

If r is small with respect to n, then $\mathcal{C}^n_r$ consists of one point, because any two such configurations are in the same $PSL(2n+2,\mathbb{C})$-orbit. Therefore, in this case $\mathfrak{T}_{\mathcal{L}}$ coincides with $\mathfrak{G}^n_r$. That is, to change the complex structure of M^n_r we need to change the corresponding involutions into a family of involutions, with the same configuration (up to conjugation), which is not conjugate to the given one.

The following result is a generalization of Theorem 5.2 in [Ka1]. This can be regarded as a restriction for a complex orbifold (or manifold) to be of the form $\Omega(\Gamma)/\Gamma$ (or $\Omega(\Gamma)/\check{\Gamma}$).

5.7 Proposition. *If $r>2$, then the compact complex manifolds and orbifolds $\Omega(\Gamma)/\check{\Gamma}$ and $\Omega(\Gamma)/\Gamma$, obtained in theorem 5.4, have no non-constant meromorphic functions.*

The following result is proved in [SV] using theorem 4.6.i) above and an argument similar to the one of Marstrand in [Mr]. This result is essential for our proof of Theorem 5.11 below.

5.8 Proposition. *Let $r>2$, $0<\lambda<(r-1)^{-1}$ and let Γ and $\check{\Gamma}$ be as in (4.5). Then, for every $\delta>0$, the Hausdorff dimension of $\Lambda(\Gamma)=\Lambda(\check{\Gamma})$ is less than $2n+1+\delta$, i.e., the* **transverse** *Hausdorff dimension of $\Lambda(\Gamma)=\Lambda(\check{\Gamma})$ is less than $1+\delta$.*

Next we will apply the previous estimates to compute the versal deformations of manifolds obtained from complex Schottky groups as in (5.8), whose limit sets have small Hausdorff dimension.

We first recall [Kod] that given a compact complex manifold X, a deformation of X consists of a triple $(\mathcal{X},\mathcal{B},\omega)$, where $\mathcal{X}$ and $\mathcal{B}$ are complex analytic spaces and $\omega:\mathcal{X}\to\mathcal{B}$ is a surjective holomorphic map such that $\omega^{-1}(t)$ is a complex manifold for all $t\in\mathcal{B}$ and $\omega^{-1}(t_0)=X$ for some t_0, which is called the reference point. It is known [Kur] that given X, there is always a deformation $(\mathcal{X},\mathfrak{K}_X,\omega)$

which is *universal*, in the sense that every other deformation is induced from it (see also [Kod]). The space $\mathfrak{K}_X$ is the Kuranishi space of *versal* deformations of X. If we let $\Theta := \Theta_X$ be the sheaf of germs of local holomorphic vector fields on X, then every deformation of X determines, via differentiation, an element in $H^1(X,\Theta)$, so $H^1(X,\Theta)$ is called the *space of infinitesimal deformations* of X ([Kod], Ch. 4). Furthermore ([Kod, Th. 5.6]), if $H^2(X,\Theta)=0$, then the Kuranishi space $\mathfrak{K}_X$ is smooth at the reference point t_0 and its tangent space at t_0 is canonically identified with $H^1(X,\Theta)$. In particular, in this case *every infinitesimal deformation* of X comes from an actual deformation, and vice-versa, every deformation of the complex structure on X, which is near the original complex structure, comes from an infinitesimal deformation.

The following lemma is an immediate application of (5.8) and Harvey's Theorem 1 in [Ha], which generalises the results of Scheja. The idea is that if the Hausdorff dimension of the limit set is sufficiently small, then its contribution to cohomology is not significant up to certain dimension. We refer to [SV] for a detailed proof.

5.9 Lemma. *Let $r>2$, $0<\lambda<(r-1)^{-1}$, let $\check{\Gamma}$ be as in proposition 5.8 and let $\Omega := \Omega(\Gamma)\subset P_{\mathbb{C}}^{2n+1}$ be its region of discontinuity . Then one has:*

$$H^j(\Omega, i^*(\Theta_{P_{\mathbb{C}}^{2n+1}})) \cong H^j(P_{\mathbb{C}}^{2n+1}, \Theta_{P_{\mathbb{C}}^{2n+1}})\,,\ for\ 0\le j<n\,,$$

where i is the inclusion of Ω in $P_{\mathbb{C}}^{2n+1}$. Hence, if $n>1$, then one has:

$$H^0(\Omega, i^*(\Theta_{P_{\mathbb{C}}^{2n+1}})) \cong \mathfrak{sl}(2n+2,\mathbb{C})\quad and\quad H^j(\Omega, i^*(\Theta_{P_{\mathbb{C}}^{2n+1}}))\cong 0\,,$$

for all $0<j<n$, where $\mathfrak{sl}(2n+2,\mathbb{C})$ is the Lie algebra of $PSL(2n+2,\mathbb{C})$, and it is being considered throughout this section as an additive group.

We let $M:=\Omega/\check{\Gamma}$, where $\check{\Gamma}$ is as above. We notice that Ω is simply connected when $n>1$, so that Ω is the universal covering $\widetilde{M}$ of M. Let $p:\widetilde{M}\to M$ be the covering projection; since $\check{\Gamma}$ acts freely on Ω, this projection is actually given by the group action. Let Θ_M be the sheaf of germs of local holomorphic vector fields on M and let $\widetilde{\Theta}$ be the pull-back of Θ to $\widetilde{M}$ under the covering p; $\widetilde{\Theta}$ is the sheaf $i^*(\Theta_{P_{\mathbb{C}}^{2n+1}})$ on $\widetilde{M}=\Omega$.

5.10 Lemma. *If $n>2$, then for $0\le j\le 2$ we have:*

$$H^j(M,\Theta_M)\cong H^j_\rho(\check{\Gamma},\mathfrak{sl}(2n+2,\mathbb{C}))\,,$$

where $\mathfrak{sl}(2n+2,\mathbb{C}))$ is considered as a $\check{\Gamma}$-left module via the representation

$$\rho:\check{\Gamma}\to Aut\,(\mathfrak{sl}(2n+2,\mathbb{C})))$$

given by:

$$\rho(\gamma)(v) = dT_g\circ v\circ {T_g}^{-1}\,,\quad v\in\mathfrak{sl}(2n+2,\mathbb{C})\ ,$$

where T_g is the action of $g \in \check{\Gamma}$ on $P^{2n+1}_{\mathbb{C}}$.

Proof. If $n > 2$, then (5.9) and Mumford's formula (c) in [Mu], pag 23, (see also Grothendieck [Gr], Chapter V) imply that there exists an isomorphism

$$\phi : H^j_\rho(\check{\Gamma}, H^0(\Omega, \widetilde{\Theta})) \to H^j(M, \Theta_M),$$

for $0 \le j \le 2$, where $H^0(\Omega, \widetilde{\Theta})$ is the vector space of holomorphic vector fields on the universal covering $\widetilde{M} = \Omega \subset P^{2n+1}_{\mathbb{C}}$ of M.

Now, by [Ha], Theorem 1.i), every holomorphic vector field in $\Omega(\Gamma)$, extends to a holomorphic vector field defined in all of $P^{2n+1}_{\mathbb{C}}$. Therefore,

$$H^0(\Omega, \widetilde{\Theta}) = H^0(P^{2n+1}_{\mathbb{C}}, \Theta_{P^{2n+1}_{\mathbb{C}}}) = \mathfrak{sl}(2n+2, \mathbb{C}). \quad \square$$

We recall that $\check{\Gamma}$ is a free group of rank $r-1$; let $g_1, \cdots, g_{r-1}$ be generators of $\check{\Gamma}$. By [HS], page 195 Corollary 5.2, applied to $\check{\Gamma}$, we obtain:

$$H^1_\rho(\check{\Gamma}, \mathfrak{sl}(2n+2, \mathbb{C})) \cong \mathfrak{sl}(2n+2, \mathbb{C}) \times \cdots \times \mathfrak{sl}(2n+2, \mathbb{C})/Im\,(\psi),$$

where

$$\psi : \mathfrak{sl}(2n+2, \mathbb{C}) \to \mathfrak{sl}(2n+2, \mathbb{C}) \times \cdots \times \mathfrak{sl}(2n+2, \mathbb{C})$$

is given by $\psi(v) = (g_1(v) - v, \cdots, g_{r-1}(v) - v)$. By Jordan's theorem, there can not exist a linear vector field v in $P^{2n+1}_{\mathbb{C}}$ which is invariant by $g_1, \cdots, g_{r-1}$ and is tangent to a $\check{\Gamma}$-invariant hyperplane Π. Hence ψ is actually injective and one has [SV],

$$dim_{\mathbb{C}}\, H^1(\Omega, \widetilde{\Theta}) = dim_{\mathbb{C}}\, [\mathfrak{sl}(2n+2, \mathbb{C})^{r-2}] = (r-2)\left((2n+2)^2 - 1\right).$$

By [HS], page 197, Corollary 5.6 we have $H^2_\rho(\check{\Gamma}, \mathfrak{sl}(2n+2, \mathbb{C})) = 0$. Hence, by 5.10 above, one obtains,

$$H^2(M, \Theta_M) \cong H^2_\rho(\check{\Gamma}, \mathfrak{sl}(2n+2, \mathbb{C})) = 0.$$

Thus we arrive to the following theorem:

5.11 Theorem. *Let $n, r > 2$ and let λ be an arbitrary scalar such that $0 < \lambda < (r-1)^{-1}$. Let Γ be a Schottky group as in (4.5.iii), so that the (Fubini-Study) distance from $\gamma(x)$ to the limit set Λ decreases faster that $C\lambda^k$ for every point $x \in P^{2n+1}_{\mathbb{C}}$ and any $\gamma \in \Gamma$ of word-length k (where C is some positive constant). Let $\check{\Gamma}$ be the index-two subgroup of Γ consisting of words of even length. Let Ω be the region of discontinuity of Γ, $M := \Omega/\check{\Gamma}$, and let $\mathfrak{K}^n_r$ denote the Kuranishi space of versal deformations of M, with reference point $t_0 \in \mathfrak{K}^n_r$ corresponding to M. Then, we have:*

$$H^1(M, \Theta_M) \cong H^1_\rho(\check{\Gamma}, \mathfrak{sl}(2n+2, \mathbb{C})) \cong \mathbb{C}^{(r-2)((2n+2)^2-1)},$$

and

$$H^2(M, \Theta_M) = 0\,.$$

Hence $\mathfrak{K}_r^n$ *is non-singular at* t_0*, of complex dimension* $(r-2)((2n+2)^2-1)$*, and every small deformation of* M *is obtained by a small deformation of* $\check{\Gamma}$ *as a subgroup of* $PSL(2n+2,\mathbb{C})$*, unique up to conjugation.*

Although we only considered $n > 2$ above, the last theorem remains valid for $n = 0, 1$. In fact, if $n = 0$ and $r > 2$, we have the classical Schottky groups. The manifold $\Omega/\check{\Gamma}$ is a compact Riemann surface of genus $r-1$. It is well known that in this case the moduli space has dimension $3(r-1)-3 = 3(r-2)$, which, of course, coincides with the formula above. When $n = 1$ and $r > 2$ the manifolds $\Omega/\check{\Gamma}$ are Pretzel twistor spaces of genus $g = r-1$, in the sense of Penrose [Pe2]. The theorem above gives that the dimension of the moduli space of this manifold is $15g - 15$, which coincides with Penrose's calculations in page 251 of [Pe2].

Bibliography

[A1] L. V. Ahlfors, *Möbius transformations and Clifford numbers*, Differential Geometry and Complex Analysis (I. Chavel and H. M. Farkas, eds.), Springer-Verlag, New York, 1985, pp. 65–74.

[A2] L. V. Ahlfors, *Möbius transformations in several dimensions*, Lecture notes, School of Mathematics, University of Minnesota, Minneapolis, 1981.

[A3] L. V. Ahlfors, *Finitely generated Kleinian groups*, Am. J. Math. **86** (1964), 415-429.

[At] M. F. Atiyah, *Geometry of Yang-Mills Fields*, Lezioni Ferminiane Accademia Nazionale dei Lincei, Scuola Normale Superiore, Pisa, Italia, 1979.

[AHS] M. F. Atiyah, N. J. Hitchin, I. M. Singer, *Self duality in four dimensional Riemannian geometry*, Proc. Royal Soc. London, serie A (Math. and Phys. Sciences) **362 , No. 1711** (1978), 425-461.

[Bl] A. Blanchard, *Sur les variétés analytiques complexes*, Ann. Sci. ENS **63** (1958), 157-202.

[BM] A. S. Besicovitch, P. A. P. Moran, *The measure of product and cylinder sets*, J. London Math. Soc. **20** (1945), 110-120.

[Bo] R. Bowen, *Hausdorff dimension of quasi-circles*, Publ. Math. IHES **50** (1979), 11–25.

[BR] F. E. Burstall, J. H. Rawnsley, *Twistor Theory for Riemannian Symmetric Spaces*, vol. 1424, Springer Verlag, Lect. Notes in Maths., 1990.

[BS] E. Bedford, J. Smillie, *External rays in the dynamics of polynomial automorphisms of* $\mathbb{C}^2$, preprint, to appear in the AMS proceedings of a conference in Pohang, Korea.

[Br] R. L. Bryant, *Conformal and minimal immersions of compact surfaces into the 4-sphere*, J. Diff. Geom. **17** (1982), 455-473.

[C1] E. Calabi, *Quelques applications de l'analyse complexe aux surfaces d'aire minima*, Topics in Complex Manifolds (H. Rossi, ed.), Les Presses de l'Université de Montreal, 1967, pp. 59–81.

[C2] E. Calabi, *Minimal immersions of surfaces in Euclidean spheres*, J. Diff. Geom. **1** (1967), 111-125.

[DM] P. Deligne, G. D. Mostow, *Commensurabilities among lattices in* $PU(1,n)$, Ann. Math. Study 132 (1993), Princeton Univ. Press.

[DV] M. Dubois-Violette, *Structures complexes au-dessus des variétés, applications*, in "Mathématique et Physique", edit L. Boutet et al, Progress in Maths. 37, Birkhäuser, 1983.

[Eh] C. Ehresmann, *Sur les espaces fibrés differentiables*, C. R. Acad. Sci. Paris **224** (1947), 1611-1612.

[FSp] L. Flaminio, R. J. Spatzier, *Geometrically finite groups, Patterson-Sullivan measures and Ratner's rigidity theorem*, Inventiones Math. **99** (1991), 621–626.

[Gr] A. Grothendieck, *Sur quelques points d'algèbre homologique*, Tohoku Math. J., II. Ser. 9 (1957), 119-221.

[Gu] R. Gunning, *On uniformization of complex manifolds; the role of connections*, Princeton Univ. Press, Math. Notes 22, 1978.

[Ha] R. Harvey, *Removable singularities of cohomology classes*, Amer. Journal of Math. **96 No. 3.** (1974), 498–504.

[HS] P. J. Hilton, U. Stammbach, *A Course in Homological Algebra*, vol. 4, Graduate Texts in Mathematics, Springer-Verlag. New York - Heidelberg - Berlin, 1971.

[Hi] N. J. Hitchin, *Kählerian twistor spaces*, Proc. London Math. Soc. **43** (1981), 133-150.

[Ht] R. Hartshorne, *Algebraic geometry*, vol. 52, Graduate Texts in Mathematics, Springer-Verlag. New York - Heidelberg - Berlin, 1984.

[In] Y. Inoue, *Twistor spaces of even dimensional Riemannian manifolds*, J. Math. Kyoto Univ. **32** (1992), 101-134.

[Ka1] M. Kato, *On compact complex 3-folds with lines*, Japanese J. Math. **11** (1985), 1-58.

[Ka2] M. Kato, *Factorization of compact complex 3-folds which admit certain projective structures*, Tôhoku Math. J. **41** (1989), 359-397.

[Ka3] M. Kato, *Compact Complex 3-folds with Projective Structures; The infinite Cyclic Fundamental Group Case*, Saitama Math. J. **4** (1986), 35-49.

[Ka4] M. Kato, *Compact Quotient Manifolds of Domains in a Complex 3-Dimensional Projective Space and the Lebesgue Measure of Limit Sets*, Tokyo J. Math. **19** (1996), 99-119.

[Ki] A. A. Kirilov, *Elements of the Theory of Representations*, vol. **220**, Grundlehren der mathematischen Wissenschaften, Springer-Verlag, 1976.

[Ko] S. Kobayashi, *Transformation groups in differential geometry*, Springer Verlag, 1972.

[KO] S. Kobayashi, T. Ochiai, *Holomorphic Projective Structures on Compact Complex Surfaces (I and II)*, Math. Ann. **249 (255)**, 1980 (81), 75-94 (519-521).

[Kod] K. Kodaira, *Complex Manifolds and Deformation of Complex Structures*, Springer Verlag, 1985.

[Kr] I. Kra, *On Lifting Kleinian Groups to $SL(2,\mathbb{C})$*, Differential Geometry and Complex Analysis (I. Chavel and H. M. Farkas, eds.), Springer-Verlag, New York, 1985, pp. 181–194.

[Ku1] R. S. Kulkarni, *Groups with domains of discontinuity*, Math. Ann. **237** (1978), 253-272.

[Ku2] R. S. Kulkarni, *Conformal structures and Möbius structures*, Aspects of Mathematics, edited by R.S. Kulkarni and U. Pinkhall, Max Planck Institut fur Mathematik, Vieweg (1988), 1–39.

[Kur] M. Kuranishi, *On the locally complete families of complex analytic structures*, Ann. of Maths. **75** (1962), 536-577.

[Ma1] B. Maskit, *Kleinian groups*, vol. 287, Springer Verlag, Grundelehren math. Wissenschaften **287**, 1980.

[Ma2] B. Maskit, *A characterization of Schottky groups*, J. d'Analyse Math. 19 (1967), 227–230.

[MFK] D. Mumford, J. Fogarty, F. Kirwan, *Geometric invariant theory. 3rd enl. ed.*, Ergebnisse der Mathematik und ihrer Grenzgebiete. 3. Folge. 34. Berlin: Springer-Verlag, 1993.

[Mr] J. M. Marstrand, *The dimension of cartesian product sets*, Proc. Cambridge Phil. Soc. **50** (1954), 198-202.

[Mu] D. Mumford, *Abelian Varieties*, Oxford University Press (Tata Lectures), 1974.

[Ni] P. J. Nicholls, *The Ergodic Theory of Discrete Groups*, London Math. Soc. Lecture Notes Series # **143**, Cambridge University Press, 1989.

[No] M. V. Nori, *The Schottky groups in higher dimensions*, Proceedings of Lefschetz Centennial Conference, Mexico City, AMS Contemporary Maths. **58, part I** (1986), 195-197.

[OR] N. R. O'Brien, J. H. Rawnsley, *Twistor spaces*, Ann. Global Anal. Geom. **3** (1985), 29-58.

[Pa] S. J. Patterson, *Lectures on measures on limit sets of Kleinian groups*, Analytic and geometric aspects of hyperbolic space, London Math. Soc. Lecture Notes **111**, 1987, pp. 281–323.

[Pe1] R. Penrose, *The twistor programme*, Rep. Math. Phys. **12** (1977), 65-76.

[Pe2] R. Penrose, *Pretzel twistor spaces*, in "Further advances in twistor theory", ed. L.J. Mason and L. P. Hughston, Pitman Research Notes in Maths. vol. 231 (1990), 246-253.

[Ra] J. Ratcliffe, *The foundations of hyperbolic manifolds*, Graduate texts in Math.**149**, Springer Verlag, 1994.

[SV] J. Seade, A. Verjovsky, *Higher dimensional complex Kleinian groups*, Preprint (1998).

[Su1] D. P. Sullivan, *The density at infinity of a discrete group of hyperbolic motions*, Publ. Math. IHES **50** (1979), 171–202.

[Su2] D. P. Sullivan, *Entropy, Hausdorff measures old and new, and limit sets of geometrically-finite Kleinian groups*, Acta Math. **153** (1984), 259–277.

[Tu] P. Tukia, *The Hausdorff dimension of the limit set of a geometrically-finite Kleinian group*, Acta Math. **152** (1984), 127–140.

[We] R. O., Jr. Wells, *Differential Analysis on Complex Manifolds*, Prentice-Hall, Englewood Cliffs, NJ, 1973.

[Zi] R. J. Zimmer, *Ergodic theory and semisimple groups*, vol. 81, Monographs in Mathematics, Birkhäuser, Boston-Basel-Stuttgart, 1984.

J. Seade: Instituto de Matemáticas, Universidad Nacional Autónoma de México, Unidad Cuernavaca. Av. Universidad s/n, Colonia Lomas de Chamilpa, C.P. 62210, Cuernavaca, Morelos, México.

E-mail address: `jseade@matcuer.unam.mx`

A. Verjovsky: Instituto de Matemáticas, Universidad Nacional Autónoma de México, Unidad Cuernavaca. Av. Universidad s/n, Colonia Lomas de Chamilpa, C.P. 62210, Cuernavaca, Morelos, México.

E-mail address: `alberto@matcuer.unam.mx`

Contemporary Mathematics
Volume **269**, 2001

Dynamics of Singular Holomorphic Foliations on the Complex Projective Plane

Saeed Zakeri

CONTENTS

Preface

This manuscript is a revised version of my Master's thesis which was originally written in 1992 and was presented to the Mathematics Department of University of Tehran. My initial goal was to give, in a language accessible to non-experts, highlights of the 1978 influential paper of Il'yashenko on singular holomorphic foliations on $\mathbb{CP}^2$ [**I3**], providing short, self-contained proofs. Parts of the exposition in chapters 1 and 3 were greatly influenced by the beautiful work of Gómez-Mont and Ortiz-Bobadilla [**GO**] in Spanish, which contains more material, different from what we discuss here. It must be noted that much progress has been made in this area since 1992, especially in local theory (see for instance the collection [**I6**] and the references cited there). However, Hilbert's 16th Problem and the Minimal Set Problem are still unsolved.

There is a well-known connection between holomorphic foliations in dimension 2 and dynamics of iterations of holomorphic maps in dimension 1, but many believe that this connection has not been fully exploited. It seems that some experts in each area keep an eye on progress in the other, but so far there have been rather few examples of a fruitful interaction. The conference on Laminations and Foliations held in May 1998 at Stony Brook was a successful attempt to bring both groups together. As a result, many people in dynamics expressed their interest in learning

2000 *Mathematics Subject Classification.* Primary 37F75; Secondary 34M35, 34M45.

about holomorphic foliations. I hope the present manuscript will give them a flavor of the subject and will help initiate a stronger link between the young researchers in both areas.

My special thanks go to S. Shahshahani who encouraged me to study this subject years ago and supervised my Master's thesis back in Iran. I gratefully acknowledge the financial support of IPM (Tehran) during preparation of this manuscript. A series of lectures given by C. Camacho and Yu. Il'yashenko in Trieste, Italy, in 1991 and 1992 were a source of inspiration to me. I am indebted to X. Gómez-Mont and J. Milnor who read parts of this revised version and made very useful comments. Finally, I would like to thank the Dynamical Systems group at Stony Brook, especially A. Epstein, M. Lyubich and J. Milnor. Without the interest they showed, I would have never found enough motivation to revise what I had written 6 years ago.

Introduction

Consider the differential equation

$$(*)\begin{cases} \dfrac{dx}{dt} = P(x,y) \\[2ex] \dfrac{dy}{dt} = Q(x,y) \end{cases}$$

in the real plane $(x, y) \in \mathbb{R}^2$, where P and Q are relatively prime polynomials with $\max\{\deg P, \deg Q\} = n$. What can be said, Hilbert asked, about the number and the location of limit cycles of $(*)$? In particular, is it true that there are only finitely many limit cycles for $(*)$? If so, does there exist an upper bound $H(n)$, depending only on n, for the number of limit cycles of an equation of the form $(*)$? Surprisingly, the finiteness problem has been settled only in recent years, and the existence and a possible value of $H(n)$ is still unknown, even when $n = 2$!

In an attempt to answer the Hilbert's question, H. Dulac "proved" the finiteness theorem in 1926 [**D**]. Many years later, however, his "proof" turned out to be wrong. In fact, in 1982 Yu. Il'yashenko found a fundamental mistake in Dulac's argument [**I2**], and gave a correct proof for the finiteness theorem later in 1987 [**I4**].

The second major attempt along this line was started in 1956 by a seminal paper of I. Petrovskiĭ and E. Landis [**PL1**]. They had a completely different and perhaps more radical approach. They considered $(*)$ as a differential equation in the complex plane $(x, y) \in \mathbb{C}^2$, with t now being a *complex time* parameter. The integral curves of the vector field are now either singular points which correspond to the common zeros of P and Q, or complex curves tangent to the vector field which are holomorphically immersed in $\mathbb{C}^2$. This gives rise to a holomorphic foliation by complex curves with a finite number of singular points. One can easily see that this foliation extends to the complex projective plane $\mathbb{CP}^2$, which is obtained by adding a line at infinity to the plane $\mathbb{C}^2$. The trajectories of $(*)$ in the real plane are then the intersection of these complex curves with the plane $\text{Im } x = \text{Im } y = 0$.

What makes this approach particularly useful is the possibility of applying methods of several complex variables and algebraic geometry over an algebraically closed field, not available in the real case. Intuitively, the complexified equation provides enough space to go around and observe how the integral curves behave, whereas the real-plane topology of the trajectories is only the tip of a huge iceberg.

Viewing $(*)$ as a complex differential equation, Petrovskiĭ and Landis "proved" that $H(2)$ exists and in fact one can take $H(2) = 3$ [**PL1**]. Later on, they "proved" the estimates

$$H(n) \le \begin{cases} \dfrac{6n^3 - 7n^2 + n + 4}{2} & \text{if } n \text{ is even,} \\[2ex] \dfrac{6n^3 - 7n^2 - 11n + 16}{2} & \text{if } n \text{ is odd,} \end{cases}$$

thus answering Hilbert's question [**PL2**]. The result was regarded as a great achievement: Not only did they solve a difficult problem, but they introduced a truly novel method in the geometric theory of ordinary differential equations. However, it did not take long until a serious gap was discovered in their proof. Although the estimate on $H(n)$ could not be salvaged anymore, the powerful method of this work paved the way for further studies in this direction.

In 1978, Il'yashenko made a fundamental contribution to the problem. Following the general idea of Petrovskiĭ and Landis, he studied equations $(*)$ with complex polynomials P, Q from a topological standpoint without particular attention to Hilbert's question. In his famous paper [**I3**], he showed several peculiar properties of the integral curves of such equations.

From the point of view of foliation theory, it may seem that foliations induced by equations like $(*)$ form a rather tiny class among all holomorphic foliations on $\mathbb{CP}^2$. However, as long as we impose a reasonable condition on the set of singularities, it turns out that *every* singular holomorphic foliation on $\mathbb{CP}^2$ is induced by a polynomial differential equation of the form $(*)$ in the affine chart $(x, y) \in \mathbb{C}^2$. The condition on singularities is precisely what is needed in several complex variables: the singular set of the foliation must be an analytic subvariety of codimension > 1, which is just a finite set in the case of the projective plane.

Consider a closed orbit γ of a smooth vector field in the real plane. To describe the behavior of trajectories near γ, one has the simple and useful concept of the *Poincaré first return map*: Choose a small transversal Σ at some point $p \in \gamma$, choose a point $q \in \Sigma$ near p, and look at the first point of intersection with Σ of the trajectory passing through q. In this way, one obtains the germ of a smooth diffeomorphism of Σ fixing p. The iterative dynamics of this self-map of Σ reflects the global behavior of the trajectories near γ.

For a singular holomorphic foliation on $\mathbb{CP}^2$ induced by an equation of the form $(*)$, a similar notion, called the *monodromy mapping*, had already been used by Petrovskiĭ and Landis, and extensively utilized by Il'yashenko. A closed orbit should now be replaced by a non-trivial loop γ on the leaf passing through a given point p, small transversal Σ is a 2-disk, and the result of traveling over γ on the leaf passing through a point on Σ near p gives the germ of a biholomorphism of Σ fixing p, called the *monodromy mapping associated with* γ. Note that all points in the orbit of a given point on Σ under this biholomorphism lie in the same leaf. In this way, to each non-trivial loop in the fundamental group of the leaf we associate a self-map of Σ reflecting the behavior of nearby leaves as one goes around the loop. It is easily checked that composition of loops corresponds to superposition of monodromy mappings, so the fundamental group of the leaf maps homomorphically to a subgroup of the group of germs of biholomorphisms of Σ fixing p; the latter subgroup will be called the *monodromy group* of the leaf. Thus a global problem can be reduced to a large extent to the study of germs of biholomorphisms of $\mathbb{C}$ fixing (say) the origin. The group of all these germs is denoted by $\mathrm{Bih}_0(\mathbb{C})$.

Now here is the crucial observation: For "almost every" equation of the form $(*)$, the line at infinity of $\mathbb{CP}^2$ with finitely many singular points deleted is a leaf of the extended foliation. On the other hand, no leaf can be bounded in $\mathbb{C}^2$ (this is in fact more than just the Maximum Principle), so every leaf has a point of accumulation on the line at infinity. Therefore, the monodromy group of the *leaf at infinity*, which is finitely-generated since the leaf is homeomorphic to a finitely-punctured Riemann sphere, gives us much information about the global behavior of all leaves.

What Il'yashenko observed was the fact that almost all the dynamical properties of leaves have a discrete interpretation in terms of finitely-generated subgroups of $\mathrm{Bih}_0(\mathbb{C})$, the role of which is played by the monodromy group of the leaf at infinity. Therefore, he proceeded to study these subgroups and deduced theorems about the behavior of leaves of singular holomorphic foliations. The density and ergodicity theorems for these foliations are direct consequences of the corresponding results for subgroups of $\mathrm{Bih}_0(\mathbb{C})$. The density theorem asserts that for "almost every" equation $(*)$, every leaf other than the leaf at infinity is dense in $\mathbb{CP}^2$. The ergodicity theorem says that "almost every" foliation induced by an equation of the form $(*)$ is *ergodic*, which means that every measurable saturated subset of $\mathbb{CP}^2$ has zero or full measure.

Yet another advantage of the reduction to the discrete case is the possibility of studying consequences of equivalence between two such foliations. Two singular holomorphic foliations on $\mathbb{CP}^2$ are said to be *equivalent* if there exists a homeomorphism of $\mathbb{CP}^2$ which sends each leaf of the first foliation to a leaf of the second one. Such an equivalence implies the equivalence between the monodromy groups of the corresponding leaves at infinity. The latter equivalence has the following meaning: $G, G' \subset \mathrm{Bih}_0(\mathbb{C})$ are *equivalent* if there exists a homeomorphism h of some neighborhood of $0 \in \mathbb{C}$, with $h(0) = 0$, such that $h \circ f \circ h^{-1} \in G'$ if and only if $f \in G$. The correspondence $f \mapsto h \circ f \circ h^{-1}$ is clearly a group isomorphism. Under typical conditions, one can show that the equivalence between two subgroups of $\mathrm{Bih}_0(\mathbb{C})$ must in fact be holomorphic. This gives some invariants of the equivalence classes, and reveals a rigidity phenomenon. Applying these results to the monodromy groups of the leaves at infinity, one can show the existence of *moduli of stability* and the phenomenon of *absolute rigidity* for these foliations.

All the above results are proved under typical conditions on the differential equations. After all, the existence of the leaf at infinity (an algebraic leaf) plays a substantial role in all these arguments. Naturally, one would like to study foliations which do not satisfy these conditions, in particular, those which do not admit any algebraic leaf. From this point of view, the dynamics of these foliations is far from being understood. The major contributions along this line have been made by the Brazilian and French schools. In their interesting paper [**CLS1**], C. Camacho, A. Lins Neto and P. Sad study non-trivial minimal sets of these foliations and show several properties of the leaves within a non-trivial minimal set. A *minimal set* of a foliation on $\mathbb{CP}^2$ is a compact, saturated, non-empty subset of $\mathbb{CP}^2$ which is minimal with respect to these three properties. A *non-trivial minimal set* is one which is not a singular point. It follows that the existence of a non-trivial minimal set is equivalent to the existence of a leaf which does not accumulate on any singular point. The fact that non-trivial minimal sets do not exist when the foliation admits an algebraic leaf makes the problem much more challenging. The major open problem in this context is the most primitive one: "Does there exist a singular holomorphic foliation on $\mathbb{CP}^2$ having a non-trivial minimal set?"

An intimately connected question is about limit sets of the leaves of these foliations. The classical theorem of Poincaré–Bendixson classifies all possible limit sets for foliations on the 2-sphere: Given a smooth vector field on the 2-sphere with a finite number of singular points, the ω- (or α-) limit set of any point is either a singular point or a closed orbit or a chain of singular points and trajectories starting from one of these singular points ending at another. This enables us to understand the asymptotic behavior of *all* trajectories. Naturally, one is interested in proving a complex version of the Poincaré–Bendixson Theorem for foliations on $\mathbb{CP}^2$. Now the concept of the limit set is defined as follows: Take a leaf $\mathcal{L}$ and let $K_1 \subset K_2 \subset \cdots \subset K_n \subset \cdots$ be a sequence of compact subsets of $\mathcal{L}$, with $\bigcup_{n\geq 1} K_n = \mathcal{L}$. Then the *limit set* of $\mathcal{L}$ is by definition the intersection $\bigcap_{n\geq 1} \overline{\mathcal{L} \smallsetminus K_n}$. Incidentally, the limit set of every non-singular leaf is non-vacuous, since it can be shown that no non-singular leaf can be compact. Despite some results in this direction (see e.g. [**CLS2**]), the problem of classifying possible limit sets is almost untouched.

Finally, it should be mentioned how Hilbert's question for a real equation $(*)$ is interpreted in the complex language. It is easy to prove that a limit cycle of a real equation $(*)$ is a homotopically non-trivial loop on the corresponding leaf of the complexified equation. So any piece of information about the fundamental groups of leaves could be a step toward understanding the limit cycles. A non-trivial loop on a leaf of a complex equation $(*)$ is said to be a *(complex) limit cycle* if the germ of its associated monodromy is not the identity map. Il'yashenko has shown that "almost every" equation $(*)$ has a countably infinite number of homologically independent (complex) limit cycles; nevertheless this result does not have a direct bearing on Hilbert's question. A complex version of the finiteness problem may be the following: Can the fundamental group of a (typical) leaf of such foliations be infinitely-generated? If not, does there exist an upper bound, depending only on the degree of P and Q in $(*)$, for the number of generators of the fundamental groups? Such questions, as far as I know, have not yet been answered.

1. Singular Holomorphic Foliations by Curves

This chapter introduces the concept of a singular holomorphic foliation by (complex) curves on a complex manifold, which will be quite essential in subsequent chapters. We will assume that the dimension of the singular set is small enough to allow easy application of several complex variables techniques. Moreover, when the underlying manifold is $\mathbb{CP}^2$ (2-dimensional complex projective space), the very special geometry of the space allows us to apply some standard algebraic geometry to show that all such foliations are induced by polynomial 1-forms ω on $\mathbb{CP}^2$, the leaves being the solutions of $\omega = 0$. The main tools here are extension theorems of several complex variables and the rigidity properties of holomorphic line bundles on projective spaces.

1.1. Holomorphic Foliations on Complex Manifolds. Let us start with the most basic definition in this subject.

DEFINITION 1.1. Let M be a complex manifold of dimension n and $0 < m < n$. A *(non-singular) holomorphic foliation* $\mathcal{F}$ *of codimension* m on M is an analytic atlas $\mathcal{A} = \{(U_i, \varphi_i)\}_{i \in I}$ for M which is maximal with respect to the following properties:

(i) For each $i \in I$, φ_i is a biholomorphism $U_i \to A_i \times B_i$, where A_i and B_i are open polydisks in $\mathbb{C}^{n-m}$ and $\mathbb{C}^m$, respectively.

(ii) If (U_i, φ_i) and (U_j, φ_j) are in $\mathcal{A}$ with $U_i \cap U_j \neq \emptyset$, then $\varphi_{ij} := \varphi_i \circ \varphi_j^{-1} : \varphi_j(U_i \cap U_j) \to \varphi_i(U_i \cap U_j)$ has the form

$$\varphi_{ij}(z, w) = (\psi_{ij}(z, w), \eta_{ij}(w)), \tag{1.1}$$

where $(z, w) \in \mathbb{C}^{n-m} \times \mathbb{C}^m$, and ψ_{ij} and η_{ij} are holomorphic mappings into $\mathbb{C}^{n-m}$ and $\mathbb{C}^m$, respectively.

Condition (ii) may also be expressed in the following way: Using the coordinates $(z_1, \dots, z_n)$ for $\mathbb{C}^n$, the mapping $\varphi_{ij} : (z_1, \dots, z_n) \mapsto (\varphi_{ij}^1, \dots, \varphi_{ij}^n)$ is required to satisfy $\partial \varphi_{ij}^k / \partial z_l = 0$ for $n - m + 1 \leq k \leq n$ and $1 \leq l \leq n - m$.

Each $(U_i, \varphi_i) \in \mathcal{A}$ is called a *foliation chart* (or a *flow box*) for $\mathcal{F}$. Given any foliation chart (U_i, φ_i), the sets $\varphi_i^{-1}(A_i \times \{w\}), w \in B_i$, are called *plaques* of $\mathcal{F}$ in U_i. Evidently, the plaques form a partition of the U_i into connected pieces of complex submanifolds of dimension $n - m$. Each $p \in M$ lies in at least one plaque. Two points p and q are called *equivalent* if there exists a sequence $P_1, \dots, P_k$ of plaques such that $p \in P_1, q \in P_k$, and $P_i \cap P_{i+1} \neq \emptyset, 1 \leq i \leq k - 1$. The *leaf* of $\mathcal{F}$ through p, denoted by $\mathcal{L}_p$, is the equivalence class of p under this relation. Each leaf has a natural structure of a connected $(n - m)$-dimensional complex manifold which is holomorphically immersed in M. Two leaves are disjoint or else identical.

REMARK 1.2. A holomorphic foliation of codimension $(n - 1)$ on a complex manifold M of dimension n is also called a *holomorphic foliation by curves*. Its leaves are immersed Riemann surfaces. As we will discuss later, the field of complex lines tangent to the leaves of such a foliation determines a holomorphic line bundle on M.

REMARK 1.3. It can be easily checked that the above definition is equivalent to the following, which is more natural from the geometric viewpoint: A holomorphic foliation $\mathcal{F}$ of codimension m on an n-dimensional complex manifold M is a partition of M into disjoint connected subsets $\{\mathcal{L}_\alpha\}$ (called the *leaves* of $\mathcal{F}$) such that for each $p \in M$ there exists a chart (U, φ) around p and open polydisks $A \subset \mathbb{C}^{n-m}$ and $B \subset \mathbb{C}^m$ with the property that $\varphi : U \to A \times B$ maps the connected components of $\mathcal{L}_\alpha \cap U$ to the level sets $A \times \{w\}, w \in B$.

1.2. Singular Holomorphic Foliations by Curves. Now we study those foliations which are allowed to have some "tame" singularities. Recall that a subset E of a complex manifold M is called an *analytic subvariety* if each $p \in M$ has a neighborhood U on which there are holomorphic functions $f_j : U \to \mathbb{C}$, $1 \leq j \leq k$, such that $E \cap U = \{x \in U : f_j(x) = 0, 1 \leq j \leq k\}$. Evidently, every analytic subvariety of M is closed, hence $M \smallsetminus E$ is itself a complex manifold of the same dimension as M.

DEFINITION 1.4. Let M be a complex manifold. A *singular holomorphic foliation by curves* $\mathcal{F}$ on M is a holomorphic foliation by curves on $M \smallsetminus E$, where E is an analytic subvariety of M of codimension > 1. A point $p \in E$ is called a *removable singularity* of $\mathcal{F}$ if there exists a chart (U, φ) around p, compatible with the atlas $\mathcal{A}$ of $\mathcal{F}$ restricted to $M \smallsetminus E$, in the sense that $\varphi \circ \varphi_i^{-1}$ and $\varphi_i \circ \varphi^{-1}$ have

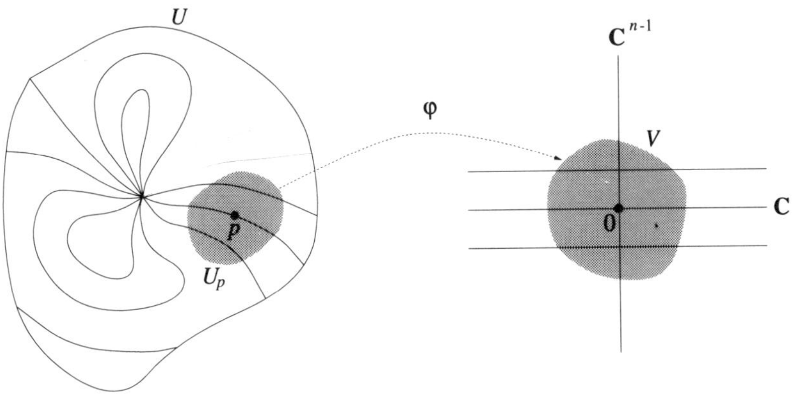

FIGURE 1. *Straightening a holomorphic vector field near a non-singular point.*

the form (1.1) for all $(U_i, \varphi_i) \in \mathcal{A}$ with $U \cap U_i \neq \emptyset$. The set of all non-removable singularities of $\mathcal{F}$ in E is called the *singular set* of $\mathcal{F}$, and is denoted by $\operatorname{sing}(\mathcal{F})$.

Naturally, as in the case of real 1-dimensional singular foliations on real surfaces, the most important examples of singular holomorphic foliations are provided by vector fields.

EXAMPLE 1.5. Let $X = \sum_{i=1}^{n} f_i \; \partial/\partial z_i$ be a holomorphic vector field on a domain $U \subset \mathbb{C}^n$. We further assume that the Jacobian $(\partial f_i/\partial z_j)_{1\leq i,j\leq n}$ has rank > 1 throughout the domain U. Then X vanishes on the analytic variety $\{z : f_1(z) = \cdots = f_n(z) = 0\}$ which has codimension > 1 (possibly the empty set). By definition, a *solution* of the differential equation $\dot{z} = X(z)$ with *initial condition* $p \in U$ is a holomorphic mapping $\eta : \mathbb{D}(0,r) \to U$ such that $\eta(0) = p$ and for every $T \in \mathbb{D}(0,r)$, $d\eta(T)/dT = X(\eta(T))$. The image $\eta(\mathbb{D}(0,r))$ is called a *local integral curve* passing through p. By the theorem of existence and uniqueness for the solutions of holomorphic differential equations [**IY**], each p has such a local integral curve passing through it, and two local integral curves through p coincide in some neighborhood of p. It follows that if $X(p) = 0$, then its integral curve will be the point p itself. If $X(p) \neq 0$, a local integral curve through p is a disk holomorphically embedded in U.

Now suppose that $X(p) \neq 0$. By the Straightening Theorem for holomorphic vector fields [**IY**], there exist neighborhoods $U_p \subset U$ of p and $V \subset \mathbb{C}^n$ of 0 and a biholomorphism $\varphi : U_p \to V$ such that $\varphi(p) = 0$ and $\varphi_*(X|_{U_p}) = \partial/\partial z_1$ (Fig. 1). Thus the connected components of the intersection of the local integral curves with U_p are mapped by φ to "horizontal" lines $\{z_2 = \text{const.}, \ldots, z_n = \text{const.}\}$. In other words, X induces a singular holomorphic foliation by curves on U, denoted by $\mathcal{F}_X$, with $\operatorname{sing}(\mathcal{F}_X) = \{X = 0\}$, whose plaques are local integral curves of X.

Conversely, every singular holomorphic foliation by curves is locally induced by a holomorphic vector field (compare Proposition 1.11 below).

Here is one property of foliations which is quite elementary and will be frequently used in subsequent arguments:

PROPOSITION 1.6. *Let $\mathcal{F}$ be a singular holomorphic foliation by curves on M. Suppose that $p \in \overline{\mathcal{L}_q}$. Then $\overline{\mathcal{L}_p} \subset \overline{\mathcal{L}_q}$.*

PROOF. There is nothing to prove if $p \in \text{sing}(\mathcal{F})$, so let p be non-singular. Let $p' \in \mathcal{L}_p$ and join p to p' by a continuous path $\gamma : [0,1] \to \mathcal{L}_p$ (thus avoiding $\text{sing}(\mathcal{F})$) such that $\gamma(0) = p$ and $\gamma(1) = p'$. Choose a partition $0 = t_0 < t_1 < \cdots < t_n = 1$ of [0,1] and foliation charts (U_i, φ_i), $0 \leq i \leq n-1$, such that $\gamma[t_i, t_{i+1}] \subset U_i$. It follows from the local picture of plaques in $\varphi_i(U_i)$ that $\gamma(t_1) \in \overline{\mathcal{L}_q}$, so by repeating this argument, $p' \in \overline{\mathcal{L}_q}$. Since p' was arbitrary, we have $\overline{\mathcal{L}_p} \subset \overline{\mathcal{L}_q}$. □

EXAMPLE 1.7. Example 1.5 has a counterpart in the context of differential forms defined on domains in $\mathbb{C}^n$. Our main example which will be seen to be quite general is the following. Let $\omega = P(x,y)dy - Q(x,y)dx$ be a holomorphic 1-form on $\mathbb{C}^2$, where P and Q are relatively prime polynomials. By definition, the singular foliation induced by ω, $\mathcal{F}_\omega : \{\omega = 0\}$, is the one induced on $\mathbb{C}^2$ by the vector field $X(x,y) = P(x,y)\partial/\partial x + Q(x,y)\partial/\partial y$ as in Example 1.5. In the language of 1-forms its leaves are obtained as follows: Take any $p \in \mathbb{C}^2$ at which P and Q are not simultaneously zero, and let $\eta : T \mapsto (x(T), y(T))$ be a holomorphic mapping on some disk $\mathbb{D}(0,r)$ which satisfies $\eta(0) = p$ and

$$P(x(T), y(T))\, y'(T) - Q(x(T), y(T))\, x'(T) = 0 \tag{1.2}$$

for all $T \in \mathbb{D}(0,r)$. The plaque through p is the image under η of some possibly smaller neighborhood of 0. Note that $\text{sing}(\mathcal{F}_\omega)$ is a finite set by Bezout's Theorem. Observe that

$$\mathcal{F}_\omega = \mathcal{F}_{f\omega} \tag{1.3}$$

for all holomorphic functions $f : \mathbb{C}^2 \to \mathbb{C}$ which are nowhere zero. It is exactly this property which allows us to extend the foliation from $\mathbb{C}^2$ to the complex projective plane $\mathbb{CP}^2$.

CONVENTION 1.8. For the rest of the manuscript, the term "Singular Holomorphic Foliation by Curves" will be abbreviated as "SHFC".

Now we make a digression to study polynomial SHFC's on $\mathbb{CP}^2$ which are obtained by extending an $\mathcal{F}_\omega$ induced by a polynomial 1-form ω on $\mathbb{C}^2$.

1.3. Geometry of $\mathbb{CP}^2$. Consider $\mathbb{C}^3 \smallsetminus \{(0,0,0)\}$ with the action of $\mathbb{C}^*$ defined by $\lambda \cdot (x_0, x_1, x_2) = (\lambda x_0, \lambda x_1, \lambda x_2)$. The orbit of (x_0, x_1, x_2) is denoted by $[x_0, x_1, x_2]$. The quotient of $\mathbb{C}^3 \smallsetminus \{(0,0,0)\}$ modulo this action (with the quotient topology) is called the *complex projective plane* $\mathbb{CP}^2$, and the natural projection $\mathbb{C}^3 \smallsetminus \{(0,0,0)\} \to \mathbb{CP}^2$ is denoted by π. $\mathbb{CP}^2$ can be made into a compact complex 2-manifold in the following way: Cover $\mathbb{CP}^2$ by three open sets

$$U_i := \{[x_0, x_1, x_2] : x_i \neq 0\}, \qquad i = 0, 1, 2 \tag{1.4}$$

and define homeomorphisms $\phi_i : \mathbb{C}^2 \to U_i$ by

$$\begin{array}{rcl} \phi_0(x,y) & = & [1, x, y] \\ \phi_1(u,v) & = & [u, 1, v] \\ \phi_2(r,s) & = & [r, s, 1]. \end{array} \tag{1.5}$$

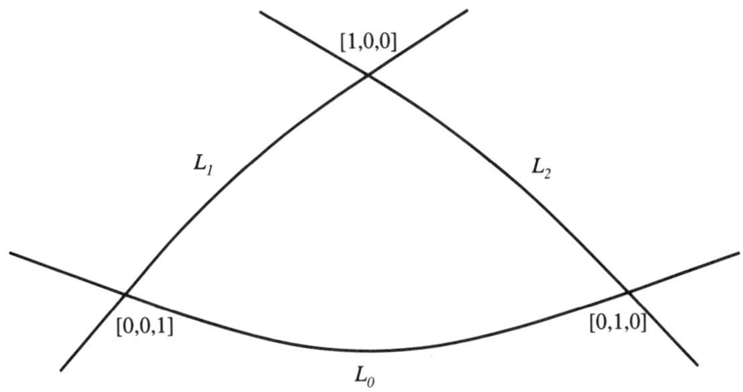

FIGURE 2. *Geometry of* $\mathbb{CP}^2$.

The change of coordinates $\phi_{ij} = \phi_j^{-1} \circ \phi_i$ are given by

$$\phi_{01}(x,y) = \phi_1^{-1} \circ \phi_0(x,y) = \left(\frac{1}{x}, \frac{y}{x}\right)$$

$$\phi_{12}(u,v) = \phi_2^{-1} \circ \phi_1(u,v) = \left(\frac{u}{v}, \frac{1}{v}\right) \tag{1.6}$$

$$\phi_{20}(r,s) = \phi_0^{-1} \circ \phi_2(r,s) = \left(\frac{s}{r}, \frac{1}{r}\right).$$

These being holomorphic, the atlas $\{(U_i, \phi_i^{-1}), i = 0,1,2\}$ determines a unique complex structure on $\mathbb{CP}^2$ for which the ϕ_i are biholomorphisms. Intuitively, $\mathbb{CP}^2$ is a one-line compactification of $\mathbb{C}^2$ for the following reason: Each (U_i, ϕ_i^{-1}) is called an *affine chart* of $\mathbb{CP}^2$. Each $L_i := \mathbb{CP}^2 \smallsetminus U_i$ has a natural structure of the Riemann sphere; for example $L_0 = \{[0,x,y] : (x,y) \in \mathbb{C}^2\}$ can be identified with $\{[x,y] : (x,y) \in \mathbb{C}^2\} \simeq \mathbb{CP}^1$ under the restriction to $\mathbb{C}^2$ of the action of $\mathbb{C}^*$. Each L_i is called the *line at infinity* with respect to the affine chart (U_i, ϕ_i^{-1}). Fig. 2 shows the relative position of the lines L_i.

It is easy to check that given any *projective line* L in $\mathbb{CP}^2$, i.e., the projection under π of any plane $ax_0 + bx_1 + cx_2 = 0$ in $\mathbb{C}^3$, one can choose a biholomorphism $\phi : \mathbb{C}^2 \to \mathbb{CP}^2 \smallsetminus L$. In this way, L may be viewed as the line at infinity with respect to some affine chart.

1.4. Algebraic Curves in $\mathbb{CP}^2$. Suppose that $P = P(x,y) = \sum a_{ij}\, x^i y^j$ is a polynomial of degree k on $(x,y) \in \mathbb{C}^2$. Using (1.6), we write P in two other affine charts as

$$P \circ \phi_{10}(u,v) = P\left(\frac{1}{u}, \frac{v}{u}\right) = u^{-k} \sum a_{ij}\, u^{k-(i+j)} v^j$$

$$P \circ \phi_{20}(r,s) = P\left(\frac{s}{r}, \frac{1}{r}\right) = r^{-k} \sum a_{ij}\, r^{k-(i+j)} s^i.$$

Set $P'(u,v) = \sum a_{ij}\, u^{k-(i+j)} v^j$ and $P''(r,s) = \sum a_{ij}\, r^{k-(i+j)} s^i$. Then the *algebraic curve* S_P in $\mathbb{CP}^2$ is defined as the compact set

$$\phi_0\{(x,y) : P(x,y) = 0\} \cup \phi_1\{(u,v) : P'(u,v) = 0\} \cup \phi_2\{(r,s) : P''(r,s) = 0\}.$$

Another way of viewing this curve is by introducing the homogeneous polynomial $\widetilde{P}$ of degree k in $\mathbb{C}^3$ as

$$\widetilde{P}(x_0, x_1, x_2) := x_0^k \, P\left(\frac{x_1}{x_0}, \frac{x_2}{x_0}\right) = \sum a_{ij} \; x_1^i x_2^j x_0^{k-(i+j)}.$$

It is then easily verified that $S_P = \pi\{(x_0, x_1, x_2) : \widetilde{P}(x_0, x_1, x_2) = 0\}$. A projective line is an algebraic curve S_P for a polynomial P of degree 1.

1.5. Extending Polynomial 1-froms on $\mathbb{CP}^2$. Consider a polynomial 1-from $\omega = Pdy - Qdx$ on $\mathbb{C}^2$ and its corresponding SHFC $\mathcal{F}_\omega$, as in Example 1.7. Using the coordinate map ϕ_0 in (1.5) one can transport $\mathcal{F}_\omega$ to U_0. To complete this picture to all of $\mathbb{CP}^2$ we have to define the foliation on L_0. This can be done as follows. First transport $\mathcal{F}_\omega$ to the affine chart (u, v). To this end, write

$$\begin{aligned}
\tilde{\omega}(u, v) := & \;\; (\phi_{10}^*\omega)(u, v) \\
= & \;\; P\left(\frac{1}{u}, \frac{v}{u}\right) d\left(\frac{v}{u}\right) - Q\left(\frac{1}{u}, \frac{v}{u}\right) d\left(\frac{1}{u}\right) \\
= & \;\; u^{-1} P\left(\frac{1}{u}, \frac{v}{u}\right) dv - u^{-2}\left\{vP\left(\frac{1}{u}, \frac{v}{u}\right) - Q\left(\frac{1}{u}, \frac{v}{u}\right)\right\} du.
\end{aligned}$$

Set $R(x, y) := yP(x, y) - xQ(x, y)$. Then

$$\tilde{\omega}(u, v) = u^{-1} P\left(\frac{1}{u}, \frac{v}{u}\right) dv - u^{-1} R\left(\frac{1}{u}, \frac{v}{u}\right) du.$$

Let k be the least positive integer such that $\omega' := u^{k+1}\tilde{\omega}$ is a polynomial 1-form on $(u, v) \in \mathbb{C}^2$. Two foliations $\mathcal{F}_{\omega'}$ and $\mathcal{F}_{\tilde{\omega}}$ are then identical on $\{(u, v) \in \mathbb{C}^2 : u \neq 0\}$ by (1.3); however, $\mathcal{F}_{\omega'}$ which is defined on all of $(u, v) \in \mathbb{C}^2$ is a well-defined extension of $\mathcal{F}_{\tilde{\omega}}$. Now transport $\mathcal{F}_{\omega'}$ by ϕ_1 to U_1. It is easily checked that the foliation induced by $(\mathcal{F}_\omega, \phi_0)$ coincides with that of $(\mathcal{F}_{\omega'}, \phi_1)$ on $U_0 \cap U_1$.

In a similar way, $\mathcal{F}_\omega$ can be transported to the affine chart (r, s) by ϕ_{20} to obtain a foliation $\mathcal{F}_{\omega''}$ induced by a polynomial 1-form ω'' on $(r, s) \in \mathbb{C}^2$. Then $\mathcal{F}_{\omega''}$ is transported to U_2 by ϕ_2.

We still denote the extended foliation on $\mathbb{CP}^2$ by $\mathcal{F}_\omega$ and frequently identify $\mathcal{F}_\omega, \mathcal{F}_{\omega'}, \mathcal{F}_{\omega''}$ with their transported companions on $\mathbb{CP}^2$. Thus, without saying explicitly, the affine charts $(x, y), (u, v)$, and (r, s) are considered as *subsets* of $\mathbb{CP}^2$ by identifying them with U_0, U_1 and U_2, respectively.

It follows from the above construction that in each affine chart, $\mathcal{F}_\omega$ is given by the integral curves of the following vector fields:

$$\begin{aligned}
&\text{In } (x, y) \in U_0, \quad X_0 = P(x, y)\frac{\partial}{\partial x} + Q(x, y)\frac{\partial}{\partial y} \\
&\text{In } (u, v) \in U_1, \quad X_1 = u^k P\left(\frac{1}{u}, \frac{v}{u}\right)\frac{\partial}{\partial u} + u^k R\left(\frac{1}{u}, \frac{v}{u}\right)\frac{\partial}{\partial v} \qquad (1.7) \\
&\text{In } (r, s) \in U_2, \quad X_2 = -r^l Q\left(\frac{s}{r}, \frac{1}{r}\right)\frac{\partial}{\partial r} + r^l R\left(\frac{s}{r}, \frac{1}{r}\right)\frac{\partial}{\partial s},
\end{aligned}$$

where k and l are the least positive integers making the above into polynomial vector fields. This shows that $\mathcal{F}_\omega$ is an SHFC on $\mathbb{CP}^2$ with $\operatorname{sing}(\mathcal{F}_\omega) = S_P \cap S_Q \cap S_R$.

1.6. Holomorphic Vector Fields on $\mathbb{CP}^2$. So far it should seem to be that leaves of the SHFC's $\mathcal{F}_\omega$ constructed on $\mathbb{CP}^2$ are realized as integral curves of holomorphic *line fields* on $\mathbb{CP}^2$ rather than *vector fields.* This distinction should have been observed when we multiplied $\tilde{\omega}$ by a power of u to cancel the pole at $u = 0$ (compare §1.5). In fact, the special geometry of the projective plane puts a severe restriction on holomorphic vector fields on $\mathbb{CP}^2$. As a result, very few SHFC's on $\mathbb{CP}^2$ can be globally described by a single holomorphic vector field.

The following proposition describes all holomorphic vector fields on $\mathbb{CP}^2$. The same characterization is true for $\mathbb{CP}^n$, as can be shown by a coordinate-free argument [**CKP**], but here we present a very elementary proof for $\mathbb{CP}^2$.

PROPOSITION 1.9. *Every holomorphic vector field on $\mathbb{CP}^2$ lifts to a linear vector field on $\mathbb{C}^3$.*

PROOF. Let X be a holomorphic vector field on $\mathbb{CP}^2$, which has the following expressions in the three affine charts U_0, U_1, and U_2:

$$\begin{aligned}
\text{In } (x,y) \in U_0 &: X_0 = f_0 \frac{\partial}{\partial x} + g_0 \frac{\partial}{\partial y} \\
\text{In } (u,v) \in U_1 &: X_1 = f_1 \frac{\partial}{\partial u} + g_1 \frac{\partial}{\partial v} \\
\text{In } (r,s) \in U_2 &: X_2 = f_2 \frac{\partial}{\partial r} + g_2 \frac{\partial}{\partial s}.
\end{aligned}$$

Since $(\phi_{01})_* X_0 = X_1$ and $(\phi_{02})_* X_0 = X_2$, we obtain

$$f_1(u,v) = -u^2 f_0\left(\frac{1}{u}, \frac{v}{u}\right) \tag{1.8}$$

$$g_1(u,v) = -uv\ f_0\left(\frac{1}{u}, \frac{v}{u}\right) + u\ g_0\left(\frac{1}{u}, \frac{v}{u}\right), \tag{1.9}$$

and

$$f_2(r,s) = -r^2 g_0\left(\frac{s}{r}, \frac{1}{r}\right) \tag{1.10}$$

$$g_2(r,s) = r f_0\left(\frac{s}{r}, \frac{1}{r}\right) - rs\ g_0\left(\frac{s}{r}, \frac{1}{r}\right). \tag{1.11}$$

Consider the power series expansions

$$f_0(x,y) = \sum_{i,j \geq 0} a_{ij}\ x^i y^j \quad \text{and} \quad g_0(x,y) = \sum_{i,j \geq 0} b_{ij}\ x^i y^j.$$

Since f_1, g_1, f_2, g_2 are holomorphic on $\mathbb{C}^2$, we have the following:

$$\begin{array}{llll}
a_{ij} = 0 & \text{if} & i+j \geq 3 & (\text{by}(1.8)) \\
b_{ij} = 0 & \text{if} & i+j \geq 3 & (\text{by}(1.9)) \\
a_{02} = b_{20} = 0 & a_{20} = b_{11} & b_{02} = a_{11} & (\text{by}(1.9))
\end{array}$$

and (1.10) and (1.11) give no new relations. Now it can be easily checked that X lifts to the linear vector field

$$\begin{aligned}\tilde{X} &= (x_0 - a_{20}x_1 - a_{11}x_2)\frac{\partial}{\partial x_0} \\ &+(a_{00}x_0 + (a_{10}+1)x_1 + a_{01}x_2)\frac{\partial}{\partial x_1} \\ &+(b_{00}x_0 + b_{10}x_1 + (b_{01}+1)x_2)\frac{\partial}{\partial x_2}\end{aligned}$$

on $\mathbb{C}^3$. Conversely, every linear vector field on $\mathbb{C}^3$ descends to a holomorphic vector field on $\mathbb{CP}^2$, and we are done. □

As a result, every holomorphic vector field on $\mathbb{CP}^2$ is seen in the affine chart (x, y) as $X_0 = f_0\ \partial/\partial x + g_0\ \partial/\partial y$, where

$$\begin{aligned} f_0(x,y) &= a_1 + a_2x + a_3y + x(Ax+By) \\ g_0(x,y) &= b_1 + b_2x + b_3y + y(Ax+By), \end{aligned} \tag{1.12}$$

for some complex constants a_i, b_i, A, B.

Our next goal is to give in detail the proof of the remarkable fact that every SHFC on $\mathbb{CP}^2$ is of the form $\mathcal{F}_\omega$ for some polynomial 1-form ω (equivalently, a polynomial vector field) on $\mathbb{C}^2$. The same proof works for every $\mathbb{CP}^n$, $n \geq 2$, with only minor modifications. We refer the reader to [**GO**] for a more general set up. Another proof for this fact can be given by methods of algebraic geometry (see [**I3**]).

The proof goes along the following lines: First we associate to each SHFC on a complex manifold M a (holomorphic) line bundle $B' \hookrightarrow TM$ over $M \smallsetminus E$ (E is an analytic subvariety of M, as in Definition 1.3). Then we show that B' may be extended to a tangent line bundle B over M (Theorem 1.12). This establishes a natural relationship between SHFC's on M and holomorphic "bundle maps" $\beta : B \to TM$. Rigidity of line bundles in the case $M = \mathbb{CP}^2$ will then be applied to show that each bundle map $\beta : B \to T\mathbb{CP}^2$ is induced by a polynomial 1-form on $\mathbb{C}^2$ (Corollary 1.17). The foundational material on holomorphic line bundles on complex manifolds used here can be found in [**GH**] or [**Kod**].

1.7. SHFC's and Line Bundles. Let $\mathcal{F}$ be an SHFC on a complex manifold M. Let $\{(U_i, \varphi_i)\}_{i\in I}$ be the collection of foliation charts on $M' := M \smallsetminus E$. By (1.1) the transition functions $\varphi_{ij} = \varphi_i \circ \varphi_j^{-1} = (\varphi_{ij}^1, \ldots, \varphi_{ij}^n)$ satisfy $\partial\varphi_{ij}^k/\partial z_1 = 0$ for $2 \leq k \leq n$. Applying the chain rule to $\varphi_{ij}^1 = \varphi_{ik}^1 \circ \varphi_{kj}^1$, we obtain

$$\frac{\partial\varphi_{ij}^1}{\partial z_1}(p) = \frac{\partial\varphi_{ik}^1}{\partial z_1}(\varphi_{kj}(p))\frac{\partial\varphi_{kj}^1}{\partial z_1}(p) \tag{1.13}$$

for every $p \in \varphi_j(U_i \cap U_j \cap U_k)$. Define $\xi_{ij} : U_i \cap U_j \to \mathbb{C}^*$ by

$$\xi_{ij}(p) := \frac{\partial\varphi_{ij}^1}{\partial z_1}(\varphi_j(p)). \tag{1.14}$$

The cocycle relation $\xi_{ij} = \xi_{ik} \cdot \xi_{kj}$ on $U_i \cap U_j \cap U_k$ follows from (1.13). Let B' be the holomorphic line bundle on M' defined by this cocycle. This line bundle has a natural holomorphic injection into TM in such a way that the image of the fiber over p under this injection coincides with the tangent line to $\mathcal{L}_p$ at p. To see this, observe that

$$B' = \bigcup_{i \in I} (U_i \times \mathbb{C})/\sim,$$

where $(p,t) \in U_i \times \mathbb{C}$ is identified under $\sim$ with $(q,t') \in U_j \times \mathbb{C}$ if and only if $p = q$ and $t = \xi_{ij}(p)t'$. For $p \in M'$, let B'_p be the fiber of B' over p and define $\beta'_i : U_i \times \mathbb{C} \to TM|_{U_i}$ by

$$\beta'_i(p,t) := t((\varphi_i^{-1})_* \frac{\partial}{\partial z_1})(p). \tag{1.15}$$

It follows then from (1.14) that if $p \in U_i \cap U_j$ and $(p,t) \sim (p,t')$, then $\beta'_i(p,t) = \beta'_j(p,t')$. Therefore, the β'_i give rise to a holomorphic bundle map $\beta' : B' \to TM$ which is injective and $\beta'(B'_p)$ is the tangent line to $\mathcal{L}_p$ at p, a subspace of T_pM.

Can B' be extended to a line bundle over all of M? The crucial point for the answer, which is affirmative, is the condition that the codimension of E is > 1. Recall the following classical theorem of F. Hartogs (see for example [**W**] for a proof):

THEOREM 1.10. *Let $U \subset \mathbb{C}^n$ be a domain and $E \subset U$ be an analytic subvariety of U of codimension > 1. Then every holomorphic (resp. meromorphic) function on $U \smallsetminus E$ can be extended to a holomorphic (resp. meromorphic) function on U.*

Using this theorem one can extend line bundles induced by SHFC's. To this end, we have to prove the following simple but remarkable proposition (compare [**GO**]). The letter E will denote an analytic subvariety of the ambient space which satisfies Definition 1.4.

PROPOSITION 1.11. *Let $U \subset \mathbb{C}^n$ be a domain and $\mathcal{F}$ be an SHFC on U. Then for each $p \in U$ there exists a holomorphic vector field X on some neighborhood U_p of p such that X is non-vanishing on $U_p \smallsetminus E$ and is tangent to the leaves of $\mathcal{F}$. This X is unique up to multiplication by a holomorphic function which is non-zero in a neighborhood of p. Moreover, $q \in U_p \cap E$ is a removable singularity of $\mathcal{F}$ if and only if $X(q) \neq 0$.*

PROOF. There is nothing to prove if $p \notin E$, so let $p \in E$ and let U_p be a connected neighborhood of p in U. Then $U'_p := U_p \smallsetminus E$ is open and connected. Each $q \in U'_p$ has a small connected neighborhood $U_q \subset U'_p$ on which there is a holomorphic vector field representing $\mathcal{F}$ on U_q. Without loss of generality we may assume that the first component of these vector fields is not identically zero over U'_p. If $(Y_1, \dots, Y_n)$ is the vector field representing $\mathcal{F}$ on U_q, then $(1, Y_2/Y_1, \dots, Y_n/Y_1)$ is a meromorphic vector field on U_q representing $\mathcal{F}$ on $U_q \smallsetminus \{\text{zeros of } Y_1\}$. Repeating this argument for each $q \in U'_p$, and noting that away from E the vector field representing $\mathcal{F}$ is uniquely determined up to multiplication by a non-vanishing holomorphic function, one concludes that there are meromorphic functions $F_2, \dots, F_n$ defined on U'_p such that $(1, F_2, \dots, F_n)$ represents $\mathcal{F}$ on $U'_p \smallsetminus \{\text{poles of the } F_i \text{ in } U'_p\}$. By Theorem 1.10, each function F_i can be extended to a meromorphic function on U_p (still denoted by F_i) since the codimension of $U_p \cap E$ is > 1. The germ of F_i can be uniquely written as $F_i = f_i/g_i$ by choosing U_p small enough, where f_i and g_i are relatively prime holomorphic functions on U_p. Then $X := (g, gF_2, \dots, gF_n)$ is a holomorphic vector field on U_p representing $\mathcal{F}$ on $U'_p \smallsetminus \{\text{zeros of } g\}$, where g

is the least common multiple of the g_i. Note that codim $\{z : X(z) = 0\} > 1$ (the components of X are relatively prime).

Now let $q \in U'_p$, and (U, φ) be a foliation chart around q. The vector field $\varphi_* X$ has the form $h\, \partial/\partial z_1$ since it is tangent to the horizontal lines $\{z_2 = \text{const.}, \dots, z_n = \text{const.}\}$ away from zeros of $g \circ \varphi^{-1}$. Since the zero set of h either is empty or has codimension 1, while the zero set of $\varphi_* X$ has codimension >1, it follows that h is nowhere zero and $X(q) \neq 0$. Thus $X(q)$ is tangent to $\mathcal{L}_q$ at q.

For the uniqueness part, let $\tilde{X}$ be another such vector field. Then $\tilde{X} = \xi X$ on U'_p, where $\xi : U'_p \to \mathbb{C}^*$ is holomorphic. Extend ξ over U_p by Theorem 1.10. This new ξ is nowhere vanishing, since its zero set, if non-empty, would have codimension 1.

Finally, let $q \in U_p \cap E$ be a removable singularity of $\mathcal{F}$. Choose a compatible chart (U, φ) around q (compare Definition 1.4) and let $\tilde{X} = \varphi_*^{-1}(\partial/\partial z_1)$. Then $\tilde{X}$ describes $\mathcal{F}$ on U, so by the above uniqueness we have $\tilde{X} = \xi X$, with ξ being a holomorphic non-vanishing function on some neighborhood of q. Thus $X(q) \neq 0$. Conversely, suppose that $X(q) \neq 0$, and let (U, φ) be a local chart around q straightening the integral curves of X, i.e., $\varphi_*(X|_U) = \partial/\partial z_1$. Then it is evident that (U, φ) is compatible with every foliation chart of $\mathcal{F}$ in U'_p. □

Now let $\mathcal{F}$ be an SHFC on M and $\beta' : B' \to TM$ be the bundle map constructed in (1.15). According to Proposition 1.11, each $p \in M$ has a neighborhood U_i and a holomorphic vector field X_i defined on U_i representing $\mathcal{F}$ on $U_i \smallsetminus E$. By the uniqueness part of Proposition 1.11, whenever $U_i \cap U_j \neq \emptyset$, we have $X_j = \xi_{ij} X_i$ on $U_i \cap U_j$, where $\xi_{ij} : U_i \cap U_j \to \mathbb{C}^*$ is holomorphic. Let B be the line bundle over M defined by the cocycle $\{\xi_{ij}\}$. As in the construction of β' in (1.15), define $\beta_i : U_i \times \mathbb{C} \to TM|_{U_i}$ by

$$\beta_i(p, t) := t X_i(p). \tag{1.16}$$

The definition of B shows that the β_i patch together to yield a well-defined bundle map $\beta : B \to TM$ for which $\beta(B_p)$ is the tangent line to $\mathcal{L}_p$ at p if $p \notin E$. Note that B is an extension of B' since $\beta^{-1} \circ \beta' : B' \to B|_{M'}$ is an isomorphism of line bundles. If $\eta : B \to TM$ is another bundle map which represents $\mathcal{F}$ away from E, then $\eta : U_i \times \mathbb{C} \to TM|_{U_i}$ satisfies $\eta(p,t) = t\lambda_i(p) X_i(p)$ for $p \in U_i \smallsetminus E$, where $\lambda_i : U_i \smallsetminus E \to \mathbb{C}$ is non-vanishing. Extend λ_i to U_i by Theorem 1.10. Note that the action of η on B_p is well-defined, so that $\lambda_i(p) = \lambda_j(p)$ if $p \in U_i \cap U_j$. Defining $\lambda : M \to \mathbb{C}$ by $\lambda|_{U_i} := \lambda_i$, we see that $\eta = \lambda \cdot \beta$. Furthermore, λ can only vanish on E, but its zero set, if non-empty, would have to have codimension 1. Therefore, λ does not vanish at all. Finally, the last part of Proposition 1.11 and (1.16) show that $p \in E$ is a removable singularity of $\mathcal{F}$ if and only if $\beta(B_p) \neq 0$.

Summarizing the above argument, we obtain

THEOREM 1.12. *The line bundle B' over $M \smallsetminus E$ associated to an SHFC $\mathcal{F}$ on M can be extended to a line bundle B over M. There exists a holomorphic bundle map $\beta : B \to TM$ for which $\beta(B_p)$ is the tangent line to $\mathcal{L}_p$ at p if $p \notin E$. This β is unique up to multiplication by a nowhere vanishing holomorphic function on M. A point $p \in E$ is a removable singularity of $\mathcal{F}$ if and only if $\beta(B_p) \neq 0$.*

REMARK 1.13. It follows from the last part of Theorem 1.12 that sing$(\mathcal{F})$ is precisely $\{p \in M : \beta(B_p) = 0\}$; in particular, after removing all the possible removable singularities of $\mathcal{F}$ in E, $\text{sing}(\mathcal{F})$ turns out to be an analytic subvariety of M of codimension > 1.

REMARK 1.14. If M is a *compact* complex manifold (in particular, if $M = \mathbb{CP}^2$), then every two bundle maps $\beta, \beta' : B \to TM$ representing $\mathcal{F}$ differ by a non-zero multiplicative constant.

REMARK 1.15. The bundle map constructed in the above theorem is unique in the sense that *two bundle maps $\beta : B \to TM$ and $\tilde{\beta} : \tilde{B} \to TM$ represent the same SHFC if and only if there exists a bundle isomorphism $\psi : B \to \tilde{B}$ with $\beta = \tilde{\beta} \circ \psi$.* To see this, first assume that both bundle maps represent $\mathcal{F}$. By the construction, we may assume that B (resp. $\tilde{B}$) is defined by $(\{U_i\}, \{\xi_{ij}\})$ (resp. $(\{V_k\}, \{\eta_{kl}\})$) and there are holomorphic vector fields X_i on U_i (resp. Y_k on V_k) satisfying $X_j = \xi_{ij} X_i$ on $U_i \cap U_j$ (resp. $Y_l = \eta_{kl} Y_k$ on $V_k \cap V_l$) and $\beta_i(p,t) := tX_i(p)$ (resp. $\tilde{\beta}_k(p,t) := tY_k(p)$). Since B and $\tilde{B}$ both represent $\mathcal{F}$, for every i, k with $U_i \cap V_k \neq \emptyset$, there is a nowhere vanishing holomorphic function λ_{ik} such that $X_i(p) = \lambda_{ik}(p) Y_k(p)$ for all $p \in (U_i \cap V_k) \smallsetminus \text{sing}(\mathcal{F})$. By Theorem 1.10, λ_{ik} can be extended to $U_i \cap V_k$. Note that the extended function cannot vanish at all, since its only possible zero set is $U_i \cap V_k \cap \text{sing}(\mathcal{F})$ which has codimension >1. Now define $\psi : B \to \tilde{B}$ by mapping the class of $(p,t) \in U_i \times \mathbb{C}$ to the class of $(p, t\lambda_{ik}(p)) \in V_k \times \mathbb{C}$. It is quite easy to see that ψ defines an isomorphism of line bundles with $\beta = \tilde{\beta} \circ \psi$.

Conversely, let $\beta : B \to TM$ represent $\mathcal{F}$. If $\tilde{B}$ is any line bundle over M isomorphic to B by $\psi : B \to \tilde{B}$, and if $\tilde{\beta} = \beta \circ \psi^{-1}$, then $\tilde{\beta} : \tilde{B} \to TM$ also represents $\mathcal{F}$.

1.8. Line Bundles and $H^1(M, \mathcal{O}^*)$. Every holomorphic line bundle B on a complex manifold M is uniquely determined by an open covering $\mathcal{U} = \{U_i\}_{i \in I}$ of M and a family $\{\xi_{ij}\}_{i,j \in I}$ of non-vanishing holomorphic functions on each $U_i \cap U_j \neq \emptyset$ satisfying the cocycle relation $\xi_{ij} = \xi_{ik} \cdot \xi_{kj}$ on $U_i \cap U_j \cap U_k$. Thus $\{\xi_{ij}\}$ may be regarded as an element of $Z^1(\mathcal{U}, \mathcal{O}^*)$, the group of Čech 1-cocycles with coefficients in the sheaf of non-vanishing holomorphic functions on M, with respect to the covering $\mathcal{U}$. Let B' be another line bundle on M defined by cocycle $\{\eta_{ij}\}$. It is not difficult to see that B' is isomorphic to B if and only if there exist non-vanishing holomorphic functions $f_i : U_i \to \mathbb{C}^*$ such that $\eta_{ij} = (f_i/f_j)\xi_{ij}$ on $U_i \cap U_j$. Interpreting $\{\xi_{ij}\}$ and $\{\eta_{ij}\}$ as elements of $Z^1(\mathcal{U}, \mathcal{O}^*)$, the last condition may be written as $\{\xi_{ij}/\eta_{ij}\} \in B^1(\mathcal{U}, \mathcal{O}^*)$, the group of Čech 1-coboundaries. We conclude that *two holomorphic line bundles on M are isomorphic if and only if they represent the same element of the Čech cohomology group $H^1(M, \mathcal{O}^*) := Z^1(M, \mathcal{O}^*)/B^1(M, \mathcal{O}^*)$.*

1.9. Line Bundles on $\mathbb{CP}^2$. Most of the results presented here are true for $\mathbb{CP}^n$, $n \geq 2$. However, we only treat the case $n = 2$ for simplicity of exposition.

Consider the short exact sequence of sheaves

$$0 \to \mathbb{Z} \xrightarrow{2\pi i} \mathcal{O} \xrightarrow{\exp} \mathcal{O}^* \to 0$$

on M. From this sequence we obtain the long exact sequence of cohomology groups

$$\cdots \to H^1(M, \mathcal{O}) \to H^1(M, \mathcal{O}^*) \xrightarrow{c_1} H^2(M, \mathbb{Z}) \to H^2(M, \mathcal{O}) \to \cdots \tag{1.17}$$

For every line bundle $B \in H^1(M, \mathcal{O}^*)$, $c_1(B) \in H^2(M, \mathbb{Z})$ is called the *first Chern class* of B. Explicitly, let $B = [\{\xi_{ij}\}]$ be defined for a covering $\mathcal{U} = \{U_i\}$ for which every U_i is connected, every $U_i \cap U_j$ is simply-connected, and every $U_i \cap U_j \cap U_k$

is connected. Then $c_1(B) = [\{c_{ijk}\}]$, where $c_{ijk} := 1/(2\pi\sqrt{-1})\{\log \xi_{jk} - \log \xi_{ik} + \log \xi_{ij}\}$ and the branches of logarithms are arbitrarily chosen [**Kod**].

There are two basic facts about the sequence (1.17) in the case $M = \mathbb{CP}^2$. First, for every $i \geq 1$ we have $H^i(\mathbb{CP}^2, \mathcal{O}) = 0$ as a consequence of the Hodge Decomposition Theorem [**GH**]. Second, $H^2(\mathbb{CP}^2, \mathbb{Z}) \simeq \mathbb{Z}$ [**GH**]. It follows that c_1 in (1.17) is an isomorphism, $H^1(\mathbb{CP}^2, \mathcal{O}^*)$ is the infinite cyclic group $\mathbb{Z}$, and every holomorphic line bundle on $\mathbb{CP}^2$ is determined up to isomorphism by its first Chern class (compare §1.8).

Now consider $\mathbb{CP}^2$ with affine charts $\{(U_i, \phi_i^{-1})\}_{i=0,1,2}$, as in §1.3. For every integer n, define a line bundle $B(n)$ on $\mathbb{CP}^2$ whose cocycle $\{\xi_{ij}\}_{i,j=0,1,2}$ is given by

$$\xi_{ij} : U_i \cap U_j \to \mathbb{C}^*, \quad i, j = 0, 1, 2$$

$$\xi_{ij}[x_0, x_1, x_2] := \left(\frac{x_j}{x_i}\right)^n. \tag{1.18}$$

It is not difficult to check that $B(-1)$, called the *canonical line bundle* over $\mathbb{CP}^2$, and its dual bundle $B(1)$ are both generators for the infinite cyclic group $H^1(\mathbb{CP}^2, \mathcal{O}^*)$. By a standard convention we define $c_1(B(1)) = 1$. It follows that $c_1(B(n)) = n$ for all integers n.

Adding up the above remarks, it follows that *every holomorphic line bundle B on $\mathbb{CP}^2$ is isomorphic to $B(n)$, where $n = c_1(B)$.*

Now suppose that $\mathcal{F}$ is an SHFC on $\mathbb{CP}^2$, and let $\beta : B \to T\mathbb{CP}^2$ be its associated bundle map given by Theorem 1.12. Choose $n = c_1(B)$ so that B is isomorphic to $B(n)$ by some $\psi : B \to B(n)$. By Remark 1.15, if $\tilde{\beta}$ is defined by $\tilde{\beta} := \beta \circ \psi^{-1}$, then $\tilde{\beta} : B(n) \to T\mathbb{CP}^2$ is a bundle map which also represents $\mathcal{F}$.

1.10. Explicit Form of SHFC's on $\mathbb{CP}^2$. In view of the remarks in the last paragraph, we are now going to determine the explicit form of every holomorphic bundle map $\beta : B(-n+1) \to T\mathbb{CP}^2$ for every integer n. (We choose $-n+1$ instead of n just to make later formulations easier.)

Once again consider $\mathbb{CP}^2$ equipped with the three affine charts $(U_i, \phi_i^{-1}), i = 0, 1, 2$. Restricting β to $U_i \times \mathbb{C}$, there exists a holomorphic vector field X_i on U_i representing the action of $\beta_i : U_i \times \mathbb{C} \to TU_i \simeq U_i \times \mathbb{C}^2$. In other words, $\beta_i(p, t) = (p, tX_i(p))$. Let $X_0 := f(x, y)\partial/\partial x + g(x, y)\partial/\partial y$ and $X_1 := \tilde{f}(u, v)\partial/\partial u + \tilde{g}(u, v)\partial/\partial v$, where $f, g, \tilde{f}, \tilde{g}$ are holomorphic on $\mathbb{C}^2$. Set $\tilde{U}_0 := U_0 \smallsetminus \{(x, y) : x = 0\}$ and $\tilde{U}_1 := U_1 \smallsetminus \{(u, v) : u = 0\}$. We have the following commutative diagram:

$$\begin{array}{ccc} \tilde{U}_0 \times \mathbb{C} & \xrightarrow{\ \xi\ } & \tilde{U}_1 \times \mathbb{C} \\ \Big\downarrow \beta_0 & & \Big\downarrow \beta_1 \\ T\tilde{U}_0 \simeq \tilde{U}_0 \times \mathbb{C}^2 & \xrightarrow{(\phi_{01})_*} & T\tilde{U}_1 \simeq \tilde{U}_1 \times \mathbb{C}^2 \end{array}$$

where $\xi(x, y, t) := (u, v, t/\xi_{01}(x, y))$. Note that by (1.18), $\xi_{01}(x, y) = \xi_{01}[1, x, y] = x^{-n+1} = u^{n-1}$ so that $\xi(x, y, t) = (u, v, u^{-n+1}t)$. It follows that $(\phi_{01})_* X_0 =$

$u^{-n+1}X_1$, or

$$\begin{aligned} \tilde{f}(u,v) &= -u^{n+1} f\left(\frac{1}{u}, \frac{v}{u}\right) \\ \tilde{g}(u,v) &= -u^{n}\left\{ v f\left(\frac{1}{u}, \frac{v}{u}\right) - g\left(\frac{1}{u}, \frac{v}{u}\right)\right\}. \end{aligned} \tag{1.19}$$

Since $\tilde{f}$ and $\tilde{g}$ are holomorphic on $\mathbb{C}^2$, (1.19) shows that if $n \leq -1$, then f and g both vanish. So let us assume that $n \geq 0$.

Suppose that $f = \sum_{k=0}^{\infty} f_k$ and $g = \sum_{k=0}^{\infty} g_k$, where f_k and g_k are the homogeneous parts of degree k of the power series expansions of f and g. It follows then from (1.19) that $f_k \equiv g_k \equiv 0$ for $k \geq n+2$, and

$$v f_{n+1}(1,v) - g_{n+1}(1,v) \equiv 0. \tag{1.20}$$

Coming back to the affine chart $(x,y) \in U_0$, we obtain from (1.20) that

$$y f_{n+1}(x,y) - x g_{n+1}(x,y) \equiv 0. \tag{1.21}$$

It is easy to see that there are no other restrictions on these homogeneous polynomials. Thus we have proved the following

THEOREM 1.16. *Let $\beta : B(-n+1) \to T\mathbb{CP}^2$ be a holomorphic bundle map. If $n \leq -1$, then $\beta \equiv 0$. If $n \geq 0$, then in the affine chart $(x,y) \in U_0$, β is given by a polynomial vector field of the form*

$$\sum_{k=0}^{n+1} f_k \frac{\partial}{\partial x} + \sum_{k=0}^{n+1} g_k \frac{\partial}{\partial y},$$

where f_k and g_k are homogeneous polynomials of degree k, and $y f_{n+1} - x g_{n+1} \equiv 0$.

Now let $\mathcal{F}$ be an SHFC on $\mathbb{CP}^2$ and let $\beta : B \to T\mathbb{CP}^2$ represent $\mathcal{F}$. Set $n = -c_1(B) + 1$. Then, by the above theorem $\mathcal{F}$ is induced by a polynomial 1-form

$$\omega = [(\sum_{k=0}^{n} f_k) + xh]\, dy - [(\sum_{k=0}^{n} g_k) + yh]\, dx,$$

in the affine chart U_0, where $h = f_{n+1}/x = g_{n+1}/y$ is a homogeneous polynomial of degree n or $h \equiv 0$. Let us assume for a moment that the latter happens, i.e., $f_{n+1} \equiv g_{n+1} \equiv 0$. Then, rewriting (1.19) gives us

$$\begin{aligned} \tilde{f}(u,v) &= -u^{n+1} \sum_{k=0}^{n} u^{-k} f_k(1,v) \\ \tilde{g}(u,v) &= -u^{n} \sum_{k=0}^{n} u^{-k} [v f_k(1,v) - g_k(1,v)]. \end{aligned} \tag{1.22}$$

This shows that $v f_n(1,v) - g_n(1,v) \not\equiv 0$ since otherwise $\tilde{f}$ and $\tilde{g}$ would have a common factor u, meaning that β would vanish on the entire line $u = 0$ (contradicting the fact that β must vanish at finitely many points). We summarize the above observations in the following

COROLLARY 1.17. *Let $\mathcal{F}$ be an SHFC on $\mathbb{CP}^2$ and $\beta : B \to T\mathbb{CP}^2$ be any holomorphic bundle map representing $\mathcal{F}$. Then $c_1(B) \leq 1$. If $n = -c_1(B)+1$, then $\mathcal{F}$ is given by a unique (up to a multiplicative constant) polynomial 1-form*

$$\omega = (f + xh)\,dy - (g + yh)\,dx \tag{1.23}$$

in the affine chart $(x, y) \in U_0$. Here

- *$f = \sum_{k=0}^{n} f_k$ and $g = \sum_{k=0}^{n} g_k$, with f_k and g_k being homogeneous polynomials of degree k,*
- *either h is a non-zero homogeneous polynomial of degree n, or if $h \equiv 0$, then $yf_n - xg_n \not\equiv 0$,*
- *the two polynomials $f + xh$ and $g + yh$ have no common factor.*

1.11. Geometric Degree of an SHFC on $\mathbb{CP}^2$. Here we show that the "cohomological degree" $n = -c_1(B) + 1$ of an SHFC $\mathcal{F}$ given by Corollary 1.17 coincides with a geometric invariant which we will call the "geometric degree" of $\mathcal{F}$. Roughly speaking, the geometric degree of $\mathcal{F}$ is the number of points at which a generic projective line is tangent to the leaves of $\mathcal{F}$. This notion allows us to give a stratification of the space of all SHFC's on $\mathbb{CP}^2$.

Let $\mathcal{F}$ be an SHFC on $\mathbb{CP}^2$ and L be any projective line such that $L \smallsetminus \operatorname{sing}(\mathcal{F})$ is not a leaf of $\mathcal{F}$. A point $p \in L$ is called a *tangency point* of $\mathcal{F}$ and L if either $p \in \operatorname{sing}(\mathcal{F})$, or $p \notin \operatorname{sing}(\mathcal{F})$ and L is the tangent line to $\mathcal{L}_p$ at p.

Let $\mathcal{F}$ be of the form $\mathcal{F}_\omega : \{\omega = Pdy - Qdx = 0\}$ in the affine chart $(x, y) \in U_0$, and $L \cap U_0$ be parametrized by $\ell(T) = (x_0 + aT, y_0 + bT)$, where $p = (x_0, y_0) = \ell(0)$. Then it is clear that p is a tangency point of $\mathcal{F}$ and L if and only if $T = 0$ is a root of the polynomial $T \mapsto bP(\ell(T)) - aQ(\ell(T))$. The *order* of tangency of $\mathcal{F}$ and L at p is defined to be the multiplicity of $T = 0$ as a root of this polynomial. Define

$$m(\mathcal{F}, L) := \sum_p (\text{order of tangency of } \mathcal{F} \text{ and } L \text{ at } p),$$

where the (finite) sum is taken over all the tangency points.

In the theorem below, we show that $m(\mathcal{F}, L)$ does not depend on L (as long as $L \smallsetminus \operatorname{sing}(\mathcal{F})$ is not a leaf), so that we can call it the *geometric degree* of $\mathcal{F}$.

THEOREM 1.18. *Let $\mathcal{F}$ be an SHFC on $\mathbb{CP}^2$ and $n = -c_1(B) + 1$, where B is the line bundle associated with $\mathcal{F}$. Let L be any projective line such that $L \smallsetminus \operatorname{sing}(\mathcal{F})$ is not a leaf. Then $m(\mathcal{F}, L) = n$. In particular, an SHFC $\mathcal{F}$ has geometric degree n if and only if $\mathcal{F}$ is induced by a 1-form ω as in (1.23) in which h is a non-zero homogeneous polynomial of degree n, or $h \equiv 0$ and $yf_n - xg_n \not\equiv 0$.*

PROOF. Let $n = -c_1(B) + 1$ so that $\mathcal{F}$ is induced by a 1-form ω as in (1.23). Take a projective line L such that $L \smallsetminus \operatorname{sing}(\mathcal{F})$ is not a leaf. Since the normal form (1.23) is invariant under projective transformations, we may assume that L is the x-axis. By (1.23), $(x, 0) \in L \cap U_0$ is a tangency point if and only if $g(x, 0) = \sum_{k=0}^{n} g_k(x, 0) = 0$. As for the point at infinity for L, consider the affine chart $(u, v) = (1/x, y/x) \in U_1$ in which L is given by the line $\{v = 0\}$. By (1.19), the foliation is described by the polynomial 1-form $\omega' = \tilde{f}\,dv - \tilde{g}\,du$, where

$$\tilde{f}(u,v) = -\sum_{k=0}^{n} u^{n+1-k} f_k(1,v) - h(1,v) \tag{1.24}$$

$$\tilde{g}(u,v) = -\sum_{k=0}^{n} u^{n-k}[v f_k(1,v) - g_k(1,v)].$$

This shows that L has a tangency point at infinity if and only if $u=0$ is a root of $\tilde{g}(u,0)=0$, or $g_n(1,0)=0$ by (1.24). To prove the theorem, we distinguish two cases:

(i) The polynomial $x \mapsto g_n(x,0)$ is not identically zero. In this case, $\sum_{k=0}^{n} g_k(x,0)$ is a polynomial of degree n in x, so there are exactly n finite tangency points on L counting multiplicities. Note that the point at infinity for L is not a tangency point since $g_n(1,0) \neq 0$. So in this case, $m(\mathcal{F}, L) = n$.

(ii) There is a largest $0 \leq j < n$ such that $x \mapsto g_j(x,0)$ is not identically zero (otherwise $\sum_{k=0}^{n} g_k(x,0)$ would be everywhere zero, so $L \smallsetminus \operatorname{sing}(\mathcal{F})$ would be a leaf). This means that $g(x,0)=0$ has exactly j roots counting multiplicities. In this case, the point at infinity for L is a tangency point of order $n-j$. In fact, (1.24) shows that $\tilde{g}(u,0) = -\sum_{k=0}^{n} u^{n-k} g_k(1,0) = -\sum_{k=0}^{j} u^{n-k} g_k(1,0)$, which has a root of multiplicity $n-j$ at $u=0$. Thus, there are n tangency points on L altogether, so again $m(\mathcal{F}, L) = n$. □

As an example, (1.12) shows that $\mathcal{F}$ *is induced by a holomorphic vector field on* $\mathbb{CP}^2$ *if and only if the geometric degree of* $\mathcal{F}$ *is* ≤ 1.

The set of all SHFC's on $\mathbb{CP}^2$ of geometric degree n is denoted by $\mathcal{D}_n$. Each $\mathcal{D}_n$ is topologized in the natural way: a neighborhood of $\mathcal{F} \in \mathcal{D}_n$ consists of all foliations of geometric degree n whose defining polynomials have coefficients close to that of $\mathcal{F}$, up to multiplication by a non-zero constant. To be more accurate, consider the complex linear space of all polynomial 1-forms ω as in (1.23). By Remark 1.14, ω and ω' define the same foliation if and only if there exists a non-zero constant λ such that $\omega' = \lambda\omega$. Therefore $\mathcal{D}_n$ can be considered as an open subset of the complex projective space $\mathbb{CP}^N$, where N is the dimension of the above linear space minus one, that is $N = 2\sum_{k=1}^{n+1} k + (n+1) - 1 = n^2 + 4n + 2$. It is not difficult to check that $\mathcal{D}_n$ is connected and dense in this projective space.

COROLLARY 1.19. *The set $\mathcal{D}_n$ of all SHFC's of geometric degree $n \geq 0$ on $\mathbb{CP}^2$ can be identified with an open, connected and dense subset of the complex projective space $\mathbb{CP}^N$, where $N = n^2 + 4n + 2$. So we can equip $\mathcal{D}_n$ with the induced topology and a natural Lebesgue measure class.*

The definition of $\mathcal{D}_n$ allows us to decompose the space $\mathcal{S}$ of all SHFC's on $\mathbb{CP}^2$ into a disjoint union $\bigcup \mathcal{D}_n$ and topologize it in a natural way. A subset $\mathcal{U}$ of $\mathcal{S}$ is open in this topology if and only if $\mathcal{U} \cap \mathcal{D}_n$ is open for every n. Hence this topology makes every $\mathcal{D}_n$ into a connected component of $\mathcal{S}$. Similarly, $\mathcal{S}$ inherits a natural Lebesgue measure class: A set $\mathcal{U} \subset \mathcal{S}$ has measure zero if and only if $\mathcal{U} \cap \mathcal{D}_n$ has measure zero in $\mathcal{D}_n$.

1.12. Line at Infinity as a Leaf. Let us find conditions on a 1-form ω which guarantee that the line at infinity $L_0 = \mathbb{CP}^2 \smallsetminus U_0$ with singular points of $\mathcal{F}_\omega$ deleted is a leaf. Consider an SHFC $\mathcal{F} \in \mathcal{D}_n$ induced by a polynomial 1-form ω as (1.23):

$\omega = (f + xh)dy - (g + yh)dx$, where $f = \sum_{k=0}^{n} f_k$, $g = \sum_{k=0}^{n} g_k$ and h is either a non-zero homogeneous polynomial of degree n, or $h \equiv 0$ but $yf_n - xg_n \not\equiv 0$.

THEOREM 1.20. *The line at infinity L_0 with singular points of $\mathcal{F} \in \mathcal{D}_n$ deleted is a leaf of $\mathcal{F}$ if and only if $h \equiv 0$.*

PROOF. This follows easily from the proof of Theorem 1.18. In fact, $L_0 \smallsetminus \operatorname{sing}(\mathcal{F})$ is a leaf if and only if the line $\{u = 0\}$ is a solution of $\omega' = \tilde{f}dv - \tilde{g}du = 0$. By (1.24), this happens if and only if $h(1, v) \equiv 0$. Since h is a homogeneous polynomial, the latter condition is equivalent to $h \equiv 0$. □

REMARK 1.21. Here is an alternative notation for polynomial 1-forms which will be used in many subsequent discussions. Let

$$\omega = (f + xh)dy - (g + yh)dx = Pdy - Qdx,$$

as in (1.23). Define

$$R(x, y) = yP(x, y) - xQ(x, y) = yf(x, y) - xg(x, y)$$

as in §1.5, and note that $\deg R \leq n + 1$. Then $\mathcal{F}|_{U_1}$ is given by $\{\omega' = 0\}$, where

$$\omega'(u, v) = u^k P\left(\frac{1}{u}, \frac{v}{u}\right) dv - u^k R\left(\frac{1}{u}, \frac{v}{u}\right) du,$$

and, as in (1.7), k is the least positive integer which makes ω' a polynomial 1-form. (Note that this representation should be identical to (1.24) up to a multiplicative constant.) If $h \not\equiv 0$, then $\deg P = n + 1$ and so $k = n + 1$. On the other hand, if $h \equiv 0$, then $yf_n - xg_n \not\equiv 0$ by Theorem 1.18 which means $\deg R = n + 1$. So again we have $k = n + 1$.

For a fixed $\mathcal{F}$, we denote $L_0 \smallsetminus \operatorname{sing}(\mathcal{F})$ by $\mathcal{L}_\infty$ whenever it is a leaf of $\mathcal{F}$. We often refer to $\mathcal{L}_\infty$ as the *leaf at infinity.*

As can be seen from the above theorem, for a foliation $\mathcal{F} \in \mathcal{D}_n$ the line at infinity $L_0 \smallsetminus \operatorname{sing}(\mathcal{F})$ is unlikely to be a leaf since this condition is equivalent to vanishing of a homogeneous polynomial. This is a consequence of the way we topologized the space $\mathcal{S}$ of all SHFC's on $\mathbb{CP}^2$ using the topologies on the $\mathcal{D}_n$. The decomposition $\bigcup \mathcal{D}_n$ is quite natural from the geometric point of view; however, it leads to a rather peculiar condition on the polynomials describing the associated 1-form (Theorem 1.18). The situation can be changed in a delicate way by choosing a different decomposition $\mathcal{S} = \bigcup \mathcal{A}_n$ which is more natural from the point of view of differential equations in $\mathbb{C}^2$ but has no longer an intrinsic geometric meaning. Elements of $\mathcal{A}_n$ are simply determined by the maximum degree of their defining polynomials.

DEFINITION 1.22. Fix the affine chart $(x, y) \in U_0$ and let $n \geq 0$. We say that an SHFC $\mathcal{F}$ belongs to the class $\mathcal{A}_n$ if it is induced by a polynomial 1-form $\omega = Pdy - Qdx$ with $\max\{\deg P, \deg Q\} = n$ and P, Q relatively prime. The number n is called the *affine degree* of $\mathcal{F}$ (with respect to U_0)

Note that $\mathcal{A}_n$ is well-defined since if $\mathcal{F}_\omega = \mathcal{F}_{\omega'}$, then $\omega' = \lambda\omega$ for some non-zero constant λ. It is important to realize that unlike the condition $\mathcal{F} \in \mathcal{D}_n$ (normal form (1.23)), whether or not $\mathcal{F} \in \mathcal{A}_n$ strongly depends on the choice of a particular affine coordinate system, and that is why we call n the "affine degree."

Consider the complex linear space of all polynomial 1-forms $\omega = Pdy - Qdx$ with $\max\{\deg P, \deg Q\} \leq n$, which has dimension $(n+1)(n+2)$. Then, as in the case of $\mathcal{D}_n$, one has

COROLLARY 1.23. *The set $\mathcal{A}_n$ of all SHFC's of affine degree $n \geq 0$ on $\mathbb{CP}^2$ can be identified with an open, connected and dense subset of the complex projective space $\mathbb{CP}^N$, where $N = n^2 + 3n + 1$. So we can equip $\mathcal{A}_n$ with the induced topology and a natural Lebesgue measure class.*

Using the decomposition of $\mathcal{S}$ into the disjoint union of the $\mathcal{A}_n$, we can define a new topology and measure class on $\mathcal{S}$ in the same way we did using the $\mathcal{D}_n$ (see the remarks after Corollary 1.19). In this new topology, each class $\mathcal{A}_n$ turns into a connected component of $\mathcal{S}$. The topologies and measure classes coming from $\bigcup \mathcal{A}_n$ and $\bigcup \mathcal{D}_n$ are significantly different. As a first indication of this difference, let $\mathcal{F} : \{\omega = Pdy - Qdx = 0\} \in \mathcal{A}_n$ and decompose $P = \sum_{k=0}^{n} P_k$ and $Q = \sum_{k=0}^{n} Q_k$ into the sum of the homogeneous polynomials P_k and Q_k of degree k. Then it easily follows from Theorem 1.18 and Theorem 1.20 that

COROLLARY 1.24. *The line at infinity $L_0 \smallsetminus$ sing $(\mathcal{F})$ is a leaf of $\mathcal{F} : \{Pdy - Qdx = 0\} \in \mathcal{A}_n$ if and only if $yP_n - xQ_n \not\equiv 0$.*

One concludes that in $\mathcal{A}_n$ it is very likely to have $L_0 \smallsetminus \mathrm{sing}(\mathcal{F})$ as a leaf, contrary to what we observed before in $\mathcal{D}_n$.

COROLLARY 1.25. *Fix the affine chart $(x, y) \in U_0$ and an SHFC $\mathcal{F} \in \mathcal{A}_n$.*

- *If the line at infinity $L_0 \smallsetminus \mathrm{sing}(\mathcal{F})$ is a leaf, then $\mathcal{F} \in \mathcal{A}_n \cap \mathcal{D}_n$ so that*
  $$\text{affine degree of } \mathcal{F} = \text{geometric degree of } \mathcal{F}.$$
- *Otherwise, $\mathcal{F} \in \mathcal{A}_n \cap \mathcal{D}_{n-1}$ so that*
  $$\text{affine degree of } \mathcal{F} = (\text{geometric degree of } \mathcal{F}) + 1.$$

The main reason for the contrast between $\mathcal{A}_n$ and $\mathcal{D}_n$ is the fact that $\dim\mathcal{A}_n < \dim\mathcal{D}_n < \dim\mathcal{A}_{n+1}$. In fact, it is not hard to see that $\mathcal{D}_n \subset \mathcal{A}_n \cup \mathcal{A}_{n+1}$ and $\mathcal{A}_n \subset \mathcal{D}_{n-1} \cup \mathcal{D}_n$. Fig. 3 is an attempt to illustrate the first property while Fig. 4 is a schematic diagram of the set-theoretic relations between these classes.

EXAMPLE 1.26. The two SHFC's

$$\mathcal{F}_1 : \{xdy - ydx = 0\} \qquad\qquad \mathcal{F}_2 : \{ydy - xdx = 0\}$$

both belong to $\mathcal{A}_1$ so they both have affine degree 1. However, $\mathcal{F}_1$ belongs to $\mathcal{D}_0$ hence has geometric degree 0, while $\mathcal{F}_2$ belongs to $\mathcal{D}_1$ and so it has geometric degree 1. Note that the line at infinity is *not* a leaf of $\mathcal{F}_1$ but it *is* a leaf of $\mathcal{F}_2$.

EXAMPLE 1.27. As another example, let us illustrate how the topologies coming from the two decompositions $\mathcal{S} = \bigcup \mathcal{A}_n = \bigcup \mathcal{D}_n$ are different. Consider the two SHFC's

$$\mathcal{F} : \{x^2dy - y^2dx = 0\},$$
$$\mathcal{F}_\varepsilon : \{(x^2 + \varepsilon xy^2)dy - (y^2 + \varepsilon y^3)dx = 0\}.$$

As $\varepsilon \to 0$, $\mathcal{F}_\varepsilon \to \mathcal{F}$ in the topology induced by $\bigcup \mathcal{D}_n$ but $\mathcal{F}_\varepsilon$ does not converge in the topology induced by $\bigcup \mathcal{A}_n$.

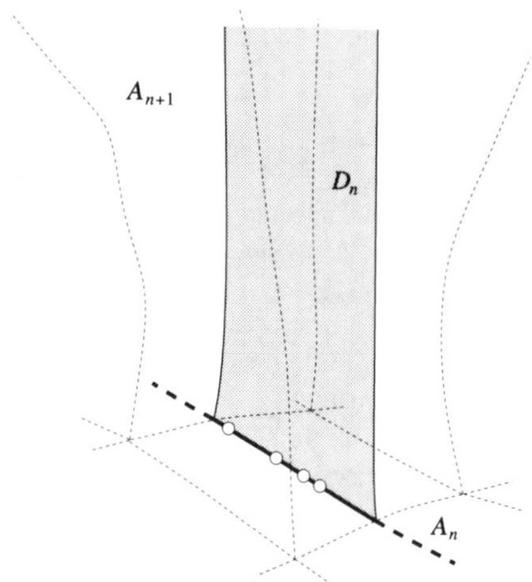

FIGURE 3. $\dim\mathcal{A}_n < \dim\mathcal{D}_n < \dim\mathcal{A}_{n+1}$.

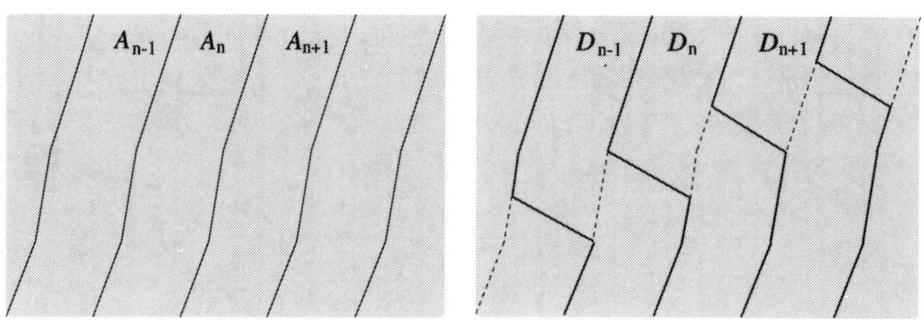

FIGURE 4. *Set-theoretic relations between* $\{\mathcal{D}_n\}$ *and* $\{\mathcal{A}_n\}$.

1.13. A Complex One-Dimensional Analogy. The following simple example may help understand the difference between the two decompositions $\{\mathcal{D}_n\}$ and $\{\mathcal{A}_n\}$: Let $\mathbf{S}$ be the space of all linear conjugacy classes of complex polynomial maps in $\mathbb{C}$ of degree at most 4 which are tangent to the identity map at the origin. This space can be naturally decomposed by the order of tangency near the fixed point at 0: For $0 \leq n \leq 2$, consider the sets $\mathbf{D}_n$ of conjugacy classes of normalized polynomials as follows:

$$\begin{aligned} \mathbf{D}_0 = &\ \langle z \mapsto z + z^4 \rangle \simeq \text{point} \\ \mathbf{D}_1 = &\ \langle z \mapsto z + z^3 + az^4 \rangle \simeq \mathbb{C} \\ \mathbf{D}_2 = &\ \langle z \mapsto z + z^2 + az^3 + bz^4 \rangle \simeq \mathbb{C}^2. \end{aligned}$$

Clearly $\mathbf{S} = \bigcup_{n=0}^{2} \mathbf{D}_n$ and this decomposition induces a topology $\tau_{\mathbf{D}}$ and a measure class $\mu_{\mathbf{D}}$ on $\mathbf{S}$. On the other hand, one can consider the following conjugacy classes determined by the degree of polynomials (i.e., by their behavior near infinity):

$$\begin{aligned} \mathbf{A}_0 = &\ \langle z \mapsto z + z^2 \rangle \simeq \text{point} \\ \mathbf{A}_1 = &\ \langle z \mapsto z + az^2 + z^3 \rangle \simeq \mathbb{C} \\ \mathbf{A}_2 = &\ \langle z \mapsto z + az^2 + bz^3 + z^4 \rangle \simeq \mathbb{C}^2. \end{aligned}$$

This gives rise to a second decomposition $\mathbf{S} = \bigcup_{n=0}^{2} \mathbf{A}_n$ hence a corresponding topology $\tau_{\mathbf{A}}$ and measure class $\mu_{\mathbf{A}}$ on $\mathbf{S}$. One has the relations

$$\begin{array}{ll} \mathbf{A}_0 \subset \mathbf{D}_2 & \mathbf{D}_0 \subset \mathbf{A}_2 \\ \mathbf{A}_1 \subset \mathbf{D}_1 \cup \mathbf{D}_2 & \mathbf{D}_1 \subset \mathbf{A}_1 \cup \mathbf{A}_2 \end{array}$$

The topologies $\tau_{\mathbf{D}}$ and $\tau_{\mathbf{A}}$ and measure classes $\mu_{\mathbf{D}}$ and $\mu_{\mathbf{A}}$ are very different. For example, $\mathbf{D}_1 \subset \mathbf{S}$ is an open set in $\tau_{\mathbf{D}}$, but it is *not* open in $\tau_{\mathbf{A}}$ since $\mathbf{D}_1 \cap \mathbf{A}_1$ is a single point. On the other hand, $\mathbf{D}_1 \subset \mathbf{A}_1 \cup \mathbf{A}_2$ has measure zero with respect to $\mu_{\mathbf{A}}$ but this is certainly not true with respect to $\mu_{\mathbf{D}}$.

1.14. Typical Properties. Certain geometric or dynamical properties often hold for "most" and not all SHFC's in $\mathcal{A}_n$ or $\mathcal{D}_n$. In these cases, we can use the Lebesgue measure class to make sense of this fact. A property $\mathcal{P}$ is said to be *typical* for elements of $\mathcal{A}_n$, or we say that a *typical* SHFC in $\mathcal{A}_n$ satisfies $\mathcal{P}$, if $\{\mathcal{F} \in \mathcal{A}_n : \mathcal{F} \text{ does not satisfy } \mathcal{P}\}$ has Lebesgue measure zero in $\mathcal{A}_n$. We can define a typical property in $\mathcal{D}_n$ in a similar way.

For example, it follows from Theorem 1.20 and Corollary 1.24 that *having the line at infinity as a leaf is not typical in $\mathcal{D}_n$ but it is typical in $\mathcal{A}_n$.*

DEFINITION 1.28. Let $\mathcal{F} \in \mathcal{A}_n$. We say that $\mathcal{F}$ has *Petrovskiĭ-Landis property* if $L_0 \smallsetminus \text{sing}(\mathcal{F})$ is a leaf of $\mathcal{F}$ and $L_0 \cap \text{sing}(\mathcal{F})$ consists of $n+1$ distinct points. The class of all such $\mathcal{F}$ is denoted by $\mathcal{A}'_n$.

If $\mathcal{F} : \{Pdy - Qdx = 0\} \in \mathcal{A}'_n$, it follows from Corollary 1.24 that $yP_n - xQ_n \not\equiv 0$. On the other hand, using the notation of Remark 1.21, if $R = yP - xQ$, we have $\deg R = n+1$ and

$$L_0 \cap \text{sing}(\mathcal{F}) = \{(0,v) : u^{n+1} R\left(\frac{1}{u}, \frac{v}{u}\right)\Big|_{u=0} = 0\}$$

in the affine chart $(u,v) \in U_1$. It follows that $u^{n+1}R(1/u, v/u)|_{u=0}$ must have $n+1$ distinct roots in v. The above two conditions on P and Q show that

COROLLARY 1.29. *A typical SHFC in $\mathcal{A}_n$ has Petrovskiĭ-Landis property.*

2. The Monodromy Group of a Leaf

Given an SHFC $\mathcal{F}$ on $\mathbb{CP}^2$ one can study individual leaves as Riemann surfaces. However, to study the so-called *transverse dynamics* of the foliation one needs a tool to describe the rate of convergence or divergence of nearby leaves. The concept of holonomy, and in particular the monodromy of a leaf, first introduced by C. Ehresmann, is the essential tool in describing the transverse dynamics near the leaf. The point is that the transverse dynamics of a leaf depends directly on its fundamental group: the smaller $\pi_1(\mathcal{L})$ is, the simpler the behavior of the leaves near $\mathcal{L}$ will be.

2.1. Holonomy Mappings and Monodromy Groups. Let $\mathcal{F}$ be an SHFC on $\mathbb{CP}^2$ and $\mathcal{L}$ be a non-singular leaf of $\mathcal{F}$. Fix $p, q \in \mathcal{L}$, and consider small transversals $\Sigma, \Sigma' \simeq \mathbb{D}$ to $\mathcal{L}$ at p, q, respectively. Let $\gamma : [0,1] \to \mathcal{L}$ be a continuous path with $\gamma(0) = p, \gamma(1) = q$. For each $z \in \Sigma$ near p one can "travel" on $\mathcal{L}_z$ "over" $\gamma[0,1]$ to reach Σ' at some point z'. To be precise, let $\{(U_i, \varphi_i)\}_{0 \leq i \leq n}$ be foliation charts of $\mathcal{F}$ and $0 = t_0 < t_1 < \cdots < t_n < t_{n+1} = 1$ be a partition of $[0,1]$ such that if $U_i \cap U_j \neq \emptyset$ then $U_i \cup U_j$ is contained in a foliation chart, and $\gamma[t_i, t_{i+1}] \subset U_i$

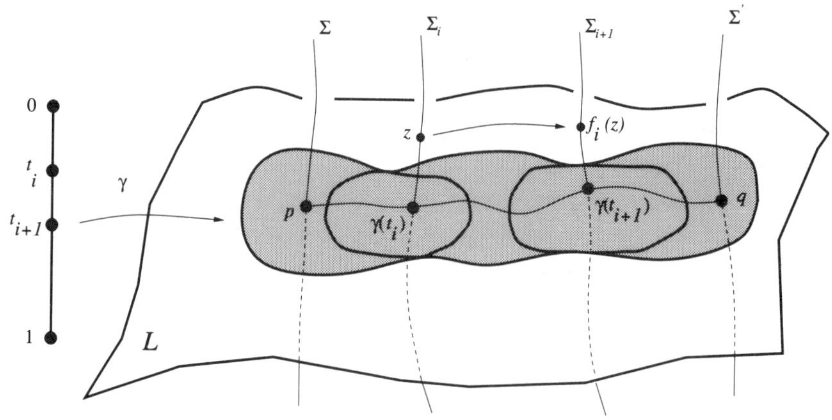

FIGURE 5. *Definition of holonomy.*

for $0 \leq i \leq n$. For each $1 \leq i \leq n$ choose a transversal $\Sigma_i \simeq \mathbb{D}$ to $\mathcal{L}$ at $\gamma(t_i)$, and set $\Sigma_0 = \Sigma$ and $\Sigma_{n+1} = \Sigma'$ (see Fig. 5). Then for each $z \in \Sigma_i$ sufficiently close to $\gamma(t_i)$ the plaque of U_i passing through z meets Σ_{i+1} in a unique point $f_i(z)$, and $z \mapsto f_i(z)$ is holomorphic, with $f_i(\gamma(t_i)) = \gamma(t_{i+1})$. It follows that the composition $f_\gamma := f_n \circ \cdots \circ f_0$ is defined for $z \in \Sigma$ near p, with $f_\gamma(p) = q$.

DEFINITION 2.1. The mapping f_γ is called the *holonomy* associated with γ.

There are several remarks about this mapping which can be checked directly from the definition (compare [**CL**]).

REMARK 2.2. f_γ is independent of the chosen transversals $\Sigma_i, 1 \leq i \leq n$, and the foliation charts U_i. Hence Σ, Σ', and γ determine the germ of f_γ at p.

REMARK 2.3. The germ of f_γ at p depends only on the homotopy class of γ rel$\{0,1\}$, that is, if η is another path joining p and q in $\mathcal{L}$ which is homotopic to γ with $\eta(0) = \gamma(0)$ and $\eta(1) = \gamma(1)$, then the germ of f_η at p coincides with that of f_γ.

REMARK 2.4. If $\gamma^{-1}(t) := \gamma(1-t)$, then $f_{\gamma^{-1}} = (f_\gamma)^{-1}$. In particular, f_γ represents the germ of a local *biholomorphism.*

REMARK 2.5. Let Σ_1 and Σ_1' be other transversals to $\mathcal{L}$ at p and q, respectively. Let $h : \Sigma \to \Sigma_1$ and $\tilde{h} : \Sigma' \to \Sigma_1'$ be projections along the plaques of $\mathcal{F}$ in a neighborhood of p and q, respectively. Then the holonomy $g_\gamma : \Sigma_1 \to \Sigma_1'$ satisfies $g_\gamma = \tilde{h} \circ f_\gamma \circ h^{-1}$.

In the special case $p = q$, we obtain a generalization of the Poincaré first return map for real vector fields.

DEFINITION 2.6. Let $\mathcal{L}$ be a non-singular leaf of an SHFC $\mathcal{F}$ on $\mathbb{CP}^2$, let $p \in \mathcal{L}$, and let Σ be a transversal to $\mathcal{L}$ at p. For each $[\gamma] \in \pi_1(\mathcal{L}, p)$, the holonomy mapping $f_\gamma : \Sigma \to \Sigma$ is called the *monodromy mapping* of $\mathcal{L}$ associated with γ (Fig. 6).

Note that by Remark 2.3, the germ of f_γ at p depends only on the homotopy class $[\gamma]$. It is quite easy to see that $[\gamma] \mapsto f_\gamma$ is a homomorphism from $\pi_1(\mathcal{L}, p)$ into the group of germs of biholomorphisms of Σ fixing p: $[\gamma \circ \eta] \mapsto f_\gamma \circ f_\eta$.

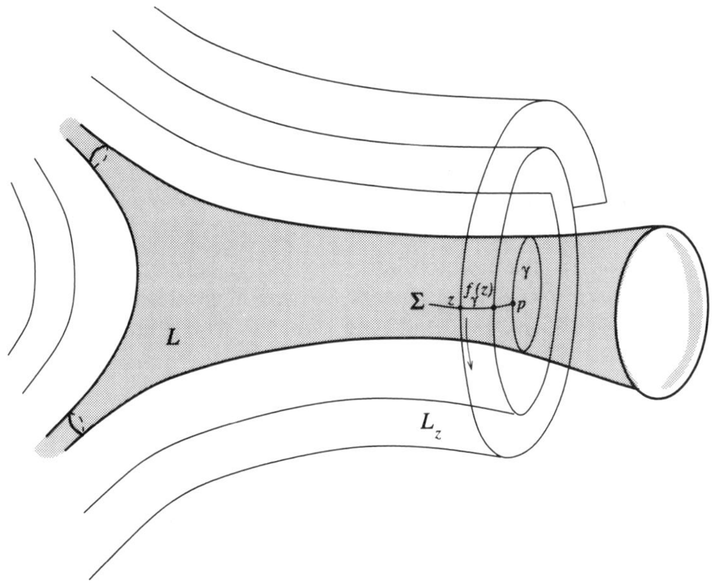

FIGURE 6. *Monodromy mapping associated with* γ.

REMARK 2.7. It should be noted that each $[\gamma] \in \pi_1(\mathcal{L}, p)$ determines only the *germ* at p of a biholomorphism f_γ of Σ, for changing γ in its homotopy class results in changing the domain of definition of f_γ.

REMARK 2.8. If the transversal Σ is replaced by another one Σ_1, then by Remark 2.5 there exists a local biholomorphism $h : \Sigma \to \Sigma_1$ fixing p such that the monodromy mapping $g_\gamma : \Sigma_1 \to \Sigma_1$ satisfies $g_\gamma = h \circ f_\gamma \circ h^{-1}$. In other words, the germ of the monodromy mapping f_γ depends only on $[\gamma]$ *up to conjugacy*. Since conjugate germs of biholomorphisms fixing p have the same iterative dynamics near p, it is not really important which transversal we choose at p.

CONVENTION 2.9. We always fix some $p \in \mathcal{L}$ as the base point for the fundamental group. We will fix some transversal Σ at p. Moreover, we will choose a coordinate on Σ in which $p = 0$. Thus, every monodromy mapping f_γ can be identified with an element of $\mathrm{Bih}_0(\mathbb{C})$, the group of germs at 0 of biholomorphisms $\mathbb{C} \to \mathbb{C}$ fixing the origin.

DEFINITION 2.10. The image of $\pi_1(\mathcal{L})$ under the monodromy mapping $[\gamma] \mapsto f_\gamma$ is called the *monodromy group* of $\mathcal{L}$, and is denoted by $G(\mathcal{L})$. Since we always fix the transversals, it can be identified with a subgroup of $\mathrm{Bih}_0(\mathbb{C})$.

Given a leaf $\mathcal{L}$ whose π_1 is finitely-generated, it is natural to fix some loops as the generators of $\pi_1(\mathcal{L})$. So we arrive at the following definition:

DEFINITION 2.11. A non-singular leaf $\mathcal{L}$ of an SHFC $\mathcal{F}$ on $\mathbb{CP}^2$ is called a *marked leaf* if $\pi_1(\mathcal{L})$ is finitely-generated and a set of loops $\{\gamma_1, \ldots, \gamma_k\}$ is given as its generators. Similarly, a finitely-generated subgroup $G \subset \mathrm{Bih}_0(\mathbb{C})$ is called a *marked subgroup* if a set of local biholomorphisms $\{f_1, \ldots, f_k\}$ is given as its generators.

Clearly, the monodromy group of a marked leaf is a marked subgroup if one chooses $\{f_{\gamma_1}, \cdots, f_{\gamma_k}\}$ as its generators.

2.2. The Monodromy Pseudo-Group of a Leaf. There is a certain difficulty in defining the orbit of a point under the action of the monodromy group $G(\mathcal{L})$ of a leaf $\mathcal{L}$. This is simply due to the fact that elements of $G(\mathcal{L})$ are *germs*, while in the definition of an orbit one should consider them as *maps*. On the other hand, given $f \in G(\mathcal{L})$, a point z may well be outside the domain of f but f may have an analytic continuation which is defined at z. Naturally, one askes to what extent this continuation really represents the same monodromy. For instance, if $\mathcal{F}$ is induced by a Hamiltonian form $\omega = dH$, then the monodromy group of the leaf at infinity $\mathcal{L}_\infty$ is abelian [**I3**]. Hence for every γ in the commutator subgroup $[\pi_1(\mathcal{L}_\infty), \pi_1(\mathcal{L}_\infty)]$ the germ of f_γ at 0 is the identity, which can be analytically continued everywhere. However, for z sufficiently far from 0, the monodromy f_γ may be undefined or it may differ from z.

Since the transverse dynamics of a leaf $\mathcal{L}$ is reflected in the orbit of points in the transversal Σ under the action of $G(\mathcal{L})$, it is quite natural to be careful about the domains of definitions. This point in addressed in the following definition:

DEFINITION 2.12. Let $G \subset \text{Bih}_0(\mathbb{C})$ be a marked subgroup with generators $\{f_1, \ldots, f_k\}$, all defined on some domain Ω around 0. The *pseudo-group* PG consists of all pairs (f, Ω_f), where $f \in G$ and Ω_f is a domain on which f is conformal, with the group operation $(f, \Omega_f) \circ (g, \Omega_g) := (f \circ g, \Omega_{f \circ g})$. The domain Ω_f is defined as follows: Let

$$f = \prod_{i=1}^{N} f_{j_i}^{\epsilon_i} \quad , \; j_i \in \{1, \ldots, k\} \; , \; \epsilon_i \in \{-1, 1\} \tag{2.1}$$

be any representation of f in terms of the generators. Any germ $\prod_{i=1}^{n} f_{j_i}^{\epsilon_i}$, with $n \leq N$, is called an *intermediate representation* of f. The domain $\Omega_{\Pi f}$ associated to the representation (2.1) is defined as the maximal starlike domain centered at 0 contained in Ω on which all the intermediate representations can be analytically continued as conformal maps, with

$$\prod_{i=1}^{n} f_{j_i}^{\epsilon_i}(\Omega_{\Pi f}) \subset \Omega, \quad \text{for all} \quad n \leq N.$$

Finally, Ω_f is defined to be the union of $\Omega_{\Pi f}$'s for all possible representations of f of the form (2.1). It is clear that f is a conformal mapping on Ω_f.

The above definition associates a *monodromy pseudo-group* $PG(\mathcal{L})$ to each marked leaf $\mathcal{L}$. The construction above allows us to define the *orbit* of $z \in \Sigma$ as $\{f(z) : (f, \Omega_f) \in PG(\mathcal{L}) \text{ and } z \in \Omega_f\}$. The basic property that we will be using is that the orbit of z under $PG(\mathcal{L})$ always lies in $\mathcal{L}_z \cap \Sigma$.

2.3. Multiplier of a Monodromy Mapping. As we will see later, the dynamics of an $f \in \text{Bih}_0(\mathbb{C})$ is essentially dominated by its derivative $f'(0)$ at the fixed point 0, also known as the *multiplier* of f at 0. Therefore it would not be surprising that the behavior of leaves near a given leaf $\mathcal{L}$ is determined to a large extent by the multipliers at 0 of the monodromy mappings f_γ for $\gamma \in \pi_1(\mathcal{L})$. Our aim here is to give a formula for $f'_\gamma(0)$ in terms of a path integral (see [**KS**] for more general formulas of this type). We will use this formula in the next section, where

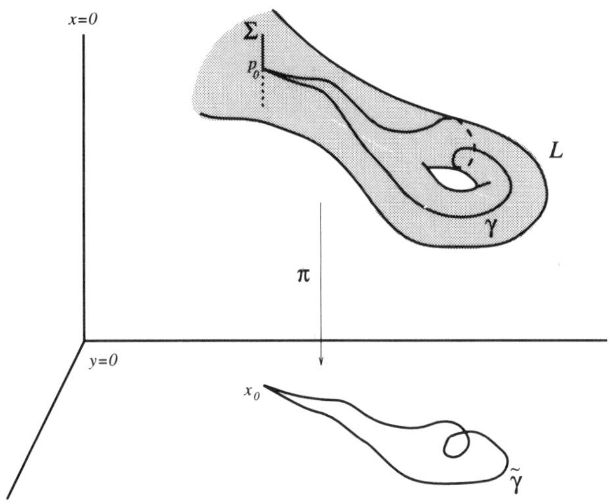

FIGURE 7. *Proof of Theorem 2.13.*

we will compute the multipliers of the monodromy mappings of the leaf at infinity of an $\mathcal{F} \in \mathcal{A}'_n$. It must be mentioned that this is quite similar to a well-known result of Reeb for real codimension 1 non-singular foliations [**Re**]. Of course a different argument is needed here since our foliations have real codimension 2. However, our proof is greatly facilitated by the fact the foliations are induced by polynomial 1-forms.

Let $\mathcal{F} : \{\omega = Pdy - Qdx = 0\} \in \mathcal{A}_n$, and suppose that $\mathcal{L}_\infty$, the line at infinity with sing($\mathcal{F}$) deleted, is a leaf of $\mathcal{F}$ (see Corollary 1.24). Fix some non-singular leaf $\mathcal{L} \neq \mathcal{L}_\infty$. It follows that $\mathcal{L}$ is completely contained in the affine chart $(x, y) \in U_0$. Without loss of generality we may assume that $E := \mathcal{L} \cap S_P$ is discrete in the leaf topology, hence countable (see §1.4 for the definition of S_P). In fact, if E is not discrete, it has a limit point z_0 in $\mathcal{L}$. Parametrizing $\mathcal{L}$ around z_0 by $T \mapsto (x(T), y(T))$ with $z_0 = (x(0), y(0))$, we conclude that $P(x(T), y(T)) \equiv 0$. By analytic continuation, $P(x, y) = 0$ for every $(x, y) \in \mathcal{L}$. Therefore we can pursue the argument with $\mathcal{L} \cap S_Q$, which is finite since P and Q are assumed to be relatively prime.

So assume that E is discrete. By integrability of ω, there exists a meromorphic 1-form $\alpha = Xdy - Ydx$, holomorphic on $\mathcal{L} \smallsetminus E$, such that $d\omega = \omega \wedge \alpha$ (since P and Q have no common factor, one can actually find *polynomial* vector fields X and Y with this property; see equation (2.3) below).

THEOREM 2.13. *Given* $\gamma \in \pi_1(\mathcal{L}, p_0)$ *one has*

$$f'_\gamma(0) = \exp\left(- \int_\gamma \alpha \right). \tag{2.2}$$

PROOF. Consider $\mathcal{L}$ as the graph of a (multi-valued) function $y = \varphi(x)$ over some region of the x-axis. Let $E = \mathcal{L} \cap S_P$ and $\tilde{E} := \{x \in \mathbb{C} : \text{There exists } y \in \mathbb{C}$ such that $(x, y) \in E\}$. Without loss of generality we can assume that the base point $p_0 = (x_0, y_0)$ is not in E. Moreover, we may replace γ by a path in its homotopy class that avoids E, if necessary. Take a "vertical" Σ transversal to $\mathcal{L}$ at p_0, and let y be the coordinate on Σ (Fig. 7). For $y \in \Sigma$ near y_0, each leaf $\mathcal{L}_y$ is the graph of the solution $\Phi(x,y)$ of $dy/dx = Q(x,y)/P(x,y)$, i.e.,

$$\frac{\partial \Phi}{\partial x}(x,y) = \frac{Q(x,\Phi(x,y))}{P(x,\Phi(x,y))}, \qquad \Phi(x_0,y) = y.$$

Define $\xi(x) := \partial\Phi/\partial y(x,y_0)$. Note that

$$\begin{aligned}
\frac{d\xi}{dx}(x) &= \frac{\partial^2\Phi}{\partial x \partial y}(x,y_0) \\
&= \frac{\partial}{\partial y}\left[\frac{Q(x,\Phi(x,y))}{P(x,\Phi(x,y))}\right](x,y_0) \\
&= \left[\frac{Q_y(x,\varphi(x))P(x,\varphi(x)) - P_y(x,\varphi(x))Q(x,\varphi(x))}{P^2(x,\varphi(x))}\right]\xi(x) \\
&=: T(x)\xi(x),
\end{aligned}$$

with $\xi(x_0) = 1$. Thus $\xi(x) = \exp(\int_{x_0}^{x} T(\tau)d\tau)$, where the path of integration avoids $\tilde{E}$. But $f'_\gamma(0)$ is the result of analytic continuation of ξ along $\tilde{\gamma}$, the projection on the x-axis of γ, so that $f'_\gamma(0) = \exp(\int_{\tilde{\gamma}} T(x)dx)$. On the other hand, the condition $d\omega = \omega \wedge \alpha$ shows that

$$YP - XQ = P_x + Q_y, \tag{2.3}$$

so that on an open neighborhood $W \subset \mathcal{L}$ of γ one has

$$\alpha|_W = (Xdy - Ydx)|_W = \left(\frac{XQ}{P} - Y\right)\Big|_W dx = -\left(\frac{P_x+Q_y}{P}\right)\Big|_{y=\varphi(x)} dx.$$

Now, compute

$$\begin{aligned}
\int_{\tilde{\gamma}} T(x)\,dx + \int_\gamma \alpha &= \int_{\tilde{\gamma}} \left[T(x) - \left(\frac{P_x+Q_y}{P}\right)\Big|_{y=\varphi(x)}\right] dx \\
&= \int_{\tilde{\gamma}} \frac{Q_yP - P_yQ - P(P_x+Q_y)}{P^2}\Big|_{y=\varphi(x)} dx \\
&= -\int_{\tilde{\gamma}} \frac{P_yQ + P_xP}{P^2}\Big|_{y=\varphi(x)} dx \\
&= -\int_{\tilde{\gamma}} \frac{\partial/\partial x(P(x,\varphi(x)))}{P}\, dx \\
&= -\int_{\tilde{\gamma}} \frac{\partial}{\partial x}\log P(x,\varphi(x))\, dx \\
&= 2\pi i n
\end{aligned}$$

for some integer n by the Argument Principle. Hence $\int_{\tilde{\gamma}} T(x)\,dx$ and $-\int_\gamma \alpha$ differ by an integer multiple of $2\pi i$, which proves the result. □

2.4. Monodromy Group of the Leaf at Infinity. Let $\mathcal{F} \in \mathcal{A}'_n$ be an SHFC on $\mathbb{CP}^2$ having Petrovskiĭ-Landis property. Recall from Definition 1.28 that this means $L_0 \smallsetminus \operatorname{sing}(\mathcal{F})$ is a leaf and $L_0 \cap \operatorname{sing}(\mathcal{F})$ consists of exactly $n+1$ points $\{p_1, \dots, p_{n+1}\}$. Having an algebraic leaf homeomorphic to the punctured Riemann sphere imposes severe restrictions on the global behavior of the leaves. Roughly speaking, for $\mathcal{F} \in \mathcal{A}'_n$ the global behavior of the leaves is essentially determined by their local behavior in some neighborhood of the leaf at infinity $\mathcal{L}_\infty$. In fact, under the assumption $\mathcal{F} \in \mathcal{A}'_n$, each non-singular leaf must accumulate on L_0 (Corollary 2.17 below). Once this is established, we can study the monodromy group of $\mathcal{L}_\infty$ to get information about the behavior of nearby leaves, which then can be transferred elsewhere.

THEOREM 2.14. *Let $X = \sum_{j=1}^n f_j \, \partial/\partial z_j$ be a holomorphic vector field on $\mathbb{C}^n$. Then every non-singular solution of the differential equation $dz/dT = X(z)$ is unbounded.*

PROOF. Fix $z_0 \in \mathbb{C}^n$ with $X(z_0) \neq 0$, and suppose by way of contradiction that the integral curve $\mathcal{L}_{z_0}$ passing through z_0 is bounded. Let $T \mapsto \eta(T)$ be a local parametrization of $\mathcal{L}_{z_0}$, with $\eta(0) = z_0$. Let $R > 0$ be the largest radius such that $\eta(T)$ can be analytically continued over $\mathbb{D}(0, R)$. If $R = +\infty$, then $\eta(T)$ will be constant by Liouville's Theorem, contrary to the assumption $X(z_0) \neq 0$. So $R < +\infty$. Recall from the Existence and Uniqueness Theorem of solutions of holomorphic differential equations that for each $T_0 \in \mathbb{C}$ and each $p \in \mathbb{C}^n$, if X is holomorphic on $\{z \in \mathbb{C}^n : |z-p| < b\}$, then there exists a local parametrization $T \mapsto \eta_p(T)$ of $\mathcal{L}_p$, with $\eta_p(T_0) = p$, defined on $\mathbb{D}(T_0, b/(M+kb))$, where $M = \sup\{|X(z)| : |z-p| < b\}$ and $k = \sup\{|dX(z)/dz| : |z-p| < b\}$ (see for example [**CoL**]). In our case, since $\mathcal{L}_{z_0}$ is bounded by the assumption, both $|X(\cdot)|$ and $|dX(\cdot)/dz|$ will be bounded on $\mathcal{L}_{z_0}$, and hence the quantity $b/(M+kb)$ has a uniform lower bound $2\delta > 0$ for all $p \in \mathcal{L}_{z_0}$. It follows that for each $T_0 \in \mathbb{D}(0, R)$ with $R - |T_0| < \delta$, we can find a local parametrization $\eta_p : \mathbb{D}(T_0, 2\delta) \to \mathbb{C}^n$ for $\mathcal{L}_{z_0}$ around $p = \eta_p(T_0)$. These local parametrizations patch together, giving an analytic continuation of η over the disk $\mathbb{D}(0, R+\delta)$. The contradiction shows that $\mathcal{L}_{z_0}$ must be unbounded. □

REMARK 2.15. The same argument in the real case gives another proof of the fact that an integral curve of a differential equation on $\mathbb{R}^n$ is either unbounded or it is bounded and parametrized by the entire real line.

COROLLARY 2.16. *Let $L \subset \mathbb{CP}^2$ be any projective line and $\mathcal{L}$ be any non-singular leaf of an SHFC $\mathcal{F}$. Then $\overline{\mathcal{L}} \cap L \neq \emptyset$.*

PROOF. Consider the affine chart (x, y) for $\mathbb{CP}^2 \smallsetminus L \simeq \mathbb{C}^2$. In this coordinate system, $L = L_0$. Note that the leaves of $\mathcal{F}|_{\mathbb{C}^2}$ are the integral curves of a (polynomial) vector field. Now the result follows from Theorem 2.14. □

COROLLARY 2.17. *Any non-singular leaf of an SHFC in $\mathcal{A}'_n$ has an accumulation point on the line at infinity.*

If the accumulation point is not singular, then the whole line at infinity is contained in the closure of the leaf by Proposition 1.6.

Knowing that each leaf accumulates on L_0, we now proceed to study the monodromy group of the leaf at infinity $\mathcal{L}_\infty$. Let $\mathcal{F} : \{\omega = P dy - Q dx = 0\} \in \mathcal{A}'_n$ and

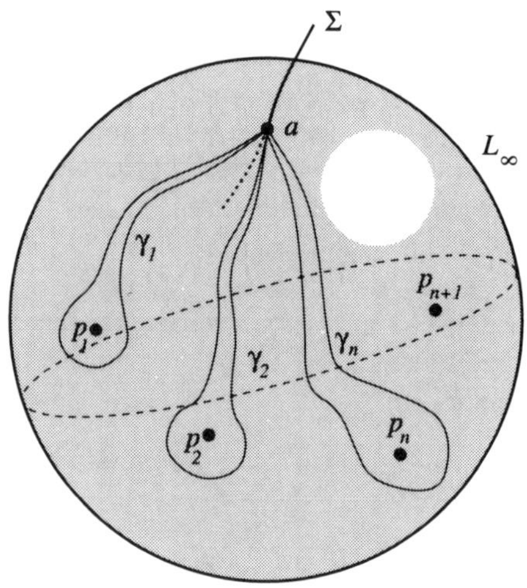

FIGURE 8. *Marking the leaf at infinity.*

$L_0 \cap \operatorname{sing}(\mathcal{F}) = \{p_1, \dots, p_{n+1}\}$. Recall from Remark 1.21 that $\mathcal{F}$ is described in the affine chart $(u, v) \in U_1$ by

$$\omega'(u,v) = u^{n+1} P\left(\frac{1}{u}, \frac{v}{u}\right) dv - u^{n+1} R\left(\frac{1}{u}, \frac{v}{u}\right) du = 0, \tag{2.4}$$

where $R(x, y) = yP(x, y) - xQ(x, y)$. The line at infinity L_0 is the closure of $\{(0, v) : v \in \mathbb{C}\}$, and $p_j := (0, a_j)$, $1 \le j \le n+1$, where the a_j are distinct roots in v of the polynomial $u^{n+1}R(1/u, v/u)|_{u=0}$.

The leaf at infinity $\mathcal{L}_\infty = L_0 \setminus \{p_1, \dots, p_{n+1}\}$ can be made into a marked leaf by choosing fixed loops $\{\gamma_1, \dots, \gamma_n\}$ as generators of $\pi_1(\mathcal{L}_\infty)$. Fixing a base point $a \notin \{p_1, \dots, p_{n+1}\}$, each γ_j goes around p_j once in the positive direction and does not encircle p_j for $i \neq j$ (Fig. 8).

DEFINITION 2.18. The monodromy mappings f_{γ_j} for $1 \le j \le n$ generate the *monodromy group of the leaf at infinity*, denoted by G_∞.

Let us compute the multiplier of each monodromy mapping f_{γ_j} in terms of P and Q. In the affine chart $(u, v) \in U_1$, the foliation is induced by the polynomial 1-form ω' of (2.4). Write

$$u^{n+1} P\left(\frac{1}{u}, \frac{v}{u}\right) =: u\tilde{P}(u,v) \qquad \text{and} \qquad u^{n+1} R\left(\frac{1}{u}, \frac{v}{u}\right) =: \tilde{R}(u,v), \tag{2.5}$$

where $\tilde{P}$ and $\tilde{R}$ are polynomials with $\tilde{R}(0, a_j) = 0$ for $1 \le j \le n+1$. Note that $\mathcal{F}|_{U_1}$ is induced by the vector field

$$X_1 = u\tilde{P}(u,v)\frac{\partial}{\partial u} + \tilde{R}(u,v)\frac{\partial}{\partial v}$$

(compare (1.7)). Let us consider the Jacobian matrix DX_1 at the singular point p_j:

$$DX_1(p_j) = \begin{pmatrix} \tilde{P}(0, a_j) & 0 \\ \tilde{R}_u(0, a_j) & \tilde{R}_v(0, a_j) \end{pmatrix}, \tag{2.6}$$

where the indices denote partial derivatives. The quotient

$$\lambda_j := \frac{\tilde{P}(0,a_j)}{\tilde{R}_v(0,a_j)} \tag{2.7}$$

of the eigenvalues of the matrix (2.6) is called the *characteristic number* of the singularity p_j. Note that since the roots of $\tilde{R}(0,v)$ are simple by assumption, we have $\tilde{R}_v(0,a_j) \neq 0$ and λ_j of (2.7) is well-defined. On the other hand, the characteristic number is evidently independent of the vector field representing $\mathcal{F}$ near p_j, since such a vector field is unique up to multiplication by a nowhere vanishing holomorphic function near p_j.

For simplicity we denote f_{γ_j} by f_j and $f_j'(0)$ by ν_j.

PROPOSITION 2.19. $\nu_j = e^{2\pi i \lambda_j}$.

PROOF. By Theorem 2.13, we have $\nu_j = \exp(-\int_{\gamma_j} \alpha)$, where α is any meromorphic 1-form which satisfies $d\omega' = \omega' \wedge \alpha$. Choose, for example, the 1-form $\alpha = -(u\tilde{P}_u + \tilde{P} + \tilde{R}_v)/\tilde{R}\, dv$. Then we have

$$\begin{aligned}
-\int_{\gamma_j} \alpha &= \int_{\gamma_j} \left.\frac{u\tilde{P}_u + \tilde{P} + \tilde{R}_v}{\tilde{R}}\right|_{u=0} dv \\
&= \int_{\gamma_j} \frac{\tilde{P}(0,v) + \tilde{R}_v(0,v)}{\tilde{R}(0,v)}\, dv \\
&= 2\pi i \operatorname{Res}\left[\frac{\tilde{P}(0,v) + \tilde{R}_v(0,v)}{\tilde{R}(0,v)}; a_j\right] \\
&= 2\pi i(\lambda_j + 1),
\end{aligned}$$

so that $\exp(-\int_{\gamma_j} \alpha) = e^{2\pi i \lambda_j}$. □

REMARK 2.20. Since $f_1 \circ \cdots \circ f_{n+1} = \mathrm{id}$, one has $\nu_1 \cdots \nu_{n+1} = 1$ so that $\sum_{j=1}^{n+1} \lambda_j$ is an integer by the above proposition. This integer turns out to be 1 by the following argument. By (2.7), λ_j is the residue at a_j of the meromorphic function $\tilde{P}(0,v)/\tilde{R}(0,v)$ on $L_0 \simeq \mathbb{CP}^1$. If $\tilde{R}(0,v) = c\prod_{j=1}^{n+1}(v - a_j)$, then c is the coefficient of y^{n+1} in $R(x,y)$ by (2.5), hence it is the coefficient of y^n in $P(x,y)$ since $R = yP - xQ$. So again by (2.5) $\tilde{P}(0,v)$ is a polynomial in v with leading term cv^n. It follows that the residue at infinity of $\tilde{P}(0,v)/\tilde{R}(0,v)$ is -1. Hence $\sum_{j=1}^{n+1} \lambda_j - 1 = 0$ by the Residue Theorem.

2.5. Equivalence of Foliations and Subgroups of $\mathbf{Bih}_0(\mathbb{C})$. For singular smooth 1-dimensional foliations on real manifolds, one can speak of topological or C^k *equivalences*, or topological or C^k *conjugacies*. In the case of an equivalence, one is concerned only about the topology of the leaves, but in the case of a conjugacy, the actual parametrization of the leaves is also relevant. Of course the latter makes sense only when the foliations are described by single smooth vector fields on the ambient space.

In the complex analytic case, we can still think of equivalences as well as conjugacies between SHFC's. But again the notion of conjugacy requires the existence of holomorphic vector fields defined on the ambient space representing the foliations.

One can develop a theory of conjugacies for holomorphic vector fields defined on non-compact complex manifolds such as open subsets of $\mathbb{C}^n$. However, the study of conjugacies for holomorphic vector fields on compact complex manifolds is certainly less interesting, as the possible examples are rather rare and often trivial (compare Proposition 1.9).

Based on this observation, the most natural notion of equivalence between SHFC's on $\mathbb{CP}^2$ seems to be the following:

DEFINITION 2.21. Two SHFC's $\mathcal{F}$ and $\mathcal{F}'$ on $\mathbb{CP}^2$ are said to be *topologically* (resp. *holomorphically*) *equivalent* if there exists a homeomorphism (resp. biholomorphism) $H : \mathbb{CP}^2 \to \mathbb{CP}^2$ which maps the leaves of $\mathcal{F}$ to those of $\mathcal{F}'$.

The existence of an equivalence between two SHFC's has the following implication on the monodromy groups:

PROPOSITION 2.22. *Let $\mathcal{F}$ and $\mathcal{F}'$ be topologically (resp. holomorphically) equivalent SHFC's linked by $H : \mathbb{CP}^2 \to \mathbb{CP}^2$. Let p be a non-singular point for $\mathcal{F}$. Then the monodromy group $G(\mathcal{L}_p)$ is isomorphic to $G(\mathcal{L}'_{H(p)})$. More precisely, there exists a local homeomorphism (resp. biholomorphism) h defined on some neighborhood of $0 \in \mathbb{C}$, with $h(0) = 0$, and a group isomorphism $k : G(\mathcal{L}_p) \to G({\mathcal{L}'}_{H(p)})$ such that $h \circ f = k(f) \circ h$ for every $f \in G(\mathcal{L}_p)$.*

PROOF. Let $\gamma \in \pi_1(\mathcal{L}_p, p)$ and Σ be a transversal to $\mathcal{L}_p$ at p. Set $q = H(p), \gamma' := H_*\gamma \in \pi_1(\mathcal{L}'_q, q)$, and let Σ' be a transversal to $\mathcal{L}'_q$ at q. As in the definition of the monodromy mapping, choose foliation charts $\{(U_i, \varphi_i)\}_{0\leq i\leq n}$ and transversals Σ_i for $\mathcal{F}$, and the corresponding data $\{(U'_i, \varphi'_i)\}_{0\leq i\leq n}$ and Σ'_i for $\mathcal{F}'$, and consider the decompositions $f_\gamma = f_n \circ \cdots \circ f_0$ and $g_{\gamma'} = g_n \circ \cdots \circ g_0$ (compare §2.1). Without loss of generality we may assume that $U'_i = H(U_i)$. Since $\mathcal{F}|_{U_i}$ (resp. $\mathcal{F}'|_{U'_i}$) is a trivial foliation, one has a projection along leaves $\pi_i : U_i \to \Sigma_i$ (resp. $\pi'_i : U'_i \to \Sigma'_i$) which sends every z to the unique intersection point of the plaque of U_i (resp. U'_i) through z with Σ_i (resp. Σ'_i).

Define $h_i : \Sigma_i \to \Sigma'_i$ by $h_i := \pi'_i \circ H$. Since H is a leaf-preserving homeomorphism (resp. biholomorphism) each h_i is also a homeomorphism (resp. biholomorphism) with inverse $h_i^{-1} = \pi_i \circ H^{-1}$ (see Fig. 9). Now the definition of f_i and g_i shows that $h_{i+1} \circ f_i = g_i \circ h_i$ for $0 \leq i \leq n$. Therefore, the two relations $f_\gamma = f_n \circ \cdots \circ f_0$ and $g_{\gamma'} = g_n \circ \cdots \circ g_0$ show that $g_{\gamma'} \circ h_0 = h_0 \circ f_\gamma$. To complete the proof, note that the mapping $f_\gamma \mapsto k(f_\gamma) := h_0 \circ f_\gamma \circ {h_0}^{-1}$ is an isomorphism between $G(\mathcal{L}_p)$ and $G(\mathcal{L}'_q)$. □

The following definition is suggested by the above proposition:

DEFINITION 2.23. Two subgroups $G, G' \subset \mathrm{Bih}_0(\mathbb{C})$ are *topologically* (resp. *holomorphically*) *equivalent* if there exists a homeomorphism (resp. biholomorphism) h defined on some neighborhood of $0 \in \mathbb{C}$, with $h(0) = 0$, such that $h \circ f \circ h^{-1} \in G'$ if and only if $f \in G$.

It follows that the mapping $f \mapsto k(f) := h \circ f \circ h^{-1} : G \to G'$ is an isomorphism, and the following diagram is commutative:

$$\begin{array}{ccc} (\mathbb{C}, 0) & \xrightarrow{\ f\ } & (\mathbb{C}, 0) \\ \downarrow h & & \downarrow h \\ (\mathbb{C}, 0) & \xrightarrow{\ k(f)\ } & (\mathbb{C}, 0) \end{array} \tag{2.8}$$

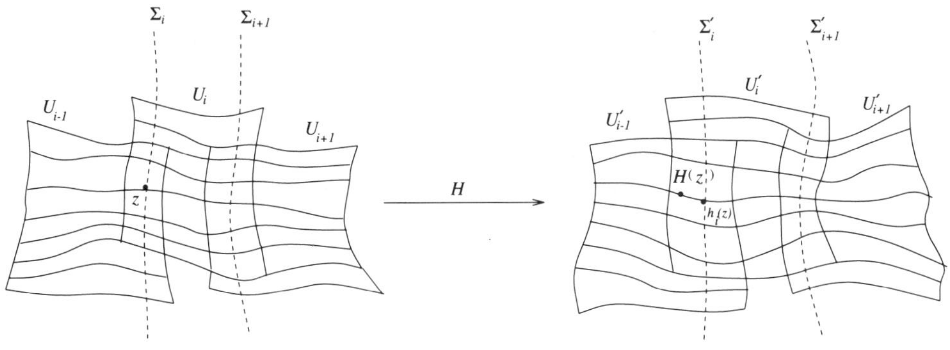

FIGURE 9. *Equivalent foliations.*

THEOREM 2.24. *If two SHFC's $\mathcal{F}$ and $\mathcal{F}'$ on $\mathbb{CP}^2$ are topologically (resp. holomorphically) equivalent, then so are the monodromy groups of the corresponding leaves.*

This theorem naturally leads us to the study of equivalent subgroups of $\mathrm{Bih}_0(\mathbb{C})$ in order to gain information about equivalent SHFC's. Below we list a few important results concerning these equivalence problems.

THEOREM 2.25. *Let $G, G' \subset \mathrm{Bih}_0(\mathbb{C})$ be two topologically equivalent marked subgroups linked by an orientation-preserving homeomorphism h such that the diagram (2.8) is commutative. Suppose that G is non-abelian, and that there exist $f_1, f_2 \in G$ such that the multiplicative subgroup of $\mathbb{C}^*$ generated by $f_1'(0), f_2'(0)$ is dense in $\mathbb{C}$. Then h is actually a biholomorphism, so that G and G' are holomorphically equivalent.*

This phenomenon is called *absolute rigidity* of subgroups of $\mathrm{Bih}_0(\mathbb{C})$. (A subgroup $G \subset \mathrm{Bih}_0(\mathbb{C})$ is called *absolutely rigid* if every subgroup which is topologically equivalent to G is holomorphically equivalent to it.) The proof of this theorem can be found in [**I3**].

Along this line, A. Shcherbakov [**Sh**] has shown the following result:

THEOREM 2.26. *A non-solvable subgroup of* $\mathrm{Bih}_0(\mathbb{C})$ *is absolutely rigid.*

For an almost complete topological and analytical classification of germs in $\mathrm{Bih}_0(\mathbb{C})$, see paper II of [**I5**].

The next result gives "moduli of stability" for typical SHFC's on $\mathbb{CP}^2$.

THEOREM 2.27. *Let $\mathcal{F}, \mathcal{F}' \in \mathcal{A}_n'$ be two SHFC's on $\mathbb{CP}^2$ having no algebraic leaves other than the leaf at infinity. Let $L_0 \cap \mathrm{sing}(\mathcal{F}) = \{p_1, \dots, p_{n+1}\}$, $L_0 \cap \mathrm{sing}(\mathcal{F}') = \{p_1', \dots, p_{n+1}'\}$, and λ_j and λ_j' be the characteristic numbers of p_j and p_j', respectively. Suppose that λ_j and λ_j' are non-zero, and $\mathcal{F}$ and $\mathcal{F}'$ are topologically equivalent by a homeomorphism $H : \mathbb{CP}^2 \to \mathbb{CP}^2$ with $H(p_j) = p_j'$. Then there exists an $\mathbb{R}$-linear transformation $A : \mathbb{C} \to \mathbb{C}$, with $A(\lambda_j) = \lambda_j'$ for $1 \leq j \leq n+1$.*

It follows in particular that the $n+1$ characteristic numbers λ_j' span a subspace of $\mathbb{C}^{n+1}$ of dimension ≤ 2. The proof of Theorem 2.27 is based on studying the equivalence of the monodromy groups of the leaves at infinity, and uses the same techniques as the proof of Theorem 2.25. It was first proved by Il'yashenko [**I3**] in the case λ_j, λ'_j are not real numbers. Later, it was generalized by V. Naishul [**N**], who presented a much more difficult argument to handle the case where the characteristic numbers are real.

Recall that $\mathcal{F} \in \mathcal{A}_n$ is *structurally stable* if there exists a neighborhood $\Omega \subset \mathcal{A}_n$ of $\mathcal{F}$ such that every SHFC in Ω is topologically equivalent to $\mathcal{F}$. Since it can be shown that the set of $\mathcal{F} \in \mathcal{A}'_n$ which do not have any algebraic leaf other than $\mathcal{L}_\infty$ is open and dense in $\mathcal{A}_n$ (see Definition 1.28 and the proof of Proposition 3.21), it follows from Theorem 2.27 that

COROLLARY 2.28. *No SHFC in $\mathcal{A}_n$ is structurally stable when $n \geq 2$.*

The following theorem, which proves a type of "absolute rigidity" for SHFC's, is a fundamental result first proved by Il'yashenko [**I3**].

THEOREM 2.29. *A typical $\mathcal{F} \in \mathcal{A}_n$ is absolutely rigid. That is, there exist neighborhoods $\Omega \subset \mathcal{A}_n$ of $\mathcal{F}$ and U of the identity mapping on $\mathbb{CP}^2$ in the uniform topology such that every SHFC in Ω which is topologically equivalent to $\mathcal{F}$ by a homeomorphism in U is holomorphically equivalent to $\mathcal{F}$.*

X. Gómez-Mont [**GO**] has generalized the above theorem to SHFC's on projective complex surfaces which have an algebraic leaf of sufficiently rich homotopy group.

3. Density and Ergodicity Theorems

In the previous chapter we noted that the behavior of leaves near the leaf at infinity $\mathcal{L}_\infty$ gives us information about their global behavior. The orbits of points under the action of the monodromy pseudo-group PG_∞ in turn give us a picture of the behavior of leaves near $\mathcal{L}_\infty$. So a natural task is to consider dynamics of germs in G_∞, that is, the iterations in a finitely-generated subgroup of $\mathrm{Bih}_0(\mathbb{C})$.

Here is a sketch of what will follow in this chapter. First we consider elements of $\mathrm{Bih}_0(\mathbb{C})$ without any attention to the relationship with the monodromy groups of SHFC's. We study the linearization of hyperbolic germs (Theorem 3.2), and approximation of a linear map by elements of a pseudo-group of germs in $\mathrm{Bih}_0(\mathbb{C})$ (Proposition 3.4) which leads us to a local density theorem (Theorem 3.5). We then consider the notion of ergodicity in $\mathrm{Bih}_0(\mathbb{C})$ and find conditions under which a finitely-generated subgroup of $\mathrm{Bih}_0(\mathbb{C})$ is ergodic (Theorem 3.15). Finally, these results will be applied to the monodromy group G_∞ of a typical $\mathcal{F} \in \mathcal{A}_n$, leading to the density theorem of M. Khudai-Veronov (Theorem 3.24) and the ergodicity theorem of Il'yashenko and Sinai (Theorem 3.26).

3.1. Linearization of Germs in $\mathbf{Bih}_0(\mathbb{C})$. Let us begin with the following definition:

DEFINITION 3.1. A germ $f \in \mathrm{Bih}_0(\mathbb{C})$ is called *linearizable* if there exists a holomorphic change of coordinate $\zeta = \zeta(z)$ near 0, with $\zeta(0) = 0$, such that

$$\zeta(f(z)) = f'(0) \cdot \zeta(z). \tag{3.1}$$

In other words, we have the following commutative diagram:

$$\begin{array}{ccc} (\mathbb{C},0) & \xrightarrow{\ f\ } & (\mathbb{C},0) \\ \big\downarrow \zeta & & \big\downarrow \zeta \\ (\mathbb{C},0) & \xrightarrow{\ f'(0)\cdot\ } & (\mathbb{C},0) \end{array}$$

It follows that the holomorphic change of coordinate ζ conjugates f with its tangent map at the fixed point 0.

If the germs f and g are conjugate, say if $\zeta \circ f \circ \zeta^{-1} = g$, then $\zeta \circ f^n \circ \zeta^{-1} = g^n$ for every $n \geq 1$, so f and g have identical iterative dynamics near 0 up to a change of coordinate. Thus the possibility of linearization can be very helpful in understanding the dynamics of iterations.

It is a remarkable fact that the possibility of linearizing a germ depends on the multiplier of the germ at the fixed point 0. In particular, it was shown by G. Koenigs that $f \in \mathrm{Bih}_0(\mathbb{C})$ is linearizable if f is *hyperbolic* in the sense that $|f'(0)| \neq 1$ [**M**].

THEOREM 3.2. *Every hyperbolic germ $f \in \mathrm{Bih}_0(\mathbb{C})$ is linearizable. The local linearization ζ is unique up to multiplication by a non-zero constant.*

PROOF. Let $\nu := f'(0)$. Without losing generality we may assume that $|\nu| < 1$, for otherwise we can consider f^{-1}. Choose a constant $c < 1$ so that $c^2 < |\nu| < c$, and let $r > 0$ be such that $|f(z)| \leq c|z|$ for $z \in \mathbb{D}(0,r)$. For every $z_0 \in \mathbb{D}(0,r)$, the orbit $\{z_n := f^n(z_0)\}_{n\geq 0}$ converges geometrically towards the origin, with $|z_n| \leq rc^n$. Choose $k > 0$ so that $|f(z) - \nu z| \leq k|z|^2$ for all $z \in \mathbb{D}(0,r)$. It follows that $|z_{n+1} - \nu z_n| \leq kr^2c^{2n}$. Set $\zeta_n(z) := f^n(z)/\nu^n$. Then $|\zeta_{n+1} - \zeta_n| \leq (kr^2/|\nu|)(c^2/|\nu|)^n$. This estimate shows that the sequence $\{\zeta_n\}$ converges uniformly on $\mathbb{D}(0,r)$ to a holomorphic map ζ. The identity $\zeta \circ f = \nu\zeta$ is immediate. Note that $\zeta'(0) = \lim \zeta_n'(0) = \lim \nu^{-n}(f^n)'(0) = 1$, so ζ is a local biholomorphism. If $\eta \circ f = \nu\eta$, then $\zeta \circ \eta^{-1}$ commutes with the linear map $z \mapsto \nu z$. Applying this condition and comparing the coefficients of the Taylor series expansions, it follows that $\zeta \circ \eta^{-1} = (\text{const.})z$, and the uniqueness follows. □

REMARK 3.3. The problem of linearizing non-hyperbolic germs is extremely difficult and has a long history. In fact, if $\nu = e^{2\pi i t}$ with $0 < t < 1$ irrational, it turns out that the possibility of linearizing a germ f with $f'(0) = \nu$ depends on the asymptotic behavior of the denominators of the continued fraction approximations to t. Part of the linearization problem was solved by A. Brjuno in the mid 60's, but the complete solution has been found only in recent years with the work of J.C. Yoccoz and R. Perez-Marco [**P**].

3.2. Approximation by Elements of a Pseudo-Group. The next proposition shows how to approximate a linear map in a suitable coordinate system by elements of a given pseudo-group PG of germs in a finitely-generated subgroup $G \subset \mathrm{Bih}_0(\mathbb{C})$. The proof is nothing but an elaboration of the following elementary fact: If $f \in \mathrm{Bih}_0(\mathbb{C})$ and $|\nu_1| < 1$, then $f'(0)z = \lim_{n\to\infty} \nu_1^{-n} f(\nu_1^n z)$.

PROPOSITION 3.4. *Let G be a marked subgroup of $\mathrm{Bih}_0(\mathbb{C})$ with generators $f_1, \cdots, f_k$, all defined on some domain Ω containing 0. Let f_1 be a hyperbolic germ and ζ be a holomorphic coordinate change linearizing f_1. Without loss of generality, assume that $|f_1'(0)| < 1$, ζ is defined on Ω, and $\zeta(\Omega) = \mathbb{D}(0,r)$ for some $r > 0$. Let DG be the tangent group of G, i.e., the multiplicative subgroup of $\mathbb{C}^*$ generated by $\nu_j := f_j'(0)$, $1 \leq j \leq k$. Then for every $\nu \in \overline{DG}$, there exists a sequence F_n in the pseudo-group PG which in coordinate ζ converges to the linear map $\zeta \mapsto \nu\zeta$ uniformly on compact subsets of $\{\zeta : |\zeta| < \min(r, r/|\nu|)\} = \zeta(\Omega \cap \nu^{-1}\Omega)$.*

By an abuse of notation, we denote by $f(\zeta)$ the germ induced by f in the coordinate ζ, where $f \in \mathrm{Bih}_0(\mathbb{C})$ and ζ is a holomorphic change of coordinate near 0.

PROOF. It suffices to consider the case where $\nu \in DG$, i.e., when $\nu = f'(0)$ for some $f \in G$. The general case will then follow by the uniformity of convergence and a standard diagonal argument. Define

$$F_n := f_1^{-n} \circ f \circ f_1^n. \tag{3.2}$$

First we claim that F_n is defined on Ω for all sufficiently large n, and $F_n(\zeta) \to \nu\zeta$ uniformly on compact subsets of $\mathbb{D}(0,r)$ as $n \to \infty$. In fact, if $f(\zeta) = \nu\zeta + \sum_{j=2}^{\infty} a_j\zeta^j$, then it easily follows from (3.2) that

$$F_n(\zeta) = \nu\zeta + \sum_{j=2}^{\infty} a_j \nu_1^{n(j-1)} \zeta^j. \tag{3.3}$$

Since $f(\zeta)$ is holomorphic on $\mathbb{D}(0,r')$ for some $0 < r' < r$, one has $\limsup_j \sqrt[j]{|a_j|} \leq 1/r'$, so for large n, $\limsup_j \sqrt[j]{|a_j||\nu_1|^{n(j-1)}} \leq 1/r$. Thus the expression on the right hand side of (3.3) has an analytic continuation to $\mathbb{D}(0,r)$, i.e., F_n is defined in Ω for large n. Now the fact that $F_n(\zeta) \to \nu\zeta$ uniformly on compact subsets of $\mathbb{D}(0,r)$ is immediate from (3.3).

What remains to be shown is that for every compact set $K \subset \Omega \cap \nu^{-1}\Omega$ there is an $N = N(K) > 0$ such that the domain Ω_{F_n} of F_n as an element of PG (see Definition 2.12) contains K for all $n > N$.

Each intermediate representation of F_n in (3.2) has the form

$$g_m := f_1^m \quad \text{or} \quad h_{mn} := f_1^{-(n-m)} \circ f \circ f_1^n, \; 0 \leq m \leq n.$$

We shall prove that for large n, $g_m(\zeta)$ and $h_{mn}(\zeta)$ have conformal extensions to $\mathbb{D}(0,r)$ and $K' := \zeta(K)$ is mapped into $\mathbb{D}(0,r)$ by them. In fact, $g_m(\zeta) = \nu_1^m\zeta$ and $h_{mn}(\zeta) \to \nu_1^m\nu\zeta$ uniformly in ζ and m as $n \to \infty$, so that $g_m(\zeta)$ and $h_{mn}(\zeta)$ have conformal extensions over $\mathbb{D}(0,r)$ for large n. Now $g_m(\zeta(K)) = \nu_1^m K' \subset \mathbb{D}(0,r)$ for all $m \geq 0$. On the other hand, let $\delta := \sup\{|\zeta| : \zeta \in K'\}$, so that $\delta < \min(r, r/|\nu|)$. Choose $0 < \epsilon < r - |\nu|\delta$, and find $N > 0$ such that $|h_{mn}(\zeta) - \nu_1^m\nu\zeta| < \epsilon$ for all $0 \leq m \leq n$ and $\zeta \in \mathbb{D}(0,r)$ whenever $n > N$. Then if $\zeta \in K'$ we have $|h_{mn}(\zeta)| < \epsilon + |\nu_1|^m|\nu|\delta < r$ for $n > N$ and all $0 \leq m \leq n$. □

THEOREM 3.5. *Let $G \subset \mathrm{Bih}_0(\mathbb{C})$ be a marked subgroup, and suppose that the tangent group DG is dense in $\mathbb{C}$. Then there exists an open neighborhood Ω of 0 such that for every $z \in \Omega \smallsetminus \{0\}$ the orbit of z under the pseudo-group PG is dense in Ω.*

PROOF. Since $\overline{DG} = \mathbb{C}$, G must contain at least one hyperbolic germ, say f_1. Let Ω and ζ be as in Proposition 3.4. By density of DG, for every $\nu \in \mathbb{C}$ there exists a sequence $\{F_n\}$ in PG such that $F_n(\zeta) \to \nu\zeta$ uniformly on compact subsets of $\zeta(\Omega \cap \nu^{-1}\Omega)$. Choose $z \in \Omega \smallsetminus \{0\}$ and let $w \in \Omega$ be arbitrary. Set $\nu := \zeta(w)/\zeta(z)$. Then $F_n(\zeta(z)) \to (\zeta(w)/\zeta(z))\,\zeta(z) = \zeta(w)$ as $n \to \infty$, and we are done. □

3.3. Ergodicity in Subgroups of $\mathrm{Bih}_0(\mathbb{C})$. Recall from ergodic theory that a measure-preserving transformation T acting on a probability space X is called *ergodic* if every T-invariant subset of X has measure 0 or 1. In what follows, for two measurable sets $A, B \subset \mathbb{C}$, the notation $A \doteq B$ means the symmetric difference $(A \smallsetminus B) \cup (B \smallsetminus A)$ has Lebesgue measure zero.

DEFINITION 3.6. Two Lebesgue measurable subsets A and B of $\mathbb{C}$ are said to be *equivalent at* 0 if there exists an open disk U around 0 such that $A \cap U \doteq B \cap U$.

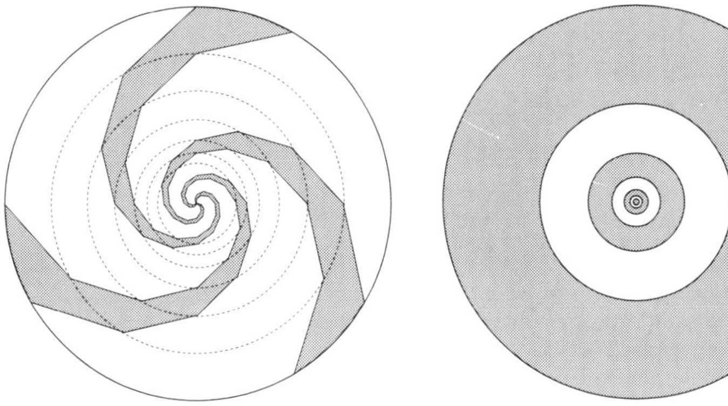

FIGURE 10. *Invariant sets A for the action $f : z \mapsto \nu z$ with $[A] \neq [\mathbb{C}]$ and $[A] \neq [\emptyset]$. Left: $|\nu| \neq 1$. Right: $|\nu| = 1$.*

The *germ* of A, denoted by $[A]$, is the equivalence class of A under this relation. Given an $f \in \mathrm{Bih}_0(\mathbb{C})$, a set A is called *f-invariant* if $[A] = [f(A)]$. Given a subgroup $G \subset \mathrm{Bih}_0(\mathbb{C})$, a set A is said to be *G-invariant* if it is f-invariant for every $f \in G$. The subgroup G is called *ergodic* if for every G-invariant set A, we have $[A] = [\mathbb{C}]$ or $[\emptyset]$.

EXAMPLE 3.7. Let $f(z) = \nu z$, $\nu \neq 0$, and G be the subgroup of $\mathrm{Bih}_0(\mathbb{C})$ generated by f. Then G is not ergodic since it is easy to construct an f-invariant measurable set A such that $[A] \neq [\mathbb{C}]$ and $[A] \neq [\emptyset]$. For example, if $|\nu| < 1$ (the case $|\nu| > 1$ is similar), take any sector B in the fundamental annulus $\{z : |\nu| < |z| \leq 1\}$ and set $A = \bigcup_{n=0}^{\infty} \nu^n B$ (see Fig. 10 left). If, on the other hand, $|\nu| = 1$, consider the invariant set $A = \bigcup_{n=0}^{\infty} \{z : 2^{-2n-1} \leq |z| \leq 2^{-2n}\}$ (see Fig. 10 right). However, we will see that if G is generated by two linear germs $f_1(z) = \nu_1 z$ and $f_2(z) = \nu_2 z$, with the tangent group $DG = \langle \nu_1, \nu_2 \rangle$ dense in $\mathbb{C}$, then G is ergodic (see Proposition 3.13).

DEFINITION 3.8. Let $A \subset \mathbb{C}$ be Lebesgue measurable. A point $z \in \mathbb{C}$ is called a *density point* of A if

$$\lim_{r \to 0} \frac{m(A \cap \mathbb{D}(z, r))}{m(\mathbb{D}(z, r))} = 1, \tag{3.4}$$

where m denotes Lebesgue measure on $\mathbb{C}$.

It is well-known that *almost every point of A is a density point of A* (see [**Ru**] for a proof).

Our next goal is to prove that under fairly general circumstances, a marked subgroup of $\mathrm{Bih}_0(\mathbb{C})$ is ergodic. The idea of the proof, due to E. Ghys, is based on a few preliminary statements, which we present below. The first one is the classical "Koebe 1/4-Theorem" (see [**Ru**]):

THEOREM 3.9. *Let $f : \mathbb{D}(z, r) \to \mathbb{C}$ be a univalent function. Then $f(\mathbb{D}(z, r))$ contains the disk $\mathbb{D}(f(z), |f'(z)|r/4)$.*

LEMMA 3.10. *Let $f(z) = \nu z$, with $0 < |\nu| < 1$, and $g \in \mathrm{Bih}_0(\mathbb{C})$. Suppose that A is both f- and g-invariant. Then there exists an open disk U around 0 such that $A \cap U \doteq \nu^{-n} g \nu^n (A) \cap U$ for all $n \geq 0$.*

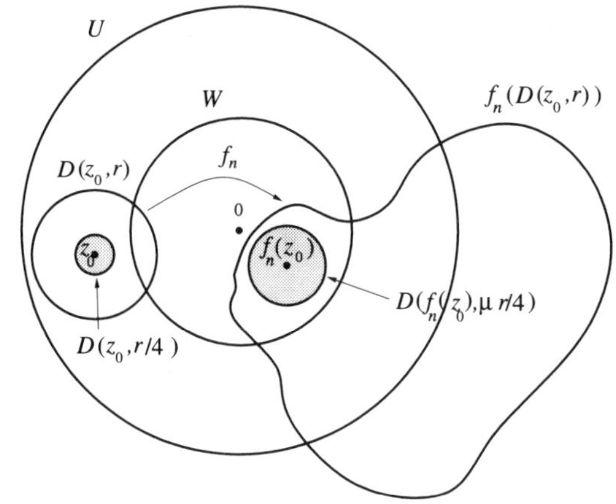

FIGURE 11. *Proof of Proposition 3.11.*

PROOF. Evidently, for each integer n, A is invariant under $f^{-n}\circ g\circ f^n$. To prove the assertion, it suffices to find a disk U that works uniformly for all sufficiently large n. Without loss of generality, in what follows we replace $\doteq$ by $=$ by modifying the inclusions up to a set of measure zero.

Choose a disk $V = \mathbb{D}(0, r_1)$ such that

$$A \cap V = (\nu A) \cap V = (\nu^{-1}A) \cap V = g(A) \cap V. \tag{3.5}$$

Set $\mu := 1/g'(0)$. Since $\nu^{-n}g^{-1}\nu^n$ converges uniformly to $z \mapsto \mu z$ in a neighborhood of 0, we can choose a disk $U = \mathbb{D}(0, r_2) \subset V$ and an $N > 0$ such that $|\nu^{-n}g^{-1}(\nu^n z)| \leq |\mu||z| + r_1/2$ for all $z \in U$ and all $n > N$. Moreover, we may choose r_2 such that $|\mu|r_2 < r_1/2$ and N large enough such that $\nu^n U \subset g(U)$ for all $n > N$.

Now suppose that $z \in A \cap U$. Then by (3.5), $w = \nu^n z \in A \cap U$. Again by (3.5) we have $w = g(x)$, where $x \in A$. If $n > N$, x is in fact in $A \cap U$ by the choice of N. If $y = \nu^{-n}x$, then $|y| = |\nu^{-n}g^{-1}(\nu^n z)| \leq |\mu||z| + r_1/2 \leq |\mu|r_2 + r_1/2 < r_1$, so that $y \in V$. Once again by (3.5) we obtain $y \in A \cap V$. Hence, $z \in \nu^{-n}g\nu^n(A) \cap U$.

The proof that $\nu^{-n}g\nu^n(A) \cap U \subset A \cap U$ for all large n is quite similar. □

PROPOSITION 3.11. *Let $f_n \in \mathrm{Bih}_0(\mathbb{C})$ be defined on some open neighborhood V of 0 for all $n \geq 1$, and $f_n \to f \in \mathrm{Bih}_0(\mathbb{C})$ uniformly on compact subsets of V. Suppose that A is f_n-invariant and that there exists a disk $U \subset V$ around 0 such that $A \cap U \doteq f_n(A) \cap U$ for all $n \geq 1$. Then A is f-invariant.*

PROOF. Without loss of generality we may assume that A has positive measure and each point of A is a density point. Choose a smaller disk $W \subset U \cap f(U) \cap f^{-1}(U)$ such that $f_n(W) \subset U$ for all n. Choose $z_0 \in A$ such that $f(z_0) \in W$. We show that $f(z_0)$ is a density point for A. This proves $f(A) \cap W \subset A \cap W$. Next, by the choice of W, $f_n^{-1}(A) \cap W = A \cap W$, so the same argument gives us $A \cap W \subset f(A) \cap W$.

So let $z_0 \in A$ with $f(z_0) \in W$. By the choice of W, we have $z_0 \in U$. Given any $\epsilon > 0$, choose $r = r(\epsilon) > 0$ so small that

(i) $\mathbb{D}(z_0, r) \subset U$,

(ii) $1 - \epsilon < \dfrac{m(A \cap \mathbb{D}(z_0, r'))}{m(\mathbb{D}(z_0, r'))} \leq 1$ for every $0 < r' \leq r$,

(iii) $|f'(z_0)| - \epsilon \leq \inf_{z \in \mathbb{D}(z_0,r)} |f_n'(z)| \leq \sup_{z \in \mathbb{D}(z_0,r)} |f_n'(z)| \leq |f'(z_0)| + \epsilon$ for all large n.

For simplicity, set $\mu := |f'(z_0)| - \epsilon$, $D := \mathbb{D}(z_0, r)$, $D' := \mathbb{D}(f(z_0), \mu r/4)$, and $D_n' := \mathbb{D}(f_n(z_0), \mu r/4)$. By Theorem 3.9, $f_n(D)$ contains D_n' if n is large enough (Fig. 11). Since $D_n' \smallsetminus A \subset f_n(D) \smallsetminus (A \cap f_n(D)) = f_n(D) \smallsetminus f_n(A \cap D)$, for large n we have

$$\begin{aligned} m(A \cap D_n') &\geq m(D_n') - m(f_n(D)) + m(f_n(A \cap D)) \\ &= m(D_n') - \int_D |f_n'|^2 \, dm + \int_{A \cap D} |f_n'|^2 \, dm \\ &\geq m(D_n') - (|f'(z_0)| + \epsilon)^2 \, m(D) + \mu^2 \, m(A \cap D) && \text{(by (iii))} \\ &\geq [1 - 16(|f'(z_0)| + \epsilon)^2 \mu^{-2} + 16(1 - \epsilon)] \, m(D_n') && \text{(by (ii))} \\ &=: \ell(\epsilon) \, m(D_n'). \end{aligned}$$

As $n \to \infty$, $m(A \cap D_n') \to m(A \cap D')$ by Lebesgue's Dominated Convergence Theorem. Therefore

$$\frac{m(A \cap D')}{m(D')} \geq \ell(\epsilon).$$

Since $\ell(\epsilon) \to 1$ as $\epsilon \to 0$, it follows that $f(z_0)$ is a density point of A. □

COROLLARY 3.12. *Under the assumptions of Proposition 3.4, every G-invariant set A is also DG-invariant.*

PROOF. Let $\nu = g'(0) \in DG$ for some $g \in G$. In the coordinate ζ one has $f_1(\zeta) = \nu_1 \zeta$, and by Proposition 3.4, $f_1^{-n} \circ g \circ f_1^n$ converges to $\zeta \to \nu\zeta$ uniformly on compact subsets of $\zeta(\Omega \cap \nu^{-1}\Omega)$. It follows from Lemma 3.10 and Proposition 3.11 that A is invariant under $\zeta \mapsto \nu\zeta$. □

The proof of the following proposition uses a standard technique in ergodic theory (see for example Appendix 11 of [**AA**]).

PROPOSITION 3.13. *Choose $\nu_j \in \mathbb{C}^*$, $1 \leq j \leq k$, and let G be the subgroup of $\mathrm{Biho}_0(\mathbb{C})$ generated by the linear germs $z \mapsto \nu_j z$, $1 \leq j \leq k$. Then G is ergodic if and only if its tangent group DG is dense in $\mathbb{C}$.*

PROOF. Suppose that DG is dense in $\mathbb{C}$. Then G contains at least one hyperbolic element $f(z) = \nu z$ with $|\nu| < 1$. Let $A \subset \mathbb{C}$ be any measurable G-invariant set. Choose a disk U around 0 such that $A \cap U \doteq (\nu A) \cap U$. Take the quotient of $U \smallsetminus \{0\}$ under the action of the group $\{f^n : n \in \mathbb{Z}\}$ generated by f, which is biholomorphic to a 2-torus $\mathbb{T}^2$. Let $\tilde{G}$ be the induced group of translations of $\mathbb{T}^2$, and $\tilde{A}$ be the induced measurable subset of $\mathbb{T}^2$. Note that $\tilde{A}$ is invariant under the action of $\tilde{G}$, and the orbit of each point in $\mathbb{T}^2$ is dense under this action.

It suffices to show that $\tilde{A}$ or $\mathbb{T}^2 \smallsetminus \tilde{A}$ has measure zero. Expand the characteristic function of $\tilde{A}$ into the Fourier series

$$\chi_{\tilde{A}}(e^{2\pi i x}, e^{2\pi i y}) = \sum_{m,n} a_{mn} \, e^{2\pi i(mx+ny)},$$

where we identify $\mathbb{T}^2$ with $\{(e^{2\pi i x}, e^{2\pi i y}) \in S^1 \times S^1\}$. Let $\tilde{f} \in \tilde{G}$ be the translation $(e^{2\pi i x}, e^{2\pi i y}) \mapsto (e^{2\pi i(x+\alpha)}, e^{2\pi i(y+\beta)})$. The $\tilde{G}$-invariance of $\tilde{A}$ shows that

$$\chi_{\tilde{A}} = \sum_{m,n} a_{mn} e^{2\pi i(m\alpha+n\beta)} \, e^{2\pi i(mx+ny)}$$

almost everywhere. Therefore, for all $m, n \in \mathbb{Z}$, $a_{mn} = a_{mn}\, e^{2\pi i(m\alpha+n\beta)}$. Since $\tilde{G}$ contains at least one irrational translation (otherwise, the orbit of each point would be finite), we conclude that $a_{mn} = 0$ for all $(m,n) \neq (0,0)$, hence $\chi_{\tilde{A}} = 0$ or 1 almost everywhere.

Conversely, let G be ergodic. Clearly G contains a hyperbolic element $f_0(z) = \nu z$, with $|\nu| < 1$. Suppose by way of contradiction that DG is not dense in $\mathbb{C}$, so that there exists an open disk $\mathbb{D}(z,r)$ such that $\overline{DG} \cap \mathbb{D}(z,r) = \emptyset$. Set $A := \bigcup_{f \in G} f(\mathbb{D}(z, r/2))$. Then A is an open, G-invariant set such that $\overline{DG} \cap A = \emptyset$. The germ of A at 0, denoted by $[A]$, is not equal to $[\emptyset]$ since every neighborhood of 0 contains $\nu^n \mathbb{D}(z, r/2)$ for large n. It follows that $[A] = [\mathbb{C}]$. The invariance of A and the fact that $|\nu| < 1$ will then show that $m(\mathbb{C} \smallsetminus A) = 0$. In particular, A is dense in $\mathbb{C}$. Let $\{w_n\}$ be a sequence in A such that $\lim w_n{=}1$. Let $w_n = \sigma_n z_n$, where $\sigma_n \in DG$ and $z_n \in \mathbb{D}(z, r/2)$. Then σ_n^{-1} belongs to $\mathbb{D}(z,r)$ for large n . This contradicts the assumption $\overline{DG} \cap \mathbb{D}(z; r) = \emptyset$. □

REMARK 3.14. The above proof shows how the notion of ergodicity for finitely-generated subgroups of $\mathrm{Bih}_0(\mathbb{C})$ containing a hyperbolic germ is related to the usual notion of ergodicity for translations of tori, justifying Definition 3.6.

THEOREM 3.15. *Let $G \subset \mathrm{Bih}_0(\mathbb{C})$ be a marked subgroup, with the tangent group DG dense in $\mathbb{C}$. Then G is ergodic.*

PROOF. Since DG is dense in $\mathbb{C}$, G must contain at least one hyperbolic element. By Corollary 3.12, every G-invariant set A is also DG-invariant. By Proposition 3.13, $[A] = [\mathbb{C}]$ or $[\emptyset]$. □

3.4. Density of Leaves of SHFC's on $\mathbb{CP}^2$. This section is devoted to the proof of a theorem of Khudai-Veronov on density of leaves of a typical SHFC on $\mathbb{CP}^2$. We first prove a version of this theorem which asserts that for a typical $\mathcal{F} \in \mathcal{A}_n$ all but a finite number of leaves are dense in $\mathbb{CP}^2$. Next, by applying a more elaborate argument, we show that among these exceptional leaves only $\mathcal{L}_\infty$ is robust. In fact, for a typical $\mathcal{F} \in \mathcal{A}_n$ *all* leaves except $\mathcal{L}_\infty$ are dense in $\mathbb{CP}^2$.

DEFINITION 3.16. Let X be a holomorphic vector field defined in some neighborhood U of $p \in \mathbb{C}^2$, and let p be an isolated singular point of X. Let σ_1 and σ_2 be the eigenvalues of the Jacobian matrix $DX(p)$. We say that p is a *non-degenerate* singularity if $\sigma_1\sigma_2 \neq 0$. p is called a *hyperbolic* singularity if it is non-degenerate and the characteristic number σ_1/σ_2 is not real.

THEOREM 3.17. *Every hyperbolic singularity is locally linearizable: There exist neighborhoods U of p and V of $0 \in \mathbb{C}^2$ and a biholomorphism $\varphi : U \xrightarrow{\simeq} V$ such that $(\varphi_* X)(x,y) = \sigma_1 x\, \partial/\partial x + \sigma_2 y\, \partial/\partial y$.*

In fact, in this two dimensional case, the linearization is possible when either $\sigma_1/\sigma_2 \notin \mathbb{R}$, or σ_1/σ_2 is positive but not an integer or the inverse of an integer (the so-called "non-resonant Poincaré case"). For a proof, see [**Ar**].

Using the above theorem, we can understand the local picture of a hyperbolic singularity since it is easy to integrate the linearized vector field.

Recall that for a holomorphic vector field X with an isolated singularity at p, a *local separatrix* through p is the image of a punctured disk $\mathbb{D}^*(0,r)$ under a holomorphic immersion η such that $d\eta(T)/dT = X(\eta(T))$ for $T \in \mathbb{D}^*(0,r)$, and $\lim_{T\to 0} \eta(T) = p$. It follows that a local separatrix of p is an invariant analytic

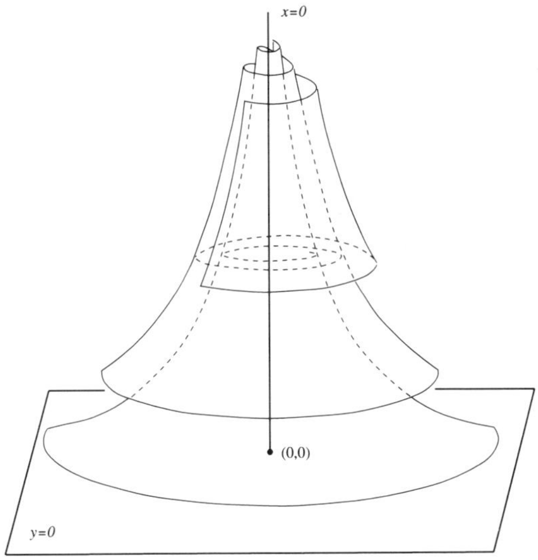

FIGURE 12. *A hyperbolic singularity at the origin of* $\mathbb{C}^2$.

curve which passes through p, perhaps with a singularity there. It is shown by C. Camacho and P. Sad that every holomorphic vector field X with an isolated singularity at $0 \in \mathbb{C}^2$ has a local separatrix through 0 [**CS**]. On the other hand, this statement is false in higher dimensions, even in $\mathbb{C}^3$ [**GL**].

COROLLARY 3.18. *Let X be a holomorphic vector field defined in a neighborhood of an isolated hyperbolic singularity $p \in \mathbb{C}^2$. Then X has exactly two local separatrices through p, and every other integral curve near p accumulates on these two separatrices.*

Fig. 12 is an attempt to visualize this situation.

PROOF. By Theorem 3.17 X can locally be transformed into $\sigma_1 x\ \partial/\partial x + \sigma_2 y\ \partial/\partial y$. The integral curve passing through (x_0, y_0) can be parametrized as $T \mapsto (x_0 e^{\sigma_1 T}, y_0 e^{\sigma_2 T})$. It follows that the punctured axes $\{x = 0\} \smallsetminus \{(0,0)\}$ and $\{y = 0\} \smallsetminus \{(0,0)\}$ are local separatrices. Since $\sigma_1/\sigma_2 \notin \mathbb{R}$, there exist sequences $\{T_n\}$ and $\{T'_n\}$ such that

$$\begin{aligned} e^{\sigma_1 T_n} &= 1 \text{ for } n = 1, 2, \dots, \qquad & e^{\sigma_2 T_n} &\to 0 \text{ as } n \to \infty, \\ e^{\sigma_2 T'_n} &= 1 \text{ for } n = 1, 2, \dots, \qquad & e^{\sigma_1 T'_n} &\to 0 \text{ as } n \to \infty. \end{aligned}$$

It follows that if $x_0 y_0 \neq 0$, the integral curve passing through (x_0, y_0) accumulates on $(x_0, 0)$ and $(0, y_0)$, hence on the axes $\{x = 0\}$ and $\{y = 0\}$ by Proposition 1.6. $\square$

Now let $\mathcal{F} \in \mathcal{A}'_n$, $L_0 \cap \operatorname{sing}(\mathcal{F}) = \{p_1, \dots, p_{n+1}\}$, and let λ_j be the characteristic number of p_j, as in (2.7). It follows that p_j is a hyperbolic singularity of X_1 (hence of any vector field representing $\mathcal{F}$ near p_j) if and only if $\lambda_j \notin \mathbb{R}$. By Proposition 2.19, the last condition is equivalent to $|\nu_j| \neq 1$, where ν_j is the multiplier at 0 of the monodromy mapping $f_j \in G_\infty$. We conclude that *p_j is a hyperbolic singularity if and only if f_j is a hyperbolic germ in* $\mathrm{Biho}_0(\mathbb{C})$.

Note that if p_j is hyperbolic, then one of the separatrices through p_j is the leaf at infinity; the other one is transversal to $\mathcal{L}_\infty$ (Fig. 13).

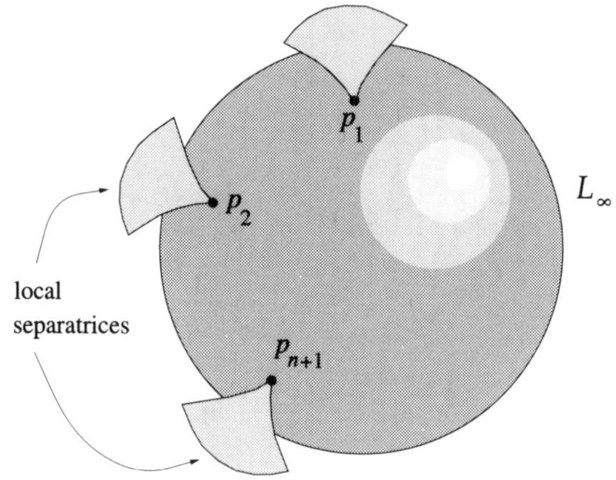

FIGURE 13. *Leaf at infinity with hyperbolic singularities.*

PROPOSITION 3.19. *A typical $\mathcal{F} \in \mathcal{A}_n$, $n \geq 2$, satisfies the following conditions:*

(i) $\mathcal{F} \in \mathcal{A}'_n$,

(ii) $|\nu_j| \neq 1$ *for all* $1 \leq j \leq n$; *in other words, all the generators of* G_∞ *are hyperbolic,*

(iii) *the tangent group* $DG_\infty = \langle \nu_1, \ldots, \nu_n \rangle$ *is dense in* $\mathbb{C}$.

PROOF. Since the union of sets of measure zero has measure zero, it suffices to prove that each condition is typical in $\mathcal{A}_n$.

The first condition is typical by Corollary 1.29. The second one is typical by (2.7): If $|\nu_j| = 1$, then $\lambda_j \in \mathbb{R}$, and this can be easily destroyed by perturbing the coefficients of P and Q.

The third condition is more subtle. By (2.7) and Proposition 2.19, it suffices to prove that for almost every $(\lambda_1, \ldots, \lambda_n) \in \mathbb{C}^n$, the multiplicative subgroup generated by $\{e^{2\pi i \lambda_1}, \ldots, e^{2\pi i \lambda_n}\}$ is dense in $\mathbb{C}$. Evidently it is enough to prove this statement for $n = 2$, since then we can take the product of the resulting subset of $\mathbb{C}^2$ by $\mathbb{C}^{n-2}$ to obtain a subset of full measure in $\mathbb{C}^n$.

We shall prove that for almost every $(\lambda_1, \lambda_2) \in \mathbb{C}^2$, the additive subgroup generated by $\{1, \lambda_1, \lambda_2\}$ is dense in $\mathbb{C}$. Suppose that (λ_1, λ_2) is chosen such that

$(*)$ no two vectors in $\{1, \lambda_1, \lambda_2\}$ are $\mathbb{R}$-dependent,

$(**)$ if $1 = a\lambda_1 + b\lambda_2$, with $a, b \in \mathbb{R}$, then b/a is irrational.

Let Λ be the lattice generated by $\{\lambda_1, \lambda_2\}$, and consider the quotient torus $\mathbb{R}^2 \xrightarrow{\pi} \mathbb{R}^2/\Lambda$. Let $L : \mathbb{R}^2 \to \mathbb{R}^2$ be a linear map with $L(\lambda_1) = (1, 0)$, $L(\lambda_2) = (0, 1)$. Then $L(1) = (a, b)$, and L induces a homeomorphism $\tilde{L} : \mathbb{R}^2/\Lambda \to \mathbb{R}^2/\mathbb{Z}^2$ such that the following diagram commutes:

$$\begin{array}{ccc} \mathbb{R}^2 & \xrightarrow{L} & \mathbb{R}^2 \\ \downarrow \pi & & \downarrow \pi' \\ \mathbb{R}^2/\Lambda & \xrightarrow{\tilde{L}} & \mathbb{R}^2/\mathbb{Z}^2 \end{array} \tag{3.6}$$

Note that the slope of $L(1)$ is irrational by $(**)$. So the sequence $\{\pi'(L(n))\}_{n \geq 0}$ is dense in $\mathbb{R}^2/\mathbb{Z}^2$. Pulling back this sequence to the torus $\mathbb{R}^2/\Lambda$ by $\tilde{L}$, it follows from diagram (3.6) that the sequence $\{\pi(n)\}_{n\geq 0}$ is dense in $\mathbb{R}^2/\Lambda$. Hence if (λ_1, λ_2) satisfies $(*)$ and $(**)$, the subgroup generated by $\{1, \lambda_1, \lambda_2\}$ is dense in $\mathbb{C}$. Finally, it is straightforward to check that $(*)$ and $(**)$ hold for almost every $(\lambda_1, \lambda_2) \in \mathbb{C}^2$. □

Now the first version of the density theorem is quite easy to prove.

THEOREM 3.20. *For a typical $\mathcal{F} \in \mathcal{A}_n$, all but at most $n+2$ non-singular leaves are dense in $\mathbb{CP}^2$.*

PROOF. Let $\mathcal{F} \in \mathcal{A}_n$ have properties (i), (ii), and (iii) of Proposition 3.19. Since all the singular points $\{p_1, \ldots, p_{n+1}\} = L_0 \cap \operatorname{sing}(\mathcal{F})$ are hyperbolic, there are exactly two local separatrices through each p_j, one of them lies in $\mathcal{L}_\infty$. Denote by $\mathcal{L}_j$ the *global separatrix* through p_j which is transversal to $\mathcal{L}_\infty$. (By the global separatrix we mean the extension of the local separatrix as a leaf.) Note that we might have $\mathcal{L}_i = \mathcal{L}_j$ even if $i \neq j$.

Now let $p \in \mathbb{CP}^2 \smallsetminus (\operatorname{sing}(\mathcal{F}) \cup \mathcal{L}_1 \cup \ldots \cup \mathcal{L}_{n+1} \cup \mathcal{L}_\infty)$. By Corollary 2.17, $\mathcal{L}_p$ has a point of accumulation on L_0. By the choice of p and Corollary 3.18, this point can be chosen on $\mathcal{L}_\infty$, hence by Proposition 1.6 the entire $\mathcal{L}_\infty$ is in the closure of $\mathcal{L}_p$. In particular, $\mathcal{L}_p$ intersects the transversal Σ to $\mathcal{L}_\infty$ at the base point a of $\pi_1(\mathcal{L}_\infty)$. By Theorem 3.5, $\mathcal{L}_p \cap \Sigma$ is dense in some neighborhood $\Omega \subset \Sigma$ of a.

Now choose any open set $U \subset \mathbb{CP}^2$ and any point $q \in U \smallsetminus (\operatorname{sing}(\mathcal{F}) \cup \mathcal{L}_1 \cup \ldots \cup \mathcal{L}_{n+1} \cup \mathcal{L}_\infty)$. By the above argument, $\mathcal{L}_q \cap \Sigma$ is dense in Ω. Therefore $q \in \overline{\mathcal{L}}_p$ by another application of Proposition 1.6. Since U was arbitrary, $\mathcal{L}_p$ is dense in $\mathbb{CP}^2$. □

The above proof shows that all non-singular leaves other than $\mathcal{L}_\infty$ and the global separatrices $\mathcal{L}_j$ are dense. The condition $p \notin \bigcup_{j=1}^{n+1} \mathcal{L}_j$ is only used to guarantee the existence of point in $\mathcal{L}_\infty \cap \overline{\mathcal{L}}_p$. If the only accumulation points of $\mathcal{L}_p$ on L_0 are singular, it turns out that $\overline{\mathcal{L}}_p$ is an algebraic curve. Since the algebraic leaves other than $\mathcal{L}_\infty$ are not typical for elements of $\mathcal{A}_n$ (Proposition 3.21 below), we can prove a stronger version of the density theorem without the assumption $p \notin \bigcup_{j=1}^{n+1} \mathcal{L}_j$ in the above proof.

PROPOSITION 3.21. *For a typical $\mathcal{F} \in \mathcal{A}_n$, the only algebraic leaf is the leaf at infinity $\mathcal{L}_\infty$.*

PROOF. The argument below is an adaptation of an idea due to I. Petrovskiĭ and E. Landis [**PL1**]. Let $\mathcal{F} : \{Pdy - Qdx = 0\} \in \mathcal{A}_n$. Suppose that the algebraic curve $S_K : \{K = 0\}$ (see §1.4) with singular points of $\mathcal{F}$ deleted is a leaf of $\mathcal{F}$, where $K = K(x, y)$ is an irreducible polynomial of degree k. Since S_K is a leaf, we have

$$\frac{\partial K}{\partial x}(x,y)P(x,y) + \frac{\partial K}{\partial y}(x,y)Q(x,y) = 0$$

whenever $K(x, y) = 0$. It follows that there exists a polynomial $\tilde{K}$ of degree at most $(n - 1)$ such that

$$\frac{\partial K}{\partial x}P + \frac{\partial K}{\partial y}Q = K\tilde{K}. \tag{3.7}$$

Conversely, if there exist polynomials $K, \tilde{K}, P$, and Q satisfying (3.7) with K irreducible and P and Q relatively prime, then $S_K \smallsetminus \operatorname{sing}(\mathcal{F})$ is an algebraic leaf of $\mathcal{F} : \{Pdy - Qdx = 0\}$.

Let E be the complex linear space of the coefficients of $K, \tilde{K}, P$, and Q, which has dimension

$$\frac{(k+1)(k+2)}{2}+\frac{n(n+1)}{2}+2\,\frac{(n+1)(n+2)}{2}=\frac{1}{2}\{(k+1)(k+2)+(3n^2+7n+4)\}.$$

If we impose (3.7) on these coefficients, we obtain equalities that define an algebraic variety S in E. Note that if a belongs to S, so does λa for every $\lambda \in \mathbb{C}^*$ by (3.7). Therefore, S projects to an algebraic variety S^* in $\mathbb{CP}^d$, with $d = \dim E - 1$. Decompose S^* as $\bigcup_{j=1}^m S_j^*$, where each S_j^* is irreducible. Let $E' \subset E$ be the subspace of coefficients of P and Q, which has dimension $n^2 + 3n + 2$. The linear projection $E \to E'$ induces a projection $\pi : \mathbb{CP}^d \to \mathbb{CP}^N$, where $N = (\dim E') - 1 = n^2+3n+1$. Each $\pi(S_j^*)$ is an algebraic variety in $\mathbb{CP}^N$. Since there are SHFC's in $\mathcal{A}_n$ which do not have any algebraic leaf other than $\mathcal{L}_\infty$, we have $\pi(S_j^*) \neq \mathbb{CP}^N$, so dim $\pi(S_j^*) \leq N-1$. Taking the union for all $j = 1, \cdots, m$, we obtain the algebraic variety $\pi(S^*)$ in $\mathbb{CP}^N$, each irreducible component of which has dimension $\leq N-1$. It follows that the Lebesgue measure of $\pi(S^*)$ in $\mathcal{A}_n$ is zero. Since points in $\pi(S^*) \cap \mathcal{A}_n$ correspond to SHFC's which have an algebraic leaf other than $\mathcal{L}_\infty$, we obtain the result. □

REMARK 3.22. By a much more difficult argument, using an index theorem of Camacho and Sad and the concept of the Milnor number of a local branch of a singular point, A. Lins Neto has shown that for $n \geq 2$ there exists an open and dense subset of $\mathcal{D}_n$ (see Corollary 1.19) consisting of SHFC's which do not have any algebraic leaf [**L**].

PROPOSITION 3.23. *Let $\mathcal{F} \in \mathcal{A}_n'$ and all points in $L_0 \cap \operatorname{sing}(\mathcal{F})$ be hyperbolic. Let $\mathcal{L}$ be a non-singular leaf of $\mathcal{F}$ such that $\overline{\mathcal{L}} \cap L_0$ consists of singular points only. Then $\mathcal{L}$ is an algebraic leaf.*

PROOF. Let $\Omega = \mathbb{CP}^2 \smallsetminus \operatorname{sing}(\mathcal{F})$. First we show that $\mathcal{L}$ is closed in Ω. Since $\overline{\mathcal{L}} \cap L_0 \subset L_0 \cap \operatorname{sing}(\mathcal{F})$, it follows that $\mathcal{L}$ coincides with the global separatrix $\mathcal{L}_j$ which is transversal to $\mathcal{L}_\infty$ at the singular point p_j on L_0. Let $p \in \overline{\mathcal{L}}$ be any non-singular point. Then, by Proposition 1.6, $L_0 \cap \overline{\mathcal{L}_p} \subset L_0 \cap \overline{\mathcal{L}} \subset L_0 \cap \operatorname{sing}(\mathcal{L})$, which shows $\mathcal{L}_p$ also coincides with $\mathcal{L}_j$. Hence $\mathcal{L}_p = \mathcal{L}$ and $p \in \mathcal{L}$.

Next we show that $\overline{\mathcal{L}}$ is an analytic subvariety of $\mathbb{CP}^2$. Let $p \in \Omega$. If $p \in \overline{\mathcal{L}}$, then $p \in \mathcal{L}$ by the above argument. Suppose that (U, φ) is a foliation chart around p, Σ is a transversal to $\mathcal{L}_p = \mathcal{L}$ at p, Σ' is another transversal to $\mathcal{L}$ at p' near $p_j \in L_0 \cap \operatorname{sing}(\mathcal{F})$, and γ is any path in $\mathcal{L}$ joining p to p' (Fig. 14). Let $f_\gamma : \Sigma \to \Sigma'$ be the associated holonomy mapping. If there exists a sequence $p_n \in \mathcal{L} \cap \Sigma$ which converges to p, then by considering the sequence $f_\gamma(p_n) \in \mathcal{L} \cap \Sigma'$ we conclude from Corollary 3.18 that $\mathcal{L}$ must have a non-singular accumulation point on $\mathcal{L}_\infty$, which contradicts our assumption. Therefore, by choosing U small enough, the only plaque of $\mathcal{L}$ in U is the one which passes through p, and evidently there exists a holomorphic function $f : U \to \mathbb{C}$ such that $f^{-1}(0) = \mathcal{L} \cap U$.

This means $\overline{\mathcal{L}} \smallsetminus \operatorname{sing}(\mathcal{F})$ is a 1-dimensional analytic subvariety of Ω. Since $\dim(\operatorname{sing}(\mathcal{F})) = 0 < 1 = \dim(\overline{\mathcal{L}} \smallsetminus \operatorname{sing}(\mathcal{F}))$, the theorem of Remmert-Stein (see for example [**GR**]) shows that $\overline{\mathcal{L}}$ is an analytic subvariety of $\mathbb{CP}^2$. Finally, every analytic subvariety of $\mathbb{CP}^2$ is algebraic by Chow's Theorem [**GH**]. □

Now, by Theorem 3.20, Proposition 3.21 and Proposition 3.23, we obtain the density theorem of Khudai-Veronov:

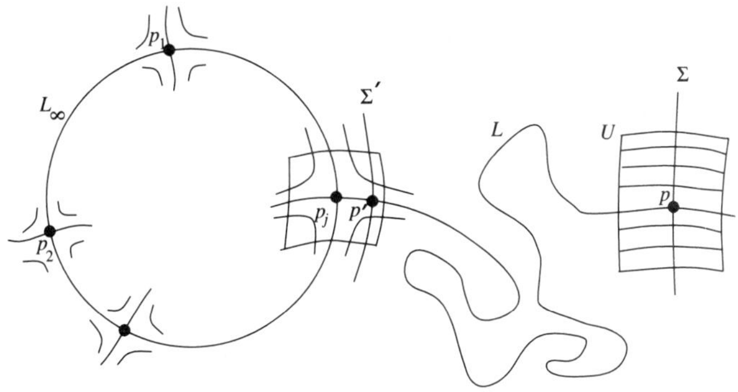

FIGURE 14. *Proof of Proposition 3.23.*

THEOREM 3.24. *For a typical $\mathcal{F} \in \mathcal{A}_n$, every leaf, except the leaf at infinity, is dense in $\mathbb{CP}^2$.*

3.5. Ergodicity of SHFC's on $\mathbb{CP}^2$. In what follows, we say that a set $A \subset \mathbb{CP}^2$ has measure zero if for every chart (U, φ) compatible with the standard smooth structure of $\mathbb{CP}^2$, the set $\varphi(A \cap U) \subset \mathbb{R}^4$ has Lebesgue measure zero. In other words, we consider the *Lebesgue measure class* on $\mathbb{CP}^2$.

DEFINITION 3.25. Let $\mathcal{F} \in \mathcal{A}_n$. The *saturation* $s(A)$ of $A \subset \mathbb{CP}^2$ is the set of all points q such that $q \in \mathcal{L}_p$ for some $p \in A$. A measurable set A is called *saturated* if $s(A) = A$ up to a set of measure zero. An SHFC $\mathcal{F}$ is called *ergodic* if for every measurable saturated set A, either A or $\mathbb{CP}^2 \smallsetminus A$ has measure zero.

Evidently if $\mathcal{F}$ is ergodic, then every non-singular leaf of $\mathcal{F}$ is either dense in $\mathbb{CP}^2$ or its closure has measure zero. Also the following observation is useful: Let $\mathcal{L}_p$ be a non-singular leaf and Σ be a transversal to $\mathcal{L}_p$ at p. Suppose that $A \subset \Sigma$ is a set of measure zero with respect to the Lebesgue measure class on $\Sigma \simeq \mathbb{D}$. Then $s(A)$ has measure zero in $\mathbb{CP}^2$. This is because $s(A)$ can be covered by a countable number of foliation charts $\{(U_i, \varphi_i)\}_{i=1}^{\infty}$ and each $s(A) \cap U_i$ has measure zero. In particular, each single leaf has measure zero (take $A = \{p\}$).

THEOREM 3.26. *Let $\mathcal{F} \in \mathcal{A}_n$ have properties (i),(ii), and (iii) of Proposition 3.19. Then $\mathcal{F}$ is ergodic. In particular, ergodicity is typical for elements of $\mathcal{A}_n$.*

PROOF. Let A be a measurable saturated subset of $\mathbb{CP}^2$. Without loss of generality, we may assume that A does not contain any global separatrix $\mathcal{L}_j$ through $p_j \in L_0 \cap \operatorname{sing}(\mathcal{F})$, for each individual leaf has measure zero. Let $p \in A$ be non-singular. Then $\mathcal{L}_p$ must accumulate on $\mathcal{L}_\infty$, so it has to intersect the transversal Σ to $\mathcal{L}_\infty$ at the base point a. By Theorem 3.15, G_∞ is ergodic. Since A is saturated, $A \cap \Sigma$ is G_∞-invariant, so there is an open disk $U \subset \Sigma$ around a such that either $A \cap U$ or $U \smallsetminus A$ has measure zero with respect to the Lebesgue measure class on Σ. It is clear that

$$s(A \cap U) = A \smallsetminus \operatorname{sing}(\mathcal{F}) \quad \text{and} \quad s(U \smallsetminus A) = \mathbb{CP}^2 \smallsetminus \{A \cup \operatorname{sing}(\mathcal{F})\}$$

up to a set of measure zero in $\mathbb{CP}^2$. By the observation before the statement of the theorem, it follows that either A or $\mathbb{CP}^2 \smallsetminus A$ has measure zero. □

4. Non-Trivial Minimal Sets

This chapter deals with a somewhat different global aspect of SHFC's on $\mathbb{CP}^2$. As will be seen, the foliations under consideration are essentially those which do not have any algebraic leaf. In particular, because of the absence of the leaf at infinity, we cannot utilize such powerful tools as the monodromy group G_∞. Recall that a typical $\mathcal{F} \in \mathcal{A}_n$ has at least one algebraic leaf (i.e., $\mathcal{L}_\infty$). Hence from the point of view of differential equations for which the decomposition into the $\mathcal{A}_n$ is more natural, the foliations we consider in this chapter almost never occur. However, from the point of view of foliation theory, for which the natural decomposition is by the $\mathcal{D}_n$, the property of having no algebraic leaf seems to be typical (see Remark 3.22).

4.1. An Open Problem. The study of limit sets of foliations and flows in the real domain has proved to be of great significance in those theories. The classical theorem of Poincaré-Bendixson asserts that for every smooth real flow on the 2-sphere, every trajectory accumulates either on a periodic orbit or a singular point (or both). It is natural to ask a similar question for SHFC's on $\mathbb{CP}^2$. Here the analogue of a periodic orbit is a compact non-singular leaf and it is not difficult to prove that no such leaves could exist (Theorem 4.10). So we naturally arrive at the following question, apparently first asked by C. Camacho:

Question. *Is there a non-singular leaf of an SHFC on $\mathbb{CP}^2$ which does not accumulate on any singular point?*

Oddly enough, the question has remained open since the mid 80's. One can formulate it in a slightly different language, commonly used in foliation theory.

DEFINITION 4.1. A *minimal set* for an SHFC on $\mathbb{CP}^2$ is a compact saturated non-empty subset of $\mathbb{CP}^2$ which is minimal with respect to these three properties. A *non-trivial minimal set* is a minimal set which is not a singular point. Throughout this chapter, $\mathcal{M}$ will always denote a non-trivial minimal set.

Minimality shows that if $p \in \mathcal{M}$, then $\overline{\mathcal{L}}_p = \mathcal{M}$. It follows that the problem of finding a non-singular leaf which does not accumulate on any singular point is equivalent to finding a non-trivial minimal set. Therefore, we can reformulate the above question as

The Minimal Set Problem. *Does there exist an SHFC on $\mathbb{CP}^2$ which has a non-trivial minimal set?*

Such a non-trivial minimal set is an example of a *Riemann surface lamination.* By definition, a Riemann surface lamination (RSL) is a compact space which locally looks like the product of the unit disk and a compact metric space (usually a Cantor set). The transition maps between various charts are required to be holomorphic in the leaf direction and only continuous in the transverse direction. Clearly every compact Riemann surface is such a space, but they form the class of *trivial* RSL's. Although there are some basic results on uniformization of RSL's (see [**Ca**]), the corresponding embedding problem is rather unexplored. A classical theorem asserts that every compact Riemann surface can be holomorphically embedded in $\mathbb{CP}^3$. The Minimal Set Problem, as Ghys has suggested [**Gh**], could be viewed as a special case of the embedding problem for RSL's: "Can a non-trivial Riemann surface lamination be holomorphically embedded in $\mathbb{CP}^2$?"

In what follows we show some basic properties of non-trivial minimal sets. This theory, due to Camacho, Lins Neto, and Sad [**CLS1**], was developed in part in the hope of arriving at a contradiction to the existence of non-trivial minimal sets.

4.2. Uniqueness of Minimal Sets. How many distinct non-trivial minimal sets, if any at all, can an SHFC on $\mathbb{CP}^2$ have?

THEOREM 4.2. *An SHFC on $\mathbb{CP}^2$ has at most one non-trivial minimal set.*

The proof of this nice fact is quite elementary, and is based on the study of the distance between two non-singular leaves and the application of the Maximum Principle for real harmonic functions (compare [**CLS1**]).

To study the distance between two leaves, we have to choose a suitable Riemannian metric on $\mathbb{CP}^2$. Consider the Hermitian metric

$$ds^2 = \frac{|dx|^2 + |dy|^2 + |xdy - ydx|^2}{(1 + |x|^2 + |y|^2)^2} \tag{4.1}$$

in the affine chart $(x, y) \in U_0$, which extends to a Hermitian metric on the entire projective plane. It is called the *Fubini-Study metric* on $\mathbb{CP}^2$ [**GH**]. We will denote by d the Riemannian distance induced by this metric. Note that the associated $(1, 1)$-form of the Fubini-Study metric is

$$\begin{aligned} \Omega &= \frac{\sqrt{-1}}{2\pi} \, \frac{dx \wedge d\bar{x} + dy \wedge d\bar{y} + (xdy - ydx) \wedge (\bar{x}d\bar{y} - \bar{y}d\bar{x})}{(1 + |x|^2 + |y|^2)^2} \\ &= \frac{\sqrt{-1}}{2\pi} \, \partial\bar{\partial} \log(1 + |x|^2 + |y|^2). \end{aligned} \tag{4.2}$$

LEMMA 4.3. *For any $p_0 = (x_0, y_0)$ and $p_1 = (x_1, y_1)$ in the affine chart $(x, y) \in U_0$, we have*

$$d(p_0, p_1) \leq \frac{|p_0 - p_1|}{(1 + \delta^2(p_0, p_1))^{1/2}},$$

where $\delta(p_0, p_1)$ is the minimum (Euclidean) distance form the origin to the line segment which joins p_0 and p_1 .

PROOF. By definition, $d(p_0, p_1) = \inf_\gamma\{\int_0^1 \|\gamma'(t)\| dt\}$, where the infimum is taken over all piecewise smooth curves $\gamma : [0, 1] \to \mathbb{CP}^2$ with $\gamma(0) = p_0, \gamma(1) = p_1$. In particular, when $\gamma(t) = (1 - t)p_0 + tp_1 =: (x(t), y(t))$, one has

$$d^2(p_0, p_1) \leq (\int_0^1 \|\gamma'(t)\| dt)^2 \leq \int_0^1 \|\gamma'(t)\|^2 dt.$$

Now we estimate $\|\gamma'\|^2$:

$$\begin{aligned}\|\gamma'(t)\|^2 &= \frac{|x'(t)|^2+|y'(t)|^2+|x(t)y'(t)-y(t)x'(t)|^2}{(1+|x(t)|^2+|y(t)|^2)^2}\\ &\leq \frac{|x'(t)|^2+|y'(t)|^2+(|x(t)|^2+|y(t)|^2)(|x'(t)|^2+|y'(t)|^2)}{(1+|x(t)|^2+|y(t)|^2)^2}\\ &= \frac{(|x'(t)|^2+|y'(t)|^2)(1+|x(t)|^2+|y(t)|^2)}{(1+|x(t)|^2+|y(t)|^2)^2}\\ &= \frac{|x_0-x_1|^2+|y_0-y_1|^2}{1+|x(t)|^2+|y(t)|^2}\\ &\leq \frac{|p_0-p_1|^2}{1+\delta^2(p_0,p_1)},\end{aligned}$$

and this completes the proof. □

COROLLARY 4.4. *Let E and F be two disjoint compact subsets of $\mathbb{CP}^2$, and $E' := E\cap U_0$ and $F' := F\cap U_0$ be both non-empty. If $\epsilon := \inf\{|p-q| : (p,q)\in E'\times F'\}$, then $\epsilon > 0$ and there exists a pair $(p,q)\in E'\times F'$ with $|p-q| = \epsilon$.*

PROOF. Let $(p_n, q_n)\in E'\times F'$ be such that $|p_n-q_n|\to\epsilon$ as $n\to\infty$. By taking subsequences, if necessary, we may assume that $p_n\to p\in E$ and $q_n\to q\in F$. If $(p,q)\in E'\times F'$, we are done. Otherwise, if $p\in E\smallsetminus E'$, one has $q\in F\smallsetminus F'$ since $|p_n-q_n|$ is bounded. Therefore $\delta(p_n,q_n)\to\infty$ as $n\to\infty$, so $d(p,q)=0$ by Lemma 4.3, which is a contradiction. □

Proof of Theorem 4.2. Let $\mathcal{M}_1$ and $\mathcal{M}_2$ be two non-trivial minimal sets of $\mathcal{F}$: $\{Pdy-Qdx=0\}$. They are necessarily disjoint by minimality. Set $\mathcal{M}_1' := \mathcal{M}_1\cap U_0$, $\mathcal{M}_2' := \mathcal{M}_2\cap U_0$ and $\epsilon := \inf\{|p-q| : (p,q)\in\mathcal{M}_1'\times\mathcal{M}_2'\}$. By Corollary 4.4, there exists $(p,q)\in\mathcal{M}_1'\times\mathcal{M}_2'$, with $|p-q|=\epsilon$. For simplicity, let $p=(0,0)$ and $q=(0,y_0)$, with $|y_0|=\epsilon$. It follows from the definition of ϵ that the y-axis is normal to $\mathcal{L}_p$ and $\mathcal{L}_q$ at p and q. Since p and q are not singular, we can parametrize the leaves by the x-parameter in a disk $\mathbb{D}(0,r)$ around the origin:

$$\mathcal{L}_p: \quad x\mapsto y_p(x),\quad y_p'(x)=\frac{Q(x,y_p(x))}{P(x,y_p(x))},\quad y_p(0)=0,$$

$$\mathcal{L}_q: \quad x\mapsto y_q(x),\quad y_q'(x)=\frac{Q(x,y_q(x))}{P(x,y_q(x))},\quad y_q(0)=y_0.$$

Define $h:\mathbb{D}(0,r)\to\mathbb{R}$ by $h(x)=\log|y_p(x)-y_q(x)|$. This is a harmonic function with a minimum at $x=0$. Therefore $h\equiv\log\epsilon$ on $\mathbb{D}(0,r)$, so that locally $\mathcal{L}_q$ is just the translation of $\mathcal{L}_p$ by y_0. By analytic continuation, this is true globally, i.e., $\mathcal{L}_q\cap U_0=(\mathcal{L}_p\cap U_0)+(0,y_0)$. By Corollary 2.16, there exists a sequence $q_n\in\mathcal{L}_q\cap U_0$ tending to infinity. The sequence $p_n := q_n-(0,y_0)\in\mathcal{L}_p\cap U_0$ also tends to infinity, so $d(p_n,q_n)\to 0$ by Lemma 4.3. This shows that $\overline{\mathcal{L}}_p\cap\overline{\mathcal{L}}_q=\mathcal{M}_1\cap\mathcal{M}_2\neq\emptyset$, which is a contradiction.

4.3. Poincaré Metric and Hyperbolicity. According to the Uniformization Theorem of Koebe-Poincaré-Riemann, every simply-connected Riemann surface is biholomorphic to one of the three standard models: The Riemann sphere $\widehat{\mathbb{C}}$, the complex plane $\mathbb{C}$, or the unit disk $\mathbb{D}$ [**Ah**]. Since every Riemann surface has a holomorphic universal covering, it follows that *every Riemann surface can be covered holomorphically by $\widehat{\mathbb{C}}$, $\mathbb{C}$, or $\mathbb{D}$, in which case it is called spherical, Euclidean, or hyperbolic, respectively.* Elementary considerations show that every spherical Riemann surface is biholomorphic to $\widehat{\mathbb{C}}$, while the Euclidean Riemann surfaces are biholomorphic to $\mathbb{C}$, or to the cylinder $\mathbb{C}/\mathbb{Z} \simeq \mathbb{C}^*$, or to a torus $\mathbb{C}/\Lambda$ for a rank 2 lattice Λ in the plane. The last two surfaces are biholomorphic to the quotients of $\mathbb{C}$ modulo the action of the groups generated by

$$\{z \mapsto z + a, \text{ for some } a \in \mathbb{C}^*\},$$

and

$$\{z \mapsto z + a, z \mapsto z + b, \text{ for some } a, b \in \mathbb{C}^* \text{ with } a/b \notin \mathbb{R}\},$$

respectively. All other Riemann surfaces are hyperbolic.

DEFINITION 4.5. The Riemannian metric

$$\rho_{\mathbb{D}} := \frac{4}{(1-|z|^2)^2}\ |dz|^2$$

is called the *Poincaré* or *hyperbolic metric* on the unit disk.

The Poincaré metric $\rho_{\mathbb{D}}$ is invariant under all biholomorphisms $\varphi : \mathbb{D} \to \mathbb{D}$. It is uniquely characterized by this property up to multiplication by a non-zero constant. Therefore, all biholomorphisms of $\mathbb{D}$ are isometries with respect to the distance induced by $\rho_{\mathbb{D}}$. The unit disk equipped with this distance is a complete metric space [**M**].

Now let X be a hyperbolic Riemann surface, with the universal covering map $\pi : \mathbb{D} \to X$. Since the Poincaré metric $\rho_{\mathbb{D}}$ is invariant under all covering transformations, $\rho_{\mathbb{D}}$ induces a well-defined Poincaré metric ρ_X on X which is invariant under all biholomorphisms $X \to X$. By the definition of ρ_X, the projection π is a local isometry.

It is a direct consequence of Schwarz Lemma that if $\varphi : X \to Y$ is a holomorphic mapping between hyperbolic Riemann surfaces, then φ decreases the Poincaré distances, i.e., for every $x, y \in X$,

$$d_Y(\varphi(x), \varphi(y)) \leq d_X(x, y),$$

where d_X and d_Y are the Riemannian distances induced by ρ_X and ρ_Y, respectively. If the equality holds for a pair (x, y), then φ will be a local isometry and a covering map [**M**].

DEFINITION 4.6. For a conformal metric $ds^2 = h^2|dz|^2$ on a Riemann surface, the *Gaussian curvature* κ is given by

$$\kappa(z) = -\frac{(\Delta \log h)(z)}{h^2(z)},$$

where, as usual, z is a local uniformizing parameter on the surface.

It follows form this definition that the *Gaussian curvature is a conformal invariant*, that is, if $\varphi : X \to Y$ is a holomorphic map between Riemann surfaces, and if ds^2 is a conformal metric on Y, then at any point $z \in X$ for which $\varphi'(z) \neq 0$, the curvature κ' at z of the pull-back metric on X is equal to the curvature κ of ds^2 at $\varphi(z)$.

It also follows from Definition 4.5 and Definition 4.6 that the Poincaré metric $\rho_{\mathbb{D}}$ on the unit disk has constant Gaussian curvature -1. The same is true for every hyperbolic surface equipped with the Poincaré metric since curvature is a conformal invariant.

The fact that hyperbolic Riemann surfaces admit a metric of strictly negative curvature is a characteristic property, as seen in the following theorem:

THEOREM 4.7. *Suppose that X is a Riemann surface that has a conformal metric whose Gaussian curvature κ satisfies $\kappa < \sigma < 0$ for some constant σ. Then X is hyperbolic.*

Indeed, this is a special case of a more general fact: A complex manifold M which admits a distance for which every holomorphic mapping $\mathbb{D} \to M$ is distance-decreasing is hyperbolic in the sense of Kobayashi. For Riemann surfaces, the usual notion of hyperbolicity is equivalent to the hyperbolicity in the sense of Kobayashi. By Ahlfors' generalized version of Schwarz Lemma, a Riemann surface X which has a conformal metric of strictly negative curvature admits a distance for which every holomorphic mapping $\mathbb{D} \to X$ is distance-decreasing (see [**Kob**], and also [**Kr**] for an elegant exposition in the case of domains in $\mathbb{C}$).

4.4. Hyperbolicity in Minimal Sets. Our next goal is to determine the conformal type of a leaf in the non-trivial minimal set. To this end, let us construct a Hermitian metric on $\mathbb{CP}^2 \smallsetminus \text{sing}(\mathcal{F})$ which induces a conformal metric of negative Gaussian curvature on each non-singular leaf of a given SHFC $\mathcal{F}$. The metric is a modification of the Fubini-Study metric (4.1).

Suppose that $\mathcal{F} : \{\omega = Pdy - Qdx = 0\} \in \mathcal{D}_n$, and let $R = yP - xQ$. Consider the following Hermitian metric on $U_0 \smallsetminus \text{sing}(\mathcal{F})$:

$$\rho := (1 + |x|^2 + |y|^2)^{n-1} \frac{|dx|^2 + |dy|^2 + |xdy - ydx|^2}{|P(x,y)|^2 + |Q(x,y)|^2 + |R(x,y)|^2}. \tag{4.3}$$

ρ extends to a Hermitian metric on $\mathbb{CP}^2 \smallsetminus \text{sing}(\mathcal{F})$. To see this, let us for example compute the extended metric on the affine chart $(u,v) \in U_1$ (compare (1.6)):

$$\begin{aligned}(\phi_{10}^*\rho)(u,v) &= \left(1 + \frac{1}{|u|^2} + \frac{|v|^2}{|u|^2}\right)^{n-1} \frac{|u|^{-4}(|du|^2 + |udv - vdu|^2 + |dv|^2)}{|P(\frac{1}{u}, \frac{v}{u})|^2 + |Q(\frac{1}{u}, \frac{v}{u})|^2 + |R(\frac{1}{u}, \frac{v}{u})|^2} \\ &= \left(1 + |u|^2 + |v|^2\right)^{n-1} \frac{|du|^2 + |dv|^2 + |udv - vdu|^2}{|\tilde{P}(u,v)|^2 + |\tilde{Q}(u,v)|^2 + |\tilde{R}(u,v)|^2},\end{aligned}$$

where $\tilde{P}(u,v) = u^{n+1}P\left(\frac{1}{u}, \frac{v}{u}\right)$, $\tilde{Q}(u,v) = u^{n+1}Q\left(\frac{1}{u}, \frac{v}{u}\right)$, and $\tilde{R}(u,v) = u^{n+1}R\left(\frac{1}{u}, \frac{v}{u}\right)$ are polynomials in u, v.

Now let $p \in U_0$ be a non-singular point of $\mathcal{F}$, and let $\eta : T \mapsto (x(T), y(T))$ be a local parametrization of $\mathcal{L}_p$ near p with $\eta(0) = p$. By (4.3) above, the induced conformal metric on $\mathcal{L}_p$ is

$$ds^2 = (1+|x(T)|^2+|y(T)|^2)^{n-1} \frac{|x'(T)|^2 + |y'(T)|^2 + |x(T)y'(T) - y(T)x'(T)|^2}{|P(\eta(T))|^2 + |Q(\eta(T))|^2 + |R(\eta(T))|^2} |dT|^2$$

$$= (1 + |x(T)|^2 + |y(T)|^2)^{n-1}|dT|^2$$

$$=: h^2(T)|dT|^2.$$

By Definition 4.6 and conformal invariance of the curvature, the Gaussian curvature of $\mathcal{L}_p$ at p is given by

$$\kappa(p) = -\frac{(\Delta \log h)(0)}{h^2(0)}.$$

Computation gives

(4.4)

$$\begin{aligned}
\kappa(p) &= \frac{-2}{(1+|p|^2)^{n-1}} \left(\frac{\partial}{\partial T}\frac{\partial}{\partial \overline{T}} \log h^2(T) \right)\Bigg|_{T=0} \\
&= \frac{-2(n-1)}{(1+|p|^2)^{n-1}} \left(\frac{\partial}{\partial T}\frac{\partial}{\partial \overline{T}} \log(1+|x(T)|^2+|y(T)|^2) \right)\Bigg|_{T=0} \\
&= \frac{-2(n-1)}{(1+|p|^2)^{n-1}} \left(\frac{|x'(T)|^2+|y'(T)|^2+|x(T)y'(T)-y(T)x'(T)|^2}{(1+|x(T)|^2+|y(T)|^2)^2} \right)\Bigg|_{T=0} \\
&= \frac{-2(n-1)}{(1+|p|^2)^{n+1}}(|P(p)|^2+|Q(p)|^2+|R(p)|^2),
\end{aligned}$$

which is strictly negative.

Now let $\mathcal{F} \in \mathcal{D}_n$ have a non-trivial minimal set $\mathcal{M}$, and $p \in \mathcal{M}$. As the above expression is a continuous function of p which extends to $\mathbb{CP}^2 \smallsetminus \operatorname{sing}(\mathcal{F})$, it follows that the Gaussian curvature of the induced metric on $\mathcal{L}_p$ is uniformly bounded from above by a negative constant. By Theorem 4.7, we obtain

THEOREM 4.8. *Every leaf in the non-trivial minimal set is a hyperbolic Riemann surface.*

EXAMPLE 4.9. We can use the preceding result to show that no SHFC $\mathcal{F}$ of geometric degree 1 can have a non-trivial minimal set. Let $\mathcal{L}$ be a leaf of $\mathcal{F}$ contained in the non-trivial minimal set $\mathcal{M}$. Choose $p \in \mathcal{L}$ and let $T \mapsto \eta(T)$ be a local parametrization of $\mathcal{L}$ near p, with $\eta(0) = p$. The germ of η can be analytically continued over the entire plane $\mathbb{C}$. To see this, observe that $\mathcal{F}$ is induced by a holomorphic vector field on $\mathbb{CP}^2$ (see the comment after Theorem 1.18). By Proposition 1.9, this vector field lifts to a linear vector field on $\mathbb{C}^3$. Since every integral curve of a linear vector field is parametrized by $\mathbb{C}$, the same must be true for $\mathcal{L}$. It follows that the result of analytic continuation of η is a single-valued function. Now $\mathcal{L}$ is hyperbolic by Theorem 4.8. Let $\pi : \mathbb{D} \to \mathcal{L}$ be the universal covering map, and lift η to obtain a holomorphic map $\tilde{\eta} : \mathbb{C} \to \mathbb{D}$ with $\pi \circ \tilde{\eta} = \eta$. By Liouville's Theorem, $\tilde{\eta}$ is constant, which is a contradiction.

4.5. Algebraic Leaves and Minimal Sets. The next theorem answers a basic question which is of special interest in the case of any foliated manifold (compare [**CLS1**]).

THEOREM 4.10. *No SHFC on $\mathbb{CP}^2$ can have a compact non-singular leaf.*

PROOF. Let $\mathcal{L}$ be a compact non-singular leaf of $\mathcal{F} : \{Pdy - Qdx = 0\}$. By Chow's Theorem [**GH**], $\mathcal{L}$ is a smooth irreducible algebraic curve in $\mathbb{CP}^2$. Note that $\mathcal{L}$ cannot be a component of both curves $S_P : \{P = 0\}$ and $S_Q : \{Q = 0\}$, since P and Q are relatively prime. Hence, for example, we may assume that P is not identically zero on $\mathcal{L}$. It follows that the intersection of $\mathcal{L}$ with S_P is a finite set $\{p_1, \dots, p_k\}$.

Consider the 1-forms

$$\alpha = \frac{\partial}{\partial y}\left(\frac{Q}{P}\right) dx \qquad \text{and} \qquad \beta = -\partial \log(1 + |x|^2 + |y|^2)$$

in the affine chart $(x, y) \in U_0$. An easy computation shows that the 1-form $\tau = \alpha+\beta$ is well-defined on $\mathbb{CP}^2 \smallsetminus S_P$. For example, in the affine chart $(u, v) \in U_1$ it is given by

$$\frac{\partial}{\partial v}\left(\frac{\tilde{R}}{\tilde{P}}\right) du - \partial \log(1 + |u|^2 + |v|^2),$$

where $\tilde{P}$ and $\tilde{R}$ are polynomials in u, v defined in §4.4. The restriction $\alpha|_{\mathcal{L}}$ has poles at the finite set $\{p_1, \dots, p_k\}$ where $\mathcal{L}$ has a vertical tangent line. Without loss of generality we assume that all the p_j are in the affine chart U_0. Furthermore, it is easy to compute the residue of $\alpha|_{\mathcal{L}}$ at p_j: If $p_j = (x_j, y_j)$ and if $y \mapsto x_j + \sum_{i=m_j}^{\infty} a_i(y - y_j)^i$ is the local parametrization of $\mathcal{L}$ near p_j with $a_{m_j} \neq 0$, then

$$\text{Res}[\alpha|_{\mathcal{L}}; p_j] = 1 - m_j.$$

Now consider small disks $D_j \subset \mathcal{L}$ around each p_j and integrate the 2-form $d\tau$ over $\mathcal{L}' = \mathcal{L} \smallsetminus \bigcup D_j$:

$$\begin{aligned} \int_{\mathcal{L}'} d\tau &= \sum \int_{\partial D_j} \tau \\ &= \sum \int_{\partial D_j} \alpha + \sum \int_{\partial D_j} \beta \\ &= (2\pi\sqrt{-1}) \sum (1 - m_j) + \sum \int_{\partial D_j} \beta. \end{aligned}$$

On the other hand, $d\tau|_{\mathcal{L}'} = d\alpha|_{\mathcal{L}'} + d\beta|_{\mathcal{L}'} = d\beta|_{\mathcal{L}'} = \bar{\partial}\beta|_{\mathcal{L}'} = (2\pi\sqrt{-1})\Omega|_{\mathcal{L}'}$, where Ω is the standard area form (4.2) coming from the Fubini-Study metric. Therefore

$$(2\pi\sqrt{-1}) \text{ area}(\mathcal{L}') = (2\pi\sqrt{-1}) \sum (1 - m_j) + \sum \int_{\partial D_j} \beta.$$

Letting D_j shrink to p_j, we get

$$\text{area}(\mathcal{L}) = \sum (1 - m_j) \leq 0,$$

which is a contradiction. □

As a result, every algebraic leaf of an SHFC $\mathcal{F}$ must have some singular points of $\mathcal{F}$ in its closure. Note that every singularity of an algebraic leaf is indeed a singular point of $\mathcal{F}$ as well.

PROPOSITION 4.11. *Let $\mathcal{M}$ be the non-trivial minimal set of an SHFC $\mathcal{F}$ on $\mathbb{CP}^2$. Then $\mathcal{M}$ intersects every algebraic curve in $\mathbb{CP}^2$.*

PROOF. Let $S_K : \{K = 0\}$ be an algebraic curve in $\mathbb{CP}^2$ of degree k. For every triple (a, b, c) of positive real numbers, define

$$\varphi(x, y) = \varphi_{a,b,c}(x, y) := \frac{|K(x, y)|^2}{(a + b|x|^2 + c|y|^2)^k}$$

which is a non-negative real-analytic function on the affine chart $(x, y) \in U_0$. Since K has degree k, φ can be extended to a real analytic function on the entire $\mathbb{CP}^2$, with $S_K = \varphi^{-1}(0)$.

Suppose that $\mathcal{M} \cap S_K = \emptyset$. Then φ attains a positive minimum on $\mathcal{M}$, i.e., there exists $p_0 \in \mathcal{M}$ such that $\varphi(p) \geq \varphi(p_0) > 0$ for all $p \in \mathcal{M}$. Define $\psi : \mathcal{M} \to \mathbb{R}$ by $\psi(p) = \log \varphi(p)$. Clearly $\psi(p) \geq \log \varphi(p_0) > -\infty$ for all $p \in \mathcal{M}$. On the other hand, ψ is superharmonic along the non-singular leaf $\mathcal{L}_{p_0}$. To see this, let $\eta : T \mapsto (x(T), y(T))$ be a local parametrization of $\mathcal{L}_{p_0}$ near p_0, with $\eta(0) = p_0$. Then $\psi(\eta(T)) > -\infty$, and

$$\begin{aligned} \Delta\psi(\eta(T)) = {} & 4\frac{\partial}{\partial T}\frac{\partial}{\partial \overline{T}}\psi(\eta(T)) \\ = {} & -4k(a + b|x(T)|^2 + c|y(T)|^2)^{-2} \\ & (ab|x'(T)|^2 + ac|y'(T)|^2 + bc|x(T)y'(T) - y(T)x'(T)|^2) \end{aligned}$$

which is negative. Since $\psi(\eta(T))$ has a minimum at $T = 0$, it follows that φ is constant on $\mathcal{L}_{p_0}$, hence on $\mathcal{M}$ since $\overline{\mathcal{L}}_{p_0} = \mathcal{M}$. Therefore, for any triple (a, b, c) of positive real numbers, there exists $\alpha > 0$ such that

$$|K(x, y)|^{2/k} = \alpha(a + b|x|^2 + c|y|^2)$$

for all $(x, y) \in \mathcal{M}$. Evidently, this is impossible. □

COROLLARY 4.12. *No SHFC on $\mathbb{CP}^2$ which has an algebraic leaf can have a non-trivial minimal set.*

PROOF. Let $\mathcal{L}$ be an algebraic leaf of an SHFC $\mathcal{F}$. By Theorem 4.10, $\overline{\mathcal{L}}$ necessarily contains a singular point of $\mathcal{F}$, say q. If $\mathcal{M}$ is a non-trivial minimal set of $\mathcal{F}$, then there exists $p \in \mathcal{M} \cap \mathcal{L}$ by Proposition 4.11. As p is non-singular, $\overline{\mathcal{L}} = \overline{\mathcal{L}}_p = \mathcal{M}$, so $q \in \mathcal{M} \cap \operatorname{sing}(\mathcal{F})$, which is a contradiction. □

Therefore, in order to find an $\mathcal{F}$ with a non-trivial minimal set, we must look for $\mathcal{F}$ in the sub-class of $\mathcal{D}_n$ consisting of SHFC's which do not admit any algebraic leaf. This sub-class is open and dense in $\mathcal{D}_n$ (see Remark 3.22).

Note that the above corollary gives another proof for the fact that no SHFC of geometric degree 1 can have a non-trivial minimal set, since one can easily see that every SHFC of geometric degree 1 has a projective line as a leaf.

We conclude with few important remarks.

REMARK 4.13. It is shown in [**CLS1**] that every leaf $\mathcal{L}$ in the non-trivial minimal set $\mathcal{M}$ has *exponential growth*. This means that if we fix some Riemannian metric on $\mathcal{L}$ and some $p \in \mathcal{L}$, then

$$\liminf_{r \to +\infty} \frac{\log(\text{area}(B_r(p)))}{r} > 0,$$

where $B_r(p)$ denotes the open ball in $\mathcal{L}$ of radius r centered at p.

REMARK 4.14. C. Bonatti, R. Langevin, and R. Moussu [**BLM**] have shown that for any non-trivial minimal set $\mathcal{M}$, there exists a leaf $\mathcal{L} \subset \mathcal{M}$ such that the monodromy group $G(\mathcal{L})$ contains a hyperbolic germ in $\text{Biho}_0(\mathbb{C})$. The real version of this theorem is a famous 1965 result of R. Sacksteder [**Sa**]: An exceptional minimal set of a transversely orientable codimension one C^2 foliation on a compact manifold contains a leaf with a hyperbolic monodromy mapping.

REMARK 4.15. Here is a related result due to A. Candel and X. Gómez-Mont [**CaG**] (see also the paper by A. Glutsyuk in [**I6**] for a generalization): Let $\mathcal{F}$ be an SHFC with hyperbolic singular points and no algebraic leaves. Then every leaf of $\mathcal{F}$ is a hyperbolic Riemann surface. In fact, a non-hyperbolic leaf gives rise to a non-trivial invariant transverse measure for $\mathcal{F}$. The support of this measure cannot intersect the leaves outside of the (possible) minimal set since then it has to be supported on the (global) separatrices by hyperbolicity of the singular points, and this means that these separatrices (with singular points added) are compact, hence algebraic, which is a contradiction. On the other hand, no invariant transverse measure can live on the minimal set by a result of [**CLS1**].

References

[**Ah**] L. Ahlfors, *Conformal Invariants: Topics in Geometric Function Theory*, McGraw-Hill, 1973.

[**An**] D. Anosov, V. Arnold (eds.), *Dynamical Systems (I)*, Encyclopedia of Mathematical Sciences, Springer-Verlag, 1988.

[**Ar**] V. Arnold, *Geometrical Methods in the Theory of Ordinary Differential Equations*, 2nd ed., Springer-Verlag, 1988.

[**AA**] V. Arnold, A. Avez, *Ergodic Problems of Classical Mechanics*, Benjamin, 1968.

[**BLM**] C. Bonatti, R. Langevin, R. Moussu, *Feuilletages de $\mathbb{CP}(n)$: de l'holonomie hyperbolique pour les minimaux exceptionnels*, Publ. Math. IHES, **75** (1992) 124-134.

[**C**] C. Camacho, *Problems on limit sets of foliations on complex projective spaces*, Proceedings of the International Congress of Mathematicians, Kyoto, Japan, 1990, 1235-1239.

[**CKP**] C. Camacho, N. Kuiper, J. Palis, *The topology of holomorphic flows with singularity*, Publ. Math. IHES, **48** (1978) 5-38.

[**CL**] C. Camacho, A. Lins Neto, *Geometric Theory of Foliations*, Birkhauser, 1985.

[**CLS1**] C. Camacho, A. Lins Neto, P. Sad, *Minimal sets of foliations on complex projective spaces*, Publ. Math. IHES, **68** (1988) 187-203.

[**CLS2**] C. Camacho, A. Lins Neto, P. Sad, *Foliations with algebraic limit sets*, Annals of Math., **136** (1992) 429-446.

[**CS**] C. Camacho, P. Sad, *Invariant varieties through singularities of holomorphic vector fields*, Annals of Math., **115** (1982) 579-595.

[**Ca**] A. Candel, *Uniformization of surface laminations*, Ann. Scient. Éc. Norm. Sup., **26** (1993) 489-516.

[**CaG**] A. Candel, X. Gómez-Mont, *Uniformization of the leaves of a rational vector field*, Ann. Inst. Fourier, **45** (1995) 1123-1133.

[**CoL**] E. Coddington, N. Levinson, *Theory of Ordinary Differential Equations*, McGraw-Hill, 1965.

[**D**] H. Dulac, *Sur les cycles limites*, Bull. Soc. Math. France, **51** (1923) 45-188.

[**Gh**] E. Ghys, *Laminations par surfaces de Riemann*, Dynamique et geometrie complexes (Lyon, 1997), Panor. Syntheses, **8** (1999) 49-95.

[**GO**] X. Gómez-Mont, L. Ortiz Bobadilla, *Systemas Dinamicos Holomorfos en Superficias*, Sociedad Matematica Mexicana, 1989.

[**GL**] X. Gómez-Mont, I. Luengo, *Germs of holomorphic vector fields in $\mathbb{C}^3$ without a separatrix*, Invent. Math., **109** (1992) 211-219.

[**GH**] P. Griffiths, J. Harris, *Principles of Algebraic Geometry*, Wiley-Interscience, New York, 1978.

[**GR**] R. Gunning, H. Rossi, *Analytic Functions of Several Complex Variables*, Prentice-Hall, 1965.

[**I1**] Yu. Il'yashenko, *Global and local aspects of the theory of complex differential equations*, Proceedings of the International Congress of Mathematicians, Helsinki, Finland, 1978, 821-826.

[I2] Yu. Il'yashenko, *Singular points and limit cycles of differential equations in the real and complex planes*, Preprint, Pushchino: Sci. Res. Comput. Cent., Acad. Sci. USSR, 1982 (in Russian).

[I3] Yu. Il'yashenko, *Topology of phase portraits of analytic differential equations in the complex projective plane*, Trudy Sem. Petrovskiĭ, **4** (1978) 84-136 (in Russian), English Translation in Sel. Math. Sov., **5** (1986) 140-199.

[I4] Yu. Il'yashenko, *Finiteness theorems for limit cycles*, Russ. Math. Surv., **42** (1987) 223.

[I5] Yu. Il'yashenko (ed.), *Nonlinear Stokes Phenomena*, Adv. Soviet Math., 14, Amer. Math. Soc., 1993.

[I6] Yu. Il'yashenko (ed.), *Differential Equations with Real and Complex Time*, Proceedings of the Steklov Institute of Mathematics, vol. 213, 1996.

[IY] Yu. Il'yashenko, S. Yakovenko, *Analytic Differential Equations*, ICTP Lecture Notes of School on Dynamical Systems, Trieste, Italy, 1991.

[J] J.P. Jouanolu, *Equations de Pfaff Algebriques*, Lecture Notes in Mathematics 708, Springer-Verlag, 1979.

[KS] B. Khanedani, T. Suwa, *First variation of holomorphic forms and some applications*, Hokkaido Math. J., **26** (1997) 323-335.

[Kob] S. Kobayashi, *Hyperbolic Manifolds and Holomorphic Mappings*, Marcel Dekker, New York, 1970.

[Kod] K. Kodaira, *Complex Manifolds and Deformation of Complex Structures*, Springer-Verlag, 1986.

[Kr] S. Krantz, *Complex Analysis: The Geometric Viewpoint*, The Carus Mathematical Monographs No. 23, Math. Assoc. Amer., 1990.

[L] A. Lins Neto, *Algebraic solutions of polynomial differential equations and foliations in dimension two*, in "Holomorphic Dynamics," Lecture Notes in Mathematics 1345, Springer-Verlag, 1988.

[M] J. Milnor, *Dynamics in One Complex Variable: Introductory Lectures*, Vieweg, 1999.

[N] V. Naishul, *Topological invariants of analytic and area-preserving mappings and their application to analytic differential equations in* $\mathbb{C}^2$ *and* $\mathbb{CP}^2$, Trans. Moscow Math. Soc., **2** (1983) 234-250.

[P] R. Perez-Marco, *Solution complète au problème de Siegel de linéarisation d'une application holomorphe au voisinage d'un point fixe (d'après J.C. Yoccoz)*, Séminaire Bourbaki, **753** (1991-1992) 1-31.

[PL1] I. Petrovskiĭ, E. Landis, *On the number of limit cycles of the equation* $dy/dx = P(x,y)/Q(x,y)$ *where* P *and* Q *are polynomials of degree 2,* Amer. Math. Society Trans., Series 2, **10** (1963) 125-176.

[PL2] I. Petrovskiĭ, E. Landis, *On the number of limit cycles of the equation* $dy/dx = P(x,y)/Q(x,y)$ *where* P *and* Q *are polynomials*, Amer. Math. Society Trans., Series 2, **14** (1964) 181-200.

[Re] G. Reeb, *Sur certaines propriétés topologiques des variétés feuilletées*, Actualités Sci. Indust., Hermann Paris, **1183** (1952) 91-154.

[Ru] W. Rudin, *Real and Complex Analysis*, 3rd ed., McGraw-Hill, 1986.

[Sa] R. Sacksteder, *Foliations and pseudo-groups*, Amer. J. Math., **87** (1965) 79-102.

[Sh] A. Shcherbakov, *Topological and analytical conjugacy of non-abelian groups of germs of conformal mappings*, Trudy Sem. Petrovskiĭ, **10** (1984) 170-196 (in Russian).

[W] H. Whitney, *Complex Analytic Varieties*, Addison-Wesley, 1972.

Department of Mathematics, University of Tehran, Iran, and Institute for Studies in Theoretical Physics and Mathematics (IPM), Tehran, Iran

Current address: Department of Mathematics, University of Pennsylvania, Philadelphia, PA 19104-6395, USA

E-mail address: `zakeri@math.upenn.edu`

Selected Titles in This Series

(Continued from the front of this publication)

244 **Leovigildo Alonso Tarrío, Ana Jeremías López, and Joseph Lipman,** Studies in duality on noetherian formal schemes and non-noetherian ordinary schemes, 1999

243 **Tsit Yuan Lam and Andy R. Magid, Editors,** Algebra, K-theory, groups, and education, 1999

242 **Bernhelm Booss-Bavnbek and Krzysztof Wojciechowski, Editors,** Geometric aspects of partial differential equations, 1999

241 **Piotr Pragacz, Michał Szurek, and Jarosław Wiśniewski, Editors,** Algebraic geometry: Hirzebruch 70, 1999

240 **Angel Carocca, Víctor González-Aguilera, and Rubí E. Rodríguez, Editors,** Complex geometry of groups, 1999

239 **Jean-Pierre Meyer, Jack Morava, and W. Stephen Wilson, Editors,** Homotopy invariant algebraic structures, 1999

238 **Gui-Qiang Chen and Emmanuele DiBenedetto, Editors,** Nonlinear partial differential equations, 1999

237 **Thomas Branson, Editor,** Spectral problems in geometry and arithmetic, 1999

236 **Bruce C. Berndt and Fritz Gesztesy, Editors,** Continued fractions: From analytic number theory to constructive approximation, 1999

235 **Walter A. Carnielli and Itala M. L. D'Ottaviano, Editors,** Advances in contemporary logic and computer science, 1999

234 **Theodore P. Hill and Christian Houdré, Editors,** Advances in stochastic inequalities, 1999

233 **Hanna Nencka, Editor,** Low dimensional topology, 1999

232 **Krzysztof Jarosz, Editor,** Function spaces, 1999

231 **Michael Farber, Wolfgang Lück, and Shmuel Weinberger, Editors,** Tel Aviv topology conference: Rothenberg Festschrift, 1999

230 **Ezra Getzler and Mikhail Kapranov, Editors,** Higher category theory, 1998

229 **Edward L. Green and Birge Huisgen-Zimmermann, Editors,** Trends in the representation theory of finite dimensional algebras, 1998

228 **Liming Ge, Huaxin Lin, Zhong-Jin Ruan, Dianzhou Zhang, and Shuang Zhang, Editors,** Operator algebras and operator theory, 1999

227 **John McCleary, Editor,** Higher homotopy structures in topology and mathematical physics, 1999

226 **Luis A. Caffarelli and Mario Milman, Editors,** Monge Ampère equation: Applications to geometry and optimization, 1999

225 **Ronald C. Mullin and Gary L. Mullen, Editors,** Finite fields: Theory, applications, and algorithms, 1999

224 **Sang Geun Hahn, Hyo Chul Myung, and Efim Zelmanov, Editors,** Recent progress in algebra, 1999

223 **Bernard Chazelle, Jacob E. Goodman, and Richard Pollack, Editors,** Advances in discrete and computational geometry, 1999

222 **Kang-Tae Kim and Steven G. Krantz, Editors,** Complex geometric analysis in Pohang, 1999

221 **J. Robert Dorroh, Giséle Ruiz Goldstein, Jerome A. Goldstein, and Michael Mudi Tom, Editors,** Applied analysis, 1999

220 **Mark Mahowald and Stewart Priddy, Editors,** Homotopy theory via algebraic geometry and group representations, 1998